Landscape Construction

Landscape Construction

DAVID SAUTER

Africa • Australia • Canada • Denmark • Japan • Mexico • New Zealand • Philippines
Puerto Rico • Singapore • Spain • United Kingdom • United States

NOTICE TO THE READER

Delmar Staff

Business Unit Director: Susan Simpfenderfer
Executive Editor: Marlene McHugh Pratt
Executive Marketing Manager: Donna Lewis
Executive Production Manager: Wendy Troeger
Developmental Editor: Andrea Edwards Myers
Production Editor: Carolyn Miller
Marketing Coordinator: Katherine M. Hans
Cover Design: Carolyn Miller

Printed in the United States of America
1 2 3 4 5 6 7 8 9 10 XXX 04 03 02 01 00 99

For more information, contact Delmar, 3 Columbia Circle, PO Box 15015, Albany, NY 12212-0515; or find us on the World Wide Web at http://www.delmar.com

Library of Congress Cataloging-in-Publication Data

Sauter, David.
    Landscape construction / David Sauter.
        p. cm.
    Includes bibliographical references and index.
    ISBN 0-8273-8427-0
    1. Landscape construction.   I. Title.
TH380.S26 1999
624—dc21   99-14917
                               CIP

Dedication

This text is dedicated to my wife Lynn,

my daughter Margaret,

my parents Ruth and Jim,

my mother-in-law Bernice,

and Murphy.

Their patience and support never ended.

Contents

SECTION 1   **BEFORE CONSTRUCTION BEGINS   1**

Introduction, 1

CHAPTER 1   **The Landscape Construction Process   2**

CHAPTER 2   **Legal Requirements   8**

CHAPTER 3   **Interpreting Construction Documents   15**

CHAPTER 4   **Safety in the Workplace   28**

CHAPTER 5   **Construction Math   33**

Preface

To change a site from an underdeveloped space to an attractive, functional, outdoor area, sound design and a skilled landscape contractor are needed. Just as the designer must perceive the visual potential and practical concerns that a site offers, the landscape contractor must possess creativity and a wide range of tangible skills to interpret and implement the designer's ideas. Such skills come only through learning and practice. Traditionally, the trade of landscape contracting has relied on apprenticeships and educational programs to teach students the art of implementation. Both of these methods provide effective hands-on schooling, but have been limited by the lack of written material necessary to form a foundation for experiential learning.

Landscape construction requires knowledge of a broad range of construction techniques. Numerous texts and manufacturer's pamphlets are available that explain one aspect of landscape construction, but the scattered nature of this information makes it difficult to grasp how one part of construction relates to another. Continuous and holistic learning is thus made difficult, leaving students with a patchy and inconsistent picture of landscape construction. These inconsistencies prevent a design from developing to its fullest potential.

The majority of "complete" landscape texts available are limited in scope, or focus on garden projects rather than construction techniques. This text attempts to correct those omissions by providing comprehensive, process-oriented coverage of the many facets and phases of landscape construction. Such information will help students develop an understanding of the construction process as well as sharpen the extensive foundation of skills required before a student can become a competent specialist. Construction basics are explained and illustrated with numerous photos and diagrams. Fundamental data related to construction techniques and materials is provided.

This text is intended for students of landscape construction and practicing professionals seeking to expand their expertise. It is prepared as a supplement to supervised classroom instruction and is not intended to be a do-it-yourself manual.

The sections of this edition are organized around the logical steps of landscape construction. The first section provides information regarding preconstruction activities, including safety, legal issues, reading construction documents, basic construction math, bidding and estimating, and basic skills. Following this initial unit are eight sections dedicated to the work typically associated with landscape construction. These sections are: site preparation; grading, site drainage, and erosion protection; site utilities; landscape retaining walls and stairs; landscape paving; wood decks and landscape structures; fences and freestanding walls; and landscape amenities. Chapters within each section detail the steps necessary to install materials covered under the section's heading.

THE PROFESSION OF LANDSCAPE CONTRACTING

When considering the importance of landscape contracting in the design and construction process, one should remember the significance of how the site is experienced. Judgments often begin with the first impression a visitor has of a site. People may not even enter a home, yet develop an opinion of the occupant by how well the exterior environment is presented. Decisions about whether to shop at a store may be made without ever seeing the merchandise, but by judging the exterior of the business. The economic or social success of many projects is determined by the landscaping. The work of the landscape contractor plays a significant role in determining that success.

As the importance of exterior image and environment has increased, so has the importance of the landscape contracting profession. Public appreciation of an attractive site has been credited with increasing business in certain markets. Advances in technologies for irrigation and issues such as xeriscaping have also contributed to industry growth. What once may have been a sideline to a retail plant sale operation or a small contracting firm has now evolved into a variety of business types and service providers known as landscape contractors.

An increase in the number of landscape contractors and the amount of work available have sparked significant changes in the industry. Basic among the types of landscape companies are the general practitioners who provide their clients a full range of construction services. The growth of the industry has also led to a variety of specialists who concentrate on a single aspect of the construction process. Specialists are common in irrigation, grading, water feature construction, and planting. Depending on the size of the market, contractors can also be found who specialize in wall construction, decorative paving, carpentry, and amenity installation.

The separation of design from construction has been reduced with the growth of design/build firms. Design/build companies prepare a design for the client and follow up with installation of the design. Many design/build firms can also provide separate services, if required, by a bidding situation. Landscape contracting services can also be obtained through many garden centers. In recognition of the value of maintenance contracts for installed landscapes, a growing number of businesses offer a "full-service" approach that includes design, building, and maintenance of a project.

With the landscape contracting industry filling a variety of roles, it is necessary to prepare professionals who can address many aspects of the building trade. This text covers these diverse technical requirements in an attempt to improve the quality of work within the profession of landscape construction. Depending on their choice of specialization, landscape contractors may be required to master one, two, or possibly all of these skills.

The varied opportunities available for practicing in the field of landscape contracting and the specific career choice may determine what education and experience are required to be successful. This text can be utilized to improve the skills of a student or contracting professional in two ways. When used as a comprehensive text, the entire process of landscape contracting can be viewed. Each section can also be used as a standalone teaching unit for introducing specialized construction techniques when appropriate. The author strongly urges you to both read and practice. When addressing a technical topic such as this, few educational techniques work better than the combination of reading, seeing, and doing.

FEATURES OF THIS EDITION

- Organization of the text around the steps of construction.
- Inclusion of a comprehensive section on preconstruction activities.
- Provision of basic math concepts and context-related math applications.
- Suggestions for project startup and increased productivity.
- Detailed descriptions of installation procedures for over forty landscape elements.

REGARDING THE USE OF THIS TEXT

- This textbook is intended to supplement supervised classroom instruction. Practice of the profession of landscape construction is filled with safety and health hazards, and use of this book without proper experience or supervision will place the reader at risk. *The reader assumes all risk when engaging in unsupervised activities.*
- All data and information contained in this text are believed to be reliable; however, no warranty of any kind, express or implied, is made with respect to the data, analysis, information, or applications contained herein; and the use of any such data, analysis, information, or applications is at the user's sole risk and expense. The author assumes no liability and expressly disclaims liability, including without limitation liability for negligence, resulting from the use of the data, analysis, information, or applications contained in this text.
- It is assumed that the readers of this text possess basic skills necessary to safely perform construction activities and operate construction equipment. Readers who are deficient in these tasks must obtain instruction in these essential areas before attempting to undertake the activities described in this book.

- It is important to understand what aspects of a project should be completed by the landscape contractor and when a professional from another trade should be called in. Throughout the text, reference is made to work tasks that should be completed by related design and construction professionals.
- Design and construction references made within this text do not supersede instructions provided by contract documents. Materials, methods, and performance required by plans and specifications must be followed.
- Historical restorations may require procedures different from those described in this text.
- Manufacturer's instructions or manuals supersede the recommendations made in this text.
- References made to proprietary materials or methods are not endorsements of those materials or methods.
- Practice of the profession will vary from region to region. Verify with local building professionals the appropriateness of recommendations made in this text. Local building officials should be contacted to verify legal requirements for the region in which any work is performed.

ABOUT THE AUTHOR

David Sauter is the dean of Business Education at Kirkwood Community College in Cedar Rapids, Iowa. Mr. Sauter has also served as a faculty member and coordinator of the Kirkwood Horticulture Department. As a professor of Horticulture, he was responsible for the development and presentation of coursework in landscape design, landscape construction, computer-aided design, computer applications in horticulture, and business management.

David Sauter completed his BS in Landscape Architecture at Iowa State University and received an MA in Higher Education from the University of Iowa. He has presented seminars for the nursery industry, master gardeners, extension service, and public service organizations. He is a registered Landscape Architect, and prior to his career in teaching, Mr. Sauter practiced as a consulting Landscape Architect in the Midwest. He has participated in landscape projects ranging from the construction of small gardens to the master plan for a national landmark site. Mr. Sauter has additional experience in plant maintenance, plant installation, landscape construction, masonry, carpentry, paving, and landscape management.

ACKNOWLEDGMENTS

The author wishes to express his appreciation to the following people, businesses, and organizations for their assistance in the preparation of this text:

Dwight Hughes, Jr.
Iowa City Landscaping
Pleasant Valley Nursery and Garden Center
Peck's Garden Center
Peck's Green Thumb Landscaping
Kirkwood Community College
King's Materials
Margaret Sauter
Wendy Johnston
Rich Kroeze
Adam Boysen
Gary Pribyl
Ron Baker
Jesse Lewis
Lou Garringer
Iowa Nursery and Landscape Association
Iowa One Call
Paul Dykstra
Melanie Schweitzer
Kinko's Copies
Aleda Kroeze Feuerbach
Stu Cole
Lucy Hershberger
Janice Carter
Interlocking Concrete Paving Institute
Prairiewoods staff
Kim Stevely
Jay Podzimek
Kim Muhl
Jeff Dierks
Nely Hernandez
Desiree Francis
Veronica Joseph
Gerry Pogue
Spring Valley Nursery
Ed Sauter
Brian Hazlett
Jim Meisheid
Dierk's Tree Moving
Borgert Paving
Kirkwood Hort Construction I class, 1997
Kirkwood Hort Construction II class, 1997
Kirkwood grounds crew
University Camera
Linn Photo
French Studios
Forevergreen Landscaping
Robin Sailor

Dayl Inglett
Joanne Moeller, FSPA
Horticulture Specialties
Scott Engledow
David Smith

Jim Hynek
Darin Chamberlin
Wilson Mendoza
Miguel Miller
H. L. Carbaugh, Jr.

The author and Delmar Publishers would also like to thank those individuals who reviewed the manuscript and offered suggestions, feedback, and assistance. Their work is greatly appreciated.

Glenn Herold
Illinois Central College
East Peoria, Illinois

David B. Nichols
University of Georgia
Athens, Georgia

Jack Ingels
Suny-Cobleskill
Cobleskill, New York

Glenn Petrick
Milwaukee Area Technical College
North Campus Center, Wisconsin

Cameron R. J. Man
Mississippi State University
Mississippi State, Mississippi

Daniel Robinson
Cal Poly Pomona
Pomona, California

Section 1

Before Construction Begins

INTRODUCTION

Before construction begins, consideration should be given to the safety, legality, and process of the task you are about to undertake. Engaging in a project without knowledge of the risks may bring forth unnecessary dangers for the crew, the public, or even the site itself. Information presented in this section will detail many activities that should be investigated before any physical work begins.

The steps of the construction process, from preconstruction activities to cleanup, are covered in Chapter 1. Chapter 2 discusses legal complexities that may influence the project, including codes, contracts, insurance requirements, and the relationship among the designer, client, and contractor. A guide to interpreting construction documents, to assist the student learning to implement design plans, is presented in Chapter 3. Chapter 4 offers important advice to help a crew reduce its risk of injury during the construction process. Essential math formulas and examples are presented in Chapter 5, to help students become competent at calculating material needs, estimating costs, and bidding before construction begins. Chapter 6 suggests methods for presenting project estimates and bids. Basic construction techniques are described in Chapter 7, including equipment operation. Accurate construction staking is addressed in Chapter 8, explaining the establishment of horizontal and vertical controls.

Although there may be a temptation to move quickly into the physical phase of a project, it is important to examine the essential tasks that should precede actual construction. Planning and preparation can reduce the common risks and minimize the potential hazards of landscape contracting.

Chapter 1

The Landscape Construction Process

The construction of landscaping should be efficient yet complementary to the other projects in progress at a site. Landscape work is full of potential conflicts, and an awareness of the activities that must be done during construction is essential to reduce potential clashes.

This chapter outlines the general process for building landscape improvements. Each major step is identified, along with the basic construction activities that will occur during the process, with an explanation for why such activities must be performed (Figure 1–1). Alternatives to the timeline are also given that accommodate variations in the industry. It should be noted that the process outlined in this chapter serves as a guideline; actual construction practices may vary from contractor to contractor, and from region to region, or may change due to the size of the project or weather conditions.

The process outline provided in Chapter 1 corresponds chronologically to the following sections and chapters of this book. The subsequent chapters will provide more detailed descriptions of what is summarized here.

PRECONSTRUCTION ACTIVITIES

Before a landscape contractor can begin work, he or she must first have a project. Bidding, or competing with other contractors for a job, is the primary way in which contractors are authorized to work on a site. The general, or prime, contractor, will sign a contract with the owner of a site to conclude the bidding process and initiate the second phase of a project: preconstruction planning. In the case of larger projects, the receipt of written authorization to proceed from the owner's representative will signal the beginning of work.

When performing landscaping work as a subcontractor, authorization to work will come from the general contractor, and the two parties may or may not sign a contract. Permission for the subcontractor to actually begin a project must be given by the general contractor. Scheduling of work is critical when the landscape contractor operates as a subcontractor for a larger firm. It is important to verify that completed work will not be disturbed by other construction activities. Restoration work typically does not qualify for additional pay, even if the work is disturbed by another contractor.

Permits, insurance, and bonding should be obtained prior to beginning any work. Orders should be placed in advance for materials that may require lead time for manufacturing, such as premanufactured bridges, amenities, lighting, special-dimensioned features, specialty paving, and wall materials.

Before construction can proceed, layout of critical dimensions and grades will be required (Figure 1–2). Anyone who is not comfortable doing this work, or is not qualified because of the critical nature of the development, should contact a registered land surveyor to provide construction staking.

SITE PREPARATION

Site preparation can vary significantly, but the performance of this step typically incorporates two basic tasks: existing site features must be protected, and existing site elements that are unwanted must be demolished and removed. Protection and demolition activities are best performed before any other phase of the project begins.

Protection of site elements could include defining an access route to construction; fencing the perimeter;

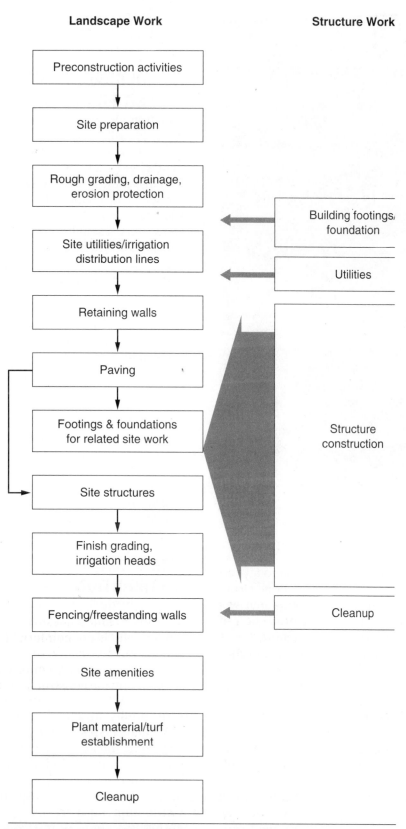

Figure 1–1 Flow chart of the landscape construction process. The right column shows how structure work relates to the landscape process.

Figure 1–2 Using a surveyor's transit to stake grades at a landscape project.

stabilizing benchmarks and baselines used to lay out the project; and protecting any plant material, structures, paving, or permanent improvements that will be part of the finished project. Proper protection of existing site features involves time and effort, but is necessary to reduce significant restoration due to construction damage. Construction documents will outline the items to be safeguarded as part of this step. For projects without construction documents, the contractor may have the responsibility of making those judgments.

Demolition involves the removal of existing site improvements that are not a part of the finished project. Disposal of the removed elements is also the contractor's responsibility. Removing old structures, worn pavement, or damaged and misplaced plant material is a common activity undertaken before new construction begins. Unlike site protection, this step is almost impossible to overlook. It can, however, do unnecessary damage to a site if not properly per-

formed. This damage will need to be restored at the expense of the contractor.

ROUGH GRADING, EROSION PROTECTION, AND SITE DRAINAGE

Rough grading to achieve the desired landform and drainage will follow the site preparation activities. A majority of the grading work will need to be completed before any further steps can be taken. Grading for large projects can be separated into distinct steps that efficiently sculpt the site and preserve resources such as topsoil. Smaller projects often condense the steps, since a small site does not lend itself well to the specialized grading activities completed in larger areas.

Initial steps in grading include the stripping of sod and vegetative cover from the areas where the grade will be changed. Once the plant material has been removed, the topsoil is scraped from the site and stockpiled for future use. Subgrade soils are then cut from areas where there is an excess and filled into areas where there are voids, (Figure 1–3). Grades established during this phase will be below what is planned for finish grades, since paving and topsoil will be placed on top of the subgrade. Grading to facilitate site drainage as well as to install drainage systems, such as tile or storm sewers, is also performed.

Grading activities are then stopped until all utilities, structures, and paving are put into place. Temporary erosion protection is installed at this time to reduce damage to the site until finish grading can occur.

SITE UTILITIES AND IRRIGATION DISTRIBUTION LINES

Depending on the size of the project, a majority of the work in this phase will probably be completed by subcontractors who specialize in installation of water, sewer, gas, electric, phone, and other communication utilities. Establishment of rough grade prior to beginning this phase is essential to maintain proper cover over utility rough-ins. Trenching and placement of major utility structures such as manholes and shutoffs should be performed and then backfilled to avoid disturbing finished elements of the landscape. Placement of conduit and direct burial wire, as well as rough-ins for site lighting, should be completed during this phase. It is also common to place major distribution lines for landscape irrigation systems during this phase of the work. It is not advisable to place valve boxes or irrigation heads until later, when all heavy traffic and finish grading is completed over the site.

Figure 1–3 Rough grading an athletic field site.

Coordination and awareness of utility installation is critical to the landscape contractor. Documentation of utility locations also aids in future construction work that requires excavation. Utilities installed in improper or unknown locations may have a significant negative impact on future plantings, landscape walls, or paving.

LANDSCAPE RETAINING WALLS AND STAIRS

Construction of landscape walls from ties or timbers, precast units, or dry-laid stone requires a stable finish grade. Once rough grade has been established over a site, construction can begin on landscape walls (Figure 1–4). In areas where utilities must pass near or under a wall, it is advisable to complete the utility and assure that the grade above the utility has been compacted prior to placing the wall. Walls that retain the subgrade for paved areas must be in place before the paving activity can take place.

LANDSCAPE PAVING

When all rough grades, utilities, and landscape walls that support paved areas are in place and heavy traffic across the site is complete, landscape paving can begin. An exception to this timing occurs when asphalt roadways are left unsurfaced, or surfaced only with a base layer, until all construction traffic has ceased. This step includes the installation of concrete paving, brick and concrete paving block, stone, and granular paving used for walks, drives, patios, and other outdoor circulation and use areas (Figure 1–5). Paving edges provide elevation guidelines for lawns and planting beds, allowing topsoil to be respread over the site. Final grade should be established after all paving operations are complete. Timing of pavement projects should be coordinated with the completion of structures.

Concrete for any fencing or landscape amenities that require footings can also be poured at this time. This facilitates the efficient delivery of materials and reduces the potential for damage to a completed site structure by paving and building activities.

Figure 1–4 Building a dry-laid, limestone, landscaping retaining wall.

SITE STRUCTURES AND WOOD CONSTRUCTION

Structures related to landscape use can be built at a variety of times during the construction process. Determining the proper time will depend on whether the structure is freestanding, such as a gazebo or trellis, or attached to a building, such as a deck. Once subgrade is established, foundations or slabs for freestanding landscape structures can be placed and erection of the structure begun. Arguments can be made to place this phase between utility construction and landscape paving, since structures may require utility connections prior to pouring of slabs and may have a paved connection to site walkways. A preferred alternative is to pour footings at the same time that landscape paving activities are underway, and to finish the structures after paving and buildings are in place.

Footings, subsurface work, and portions of the site structure that are not connected to a building may be undertaken after site utility work is complete, but finish work and connection to the building may have to be delayed. Structures that are attached to buildings will need to have building exterior work underway before construction begins. Most connections for site structures are made to the exterior sheathing or surfacing material, or are designed to integrate with building roofs. Material delivery may also influence the timing of this phase, since deliveries to buildings may require a vehicle access route near to where a structure is planned. Delivery of materials used in the construction of a site structure should be timed so that damage to the finished landscape is avoided when vehicles travel to a site.

FINISH GRADING AND COMPLETION OF IRRIGATION WORK

Following the completion of other hardscape construction activities, and prior to installation of turf areas and plantings, topsoil is removed from the stockpile and respread over the areas on which turf or plant material is placed. Finish, or fine, grading is performed after respreading of the topsoil to achieve the final grades desired by the designer. Installation of irrigation valves and heads is done after finish grading, or in some circumstances, following installation of turf areas.

FENCING AND FREESTANDING WALLS

Timing of fence and freestanding wall installation varies from project to project. Footings for freestanding walls should be installed when other footings are placed, and masonry work completed before finish grading is begun. Chain link, metal, or wood fences that require posts set in concrete should have that portion of the construction incorporated with pouring of concrete walks, footings for structures, and/or footings for amenities. Surfacing for all types of fencing will need to wait until access to the site is complete for delivery and construction vehicles. Fence surfacing can be accomplished as portions of the site become available and final grade in the area of the fence is established.

SITE AMENITIES

Amenity installation includes such work as placement of benches, flagpoles, trash receptacles, bike racks, and site lighting. These elements are the finishing touches of the hardscape, and their placement should be postponed as late as possible to protect the finishes and delicate workings. Portions of the work related to amenities, particularly footings and wiring, will need to be addressed during the landscape paving and util-

Figure 1–5 Preparing a site for landscape paving.

ity phases of project, but the final placement of the amenity should occur just before or after plant material is installed. Water elements in the landscape should be completed and their proper operation verified prior to planting. Repairs in pools and fountains require significant disturbance of the surrounding area.

PLANT MATERIAL AND TURF AREAS
This step is reserved until the end of a project to protect sensitive plant material, and because performance of any other task after this step has the potential to seriously damage plant installations and lawns. Plant installation should begin with large trees and plants in areas that are difficult to access. Smaller trees and shrubs should be installed next, followed by edgers and ground cover plantings. Post-planting care, including mulching, should be accomplished when each plant group is installed. Preparation and installation of turf areas typically begins after plants are installed. An alternative practiced by some landscape contractors is to prepare all turf areas prior to planting trees and shrubs. Any damage to turf areas is then repaired after planting is complete.

Timing of plant material installation is important to the survival of the plants and the maintenance load of the owner. Maintenance of all plants and lawns, particularly watering, consumes a great deal of time. Fall installation of plantings and lawns provides the best chance of survival and minimizes maintenance, but this timing does not always coordinate with the completion of building projects. Completion during the winter, spring or summer of the year will require a maintenance program that is instituted immediately to assure plant survival until the owner accepts the project.

CLEANUP
After the landscape installation is complete, it will be necessary to clean and touchup exterior areas where work has been completed. The time investment required to wash walkways, touch up paint on amenities, and make sure light bulbs are working is worth the effort. Cleanup is made difficult when the general contractor requires that landscaping be completed before building exteriors or corrective work can be finished. Therefore, it should be verified that all exterior work on the site is complete before applying final aesthetic treatments.

Chapter 2

Legal Requirements

Many legal controls exist that have the potential to influence landscape construction work. Few activities that the landscape contractor undertakes will be without some form of standard—either voluntary or involuntary. While these standards influence the speed and efficiency of a landscape operation, contractors' attitudes toward controls should be tempered with the knowledge that most legal requirements protect the health, safety, and welfare of the public and the contractor.

This chapter covers typical relationships, rules, contracts, standards, and entities that govern the profession of landscape contracting in many parts of the nation. National standards, or codes, provide guidance in the form of construction standards. Locally, ordinances and building departments provide requirements for location, construction, and, occasionally, style of the landscape improvements that can be installed. Local governments may also require that permits be obtained and fees paid prior to construction. Projects are often guided by contract documents, a legally binding set of instructions that provides control over the materials and methods that may be used. Insurance standards, usually expressed in limitations and restrictions incorporated into policies and contracts, also provide some level of control over landscape work.

It is important that legal requirements be verified prior to beginning a construction project. Failure to do so may mean imposition of fines or removal of parts of a project that do not comply with regulations. To verify that the locale in which you are working has regulations by which the contractor must abide, look in the government section of the phone book for county and city building departments or a zoning administrator. In some areas, look also for planning departments and forestry departments. Describe to the officials the work being done, and they should be able to direct you through the appropriate process. Ignorance of legal requirements will not relieve the contractor of any liability.

> **CAUTION**
> Legal counsel should be obtained for the interpretation of any information described herein.

RELATED INFORMATION IN OTHER CHAPTERS

Information provided in this chapter is supplemented by instructions provided elsewhere in this text. Before undertaking activities described in this chapter, read the related information in the following chapters:

- Interpreting construction documents, Chapter 3
- Estimating and bidding, Chapter 6

CONTRACTOR/CLIENT RELATIONSHIPS

Entering into the world of business requires that the contractor be aware of many methods for obtaining work, understand relationships with clients and related parties, and execute contractual relationships with clients. Because space is not available in this text to explain the detailed nature of every relationship, this chapter provides an overview of the most common relationships. The author suggests supplemental reading from these texts to obtain more information regarding these issues: *Building Contracts for Design and Construction,* Harold Hauf, Wiley-Interscience; and *Landscape Architectural Handbook,* Carpenter, Landscape Architecture Foundation.

Obtaining Landscape Contracting Work

Obtaining work is essential for the contractor to maintain a business. Methods to obtain, or procure, work in the contracting field range from informal marketing efforts to formal subcontracting arrangements with major builders. Several aspects of the contractor's work may influence the success rate in obtaining contracts, including such factors as charges for performing services, how efficiently he or she works, how quickly the services can be provided, experience in performing similar types of projects, and the quality of the contractor's work. Depending on the type of procurement method, one or more of these factors may be critical. Three basic methods for obtaining work in the landscape contracting field can be identified as direct procurement, competitive negotiation, and competitive bidding. Chapter 6 explains specific techniques for calculating and presenting costs required by these methods.

Direct Procurement. Direct procurement involves the acquisition of contracts for landscape services directly from a client. Initiation of direct procurement may be done by the contractor through advertising or marketing of services, by the client through referral from another client, or what the industry terms "walk-in" business. In any instance, the materials and services are sold to the client without direct competition from other parties.

Competitive Negotiation. Many customers prefer to "shop" for landscape services, leading to receipt of proposals from several service providers. This form of procurement often requires detailed explanations and itemization to assure the client is basing the selection of services on similar proposals. Competitive projects may be client-originated or solicited from **general contractors** (a contractor who holds the prime contract for completion of a large construction project) or from **design professionals (landscape architects,** architects, and engineers) who have projects with private clients. Competitive negotiation is a method often used by general contractors to select subcontractors or by specialized contractors who perform specific operations for a project.

Miscommunication is common in this form of obtaining services due to the client's inexperience in comparing the quality of products and services available. The situation is complicated further if the contractor has priced services without the benefit of a

design. Slight differences between interpretations of client needs can lead to substantial differences in price. Negotiation often occurs when the client wants to add or remove elements from the project, begins comparing different design ideas, or places proposing parties in competition with each other on price. The inconsistencies of this process underscore the importance of providing design services as a separate phase from installation services.

Competitive Bidding. Bidding is a formal process practiced by public clients for all but small projects and by private clients for most large projects. Competitive bidding requires that contractors offer, or bid, a sealed project cost based on a specific design. A client will open all proposals at a designated time and place. Selection of a contractor is typically based on the lowest bid submitted by a qualified contractor. Determining a qualified contractor is an activity performed by the design professional or client's counsel. Certain contractors may be disqualified from bidding based on sound legal reasons such as experience, financial insolvency, or government sanction.

Submission of bids is typically restricted to general contractors. In many projects, the landscape contractor will not be the entity submitting the bid. Landscape contractors may submit a bid or negotiated price to a general contractor, who will add that total to the other subcontractor prices and submit it as part of that contractor's bid. It is not unusual for a landscape contractor to submit prices to several different general contractors for the same project. Bidding opportunities can be obtained by reviewing public notices, contacting plan houses (such as Dodge House, a private publisher who tracks major invitations to bid), prime contractors, and design professionals. Prices prepared for bids may take different forms, as described in Chapter 6.

Relationships Between Contractor and Client

When work is secured, the contractor's relationship with the client is defined as being either a general contractor or a subcontractor. Whichever of these positions is held by the contractor will have a significant impact on control of the project. Third parties, such as design professionals acting as owner's agents and consulting construction managers, also impact the relationship between contractor and client (Figure 2–1).

General Contractor Relationships. The general contractor is responsible for completing all aspects of a project. General contractors hold the prime construction contract with the client, who is typically the owner or financier of the project. This relationship is framed by the terms of a contract spelling out the rights and

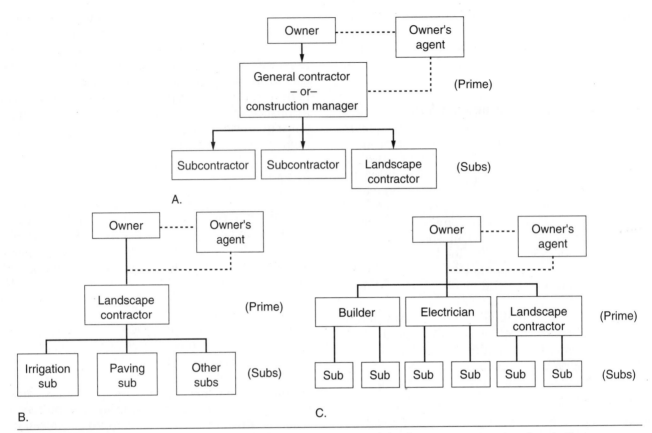

Figure 2–1 Relationship of owner and contractors in the landscape process. A. Landscape contractor as a subcontractor. B. Landscape contractor as a general contractor. C. Landscape contractor as one of several general contractors.

responsibilities of each party. On complex projects, the general contractor may be responsible for numerous suppliers and subcontractors, including those doing the landscape installation, as well as his or her own work force (Figure 2–1A). When subcontractors are used to complete portions of a project, the general contractor assumes responsibility for their work.

An alternative to a general contractor is a construction manager, who performs the responsibilities of the general contractor but has no construction work force. Construction manager duties are typically contracted to companies that specialize in procurement of materials and services, scheduling, budget management, and other construction-related project management aspects. The construction manager holds the prime contract with the project owner and arranges subcontracts and other elements required to complete the project.

In projects that cover only landscape construction elements, the landscape contractor may serve as a general contractor (Figure 2–1B). The landscape contractor is then responsible for completion of all work, including those elements that he or she may subcontract to others.

Another arrangement that is possible under the contractual arrangements for a project is the concept of multiple general contractors (Figure 2–1C). For reasons of timing, cost, or quality control, the client may choose to separate major work areas into separate contracts. Prime contracts will be offered for each of the work areas, with each general contractor being responsible for work within the appropriate trade. Coordination of activities among contracts becomes a primary concern of the general contractors, owner's agent, and other parties to the work.

Subcontractor Relationships. Subcontractors perform specialized work for a project. The subcontractor maintains a formal or informal relationship with the general contractor and has no contractual relationship to the owner or financier of the project. Subcontractors are common on large building projects that involve many building trades. Subcontracting landscape elements of a project is very common. Specialization within the landscape contracting field has also increased the possibility of having several landscape-related subcontractors for a single project, such as those who specialize in irrigation, wall construction, plant installation, and turf establishment.

Third Parties to Contractor/Client Relationships.
Communications between the prime contractor and owner will typically be referred to the owner's agent (Figure 2–1 A,B,C). In many projects, this will be the design professional. The owner's agent will serve as the clearinghouse for all communications with the owner, accepting and approving submissions and providing instructions to the general contractor. Whether landscaping work is performed by a general contractor on a small project, or by a subcontractor on a large project, any change or deviation from the plans and specifications will require approval from the owner's agent.

Forms of Contract

Execution of work in any form should be guided by a written contract. Whether simple or extensive, the contract forms a basis for a relationship between parties and provides a mechanism whereby disputes can be resolved. As mentioned in an earlier caution, seek the advice of appropriate legal counsel before preparing and entering into any contract.

Required Conditions of a Contract. Preparation of contracts of any size will require that certain conditions be contained within the contract. Contracts must contain lawful subject matter, must include an offer and exchange (a fee for services rendered), must be negotiated between competent parties, and must not be signed under duress. Contracts must also include agreement and be in proper form for the jurisdiction in which the work will be performed. In addition, the following items should be included in a contract for landscape services:[1]

- Names and addresses of the parties involved
- Date of contract preparation
- Description of the work to be accomplished (also called the scope of services) and the materials or services to be provided, including the location where it is to be accomplished
- The terms of completion, including such things as inspection, the completion date, and the penalty for not completing on time
- The terms of payment
- The signatures of both parties or their legal representatives
- The date of signing

Contract Documents. The documents used to bid and construct large projects are collectively referred to as **contract documents.** Included in a full set of contract documents for a large project would be instructions for bidding, contracts for construction, general instructions, and specific technical instructions. These preceding elements are collectively referred to

as **specifications.** Also included within the contract document definition would be the set of plans used to guide construction, provisions for change orders if construction is altered, and payment and completion documents to be used during construction. For suggestions on interpreting the drawings and specifications for construction purposes, refer to Chapter 3. All elements of this project documentation become part of a legally binding contract, whether work is performed as a general contractor or a subcontractor.

Subcontractor/Contractor Agreements. The relationship between a subcontractor and general contractor may be in the form of a contract, an informal letter agreement, or, inadvisably, through oral agreement. Formal and informal agreements should address issues such as recognition of the prime contract, work to be performed by the subcontractor, payments to the subcontractor, timing of the work, insurance and bonding required of the subcontractor, special conditions required by the project, and agreement by the parties.

Two-Party Contracts. Smaller projects may proceed without the benefit of an expansive set of contract documents. Contractors may work from written proposals or landscape drawings that have been presented to, and agreed upon by, the client. While it varies according to where the contract is prepared and the wording of the proposal, most written proposals can become legally binding if both parties sign the document. Lacking the type of specific instructions for every element of the project that contract documents provide, a great deal of communication and understanding by the parties is required to avoid disputes.

BUILDING CODES

Codes are typically prepared by national institutes or organizations that have an interest in advancing the safe application of their trade. Two of the most prominent types of building codes are structure building codes and the National Electrical Code (NEC).

Structure Building Codes

When landscape projects involve the construction of any structure to be occupied, the guidelines of the standardized building codes typically come into play. Building codes provide a wealth of information regarding the safe planning of structures. Information regarding material standards, connectors, area planning, and dimensions, and other details of building construction are included. Examples of building codes that are used in various areas of the country include BOCA (Building Officials and Code Administrators, based in Country Club Hills, Illinois), SBCCI (Southern Building Code Congress International, based in Birmingham, Alabama), UBC (Uniform Building Code), and the Standard Building Code. Building codes are adopted on a locality-by-locality basis, so check with the city or county zoning administrator or building office to determine if the codes apply to work being performed.

National Electrical Code (NEC)
Concerns over protection from fire and electrocution have led to development of guidelines found in the NEC. Electrical codes guide the planning and installation of electrical systems, including as safety precautions, circuit loading, wire sizing, connection instructions, material selection, and many other minimum safety standards. Electrical codes are available from building officials or electrical associations.

Related Building Codes
In addition to structure and electrical codes, the Standard Plumbing Code, Uniform Plumbing Code, Standard Gas Code, Uniform Mechanical Code, and BOCA National Plumbing and/or Mechanical Codes may impact the work being performed.

ORDINANCES AND DEED RESTRICTIONS
Local control over construction projects is typically expressed through the use of ordinances that control zoning and building. These regulations place stipulations on the use of land and the construction of improvements on property that lies within a legal **jurisdiction**. In large communities, these regulations can be extensive and complex, and in small communities, they may not exist at all. In absence of an ordinance and permit system, a governmental unit may require adherence to one or more of the codes mentioned in the previous section. Some local government controls of concern are listed in the following section. Contact local building officials, zoning administrators, city foresters, or homeowner association officials to determine if any regulations apply to the work being done.

Zoning Ordinance
Zoning ordinances control how land within a government's jurisdiction may be used. This control may restrict the type of structure one may place on a property, the dimensions of that structure, and the uses of the structure. Also common in a zoning ordinance are requirements for setbacks, or distances between improvements and property lines. Elements of a zon-

ing ordinance that may influence landscape construction include provisions that control fence requirements (such as around a swimming pool) and location, fence height and materials, deck location, deck materials and construction, and railing requirements.

Sign Ordinance
Signage controls may be located within a zoning ordinance, but some locales place regulations on signage in a separate ordinance. Controls on sign use, materials, location, lighting, and size are common in developed areas.

Parking Ordinance
Quantity and location of parking spaces, as well as control of access drives, can sometimes be found in an ordinance separate from the zoning ordinance.

Street Tree Ordinance
Many cities have regulations on what type of plants may be placed in city right-of-way (Figure 2–2). A

Figure 2–2 Street trees along a public boulevard.

street tree ordinance will clarify the species that are acceptable for urban planting and may offer guidelines for plant location.

Deed Restrictions

Planned communities, condominium housing complexes, historic districts, and, occasionally, communities may have restrictions that control what type of exterior improvements can be planned and built. These restrictions are intended to develop uniformity of development or to protect historical authenticity. Review plans with authorities in these situations to verify if construction work must follow established guidelines.

Building Permits and Site Plan Review

To verify compliance with ordinances, counties and communities often require that a building permit be obtained prior to beginning construction. The process of obtaining such a permit may simply involve an official who reviews ideas, or it may require a series of committee meetings in a process called site plan review. Either method will most likely require preparation of a plan that contains descriptions of the work along with elevations and dimensions. Building and zoning officials typically supervise the issuance of building permits and submissions for site plan review.

INSURANCE, BONDING, AND LICENSURE

Insurance

Considering the risk involved in construction, any contractor is advised to obtain necessary insurance to cover the perils faced when building landscapes. Risk in the landscaping industry comes primarily in the form of comprehensive liability from accidental destruction of property, employee accidents, and vehicular accidents. Natural peril coverage and errors in workmanship are also risk areas for which protection may be sought. Other activities that may require insurance protection by the contractor might include umbrella liability policies, equipment and materials coverage, and crime policies.

Contracted projects typically require that insurance be in place before work begins, and it is inadvisable to attempt any project without basic coverage. Seek the help of an insurance advisor when determining the coverage and dollar amounts that are best for a particular situation.

Bonding

Bonding is a surety tool in which a bonding agency guarantees payment in the event a contractor fails to complete his or her legal obligation. Bonding is common in large projects and is required for licensing in some localities. Bonding protects the owner of a project from losses, damages, liens, and unpaid bills on the part of the contractor. Common bonds used in construction are described in the following section. Bonding agents are typically contacted to obtain proper construction bonding.

Bid Bond. The bid bond protects the owner if a contractor withdraws his or her bid after opening. Most formal bids require that a bond in the amount of 5%–10% of the total bid be included with the bid. If the low bidder on a project withdraws, the owner may apply the bond against the expense of rebidding the project.

Performance and Payment Bonds. Successful bidders are often required to submit performance and payment bond(s) prior to beginning construction. Performance bonds provide protection to the owner in the event that the contractor fails to complete the conditions of the construction contract. Payment bonds (also known as labor and material payment bonds) guarantee payment of labor and material bills incurred by the contractor in the construction of the contracted project.

Performance and payment bonds that combine these two sureties are available, but many projects require submission of dual bonds—one covering each risk. Typically, bonding for performance and payment typically is in the amount of 100% of the contract value. If either of these conditions for which the contractor is bonded occurs, the bonding company may be contacted for assistance in completing the project and/or payments.

Contractor Licensing

Several states require that landscape and/or irrigation contractors be licensed in order to practice their trade. By licensing the professions, states can provide some level of protection for the public's health, safety, and welfare by screening out unqualified practitioners. Liscensure procedures vary from state to state, and may include completing legal documents, or exams, or providing evidence of capability to perform work. The bonding of a contractor and verification of insurance may be possible additional qualifications.

To determine whether a business will require licensure, contact the Department of Commerce in the state where practice is planned. Local zoning and building officials or local landscape association professionals may also be able to assist.

If your company is planning to engage in design/build operations, verify licensure requirements for contracting and designing. Many states require the licensure of landscape architects, and depending on the type of law, may restrict the level of design work that may be performed. Landscape architecture licensure laws are either title or practice laws. Title laws prohibit the use of the title *Landscape Architect* in any form by an unlicensed design professional. Practice laws limit the scope of work that may be designed by an unlicensed professional. In some areas, work on public and commercial projects may legally require the design skills of a licensed landscape architect rather than the **landscape designer** or contractor.

TORT ISSUES

A contractor will be required to perform activities within the boundaries of the law. While regions vary with regard to specific legal requirements, there are several general requirements that are similar without regard to locality. Failure to abide by legal requirements resulting in damage to the property of another is considered a **tort** and could lead to payment of damages to the offended party. It is to the contractor's benefit to discuss common legal requirements with industry officials, association members, and insurance and legal advisors before beginning work in the landscaping field. The next section discusses common issues that could lead to legal action.

Trespass

Trespass is an offense committed against the property of another. To be charged with trespass, physical entry without permission and commission of damage are required. Trespassing is not limited to humans physically entering a site and committing damages. Equipment, debris, drainage, or other entities under the contractor's control may enter someone else's property and cause damage.

Nuisance

A disruption that causes discomfort to another is considered a nuisance. Working in an unsafe manner or at odd hours may be considered a nuisance. Excessive noise, dust, or disruption caused by construction activities may also be construed as a nuisance if it meets the legal tests.

Negligence

Failure of a professional to act in a reasonable and prudent manner to safeguard the well-being of others leads to negligence. Failure to mark construction areas, performing overhead work without any safeguards, and other construction activities could lead to legal action if a contractor does not engage in actions or protective measures that would be reasonably expected to protect the public.

Riparian Rights

Subject to state and federal laws, landowners have rights regarding access to and use of waterways that run through or abut their properties. Changing the conditions of that waterway, possibly through irrigation or damming, may create a legal violation of those laws.

Lateral and Subjacent Rights

Construction activities that affect the lateral support of neighboring properties may lead to legal action. Landowners are entitled to maintenance of the grade at their property line. For example, if excavation for a parking lot collapses the neighboring grade, lateral support has been denied. Without prior permission, grade change activities must be completed within the boundaries of a project.

ENDNOTES

1. Ingels, *Landscaping Principles and Practices,* p 237.

Chapter 3

Interpreting Construction Documents

Large and/or complex projects require detailed instructions explaining how the project is to be built. These instructions, called plans and **specifications** (this term is often shortened to "specs"), are prepared by design professionals such as landscape architects, architects, and engineers. Plans and specifications provide the information necessary to assure that the project is constructed exactly as the owner desires.

This chapter explains the types of construction documents that may be encountered on a project and explains how to interpret the content of key documents. Each plan and specification contains a different type of information, but they all work in concert to provide necessary information. Even if bidding only landscape work, the landscape contractor should obtain a complete set of construction drawings and specifications at the beginning of the project to verify that other project activities will not change or influence the bid. Subcontractors often receive only their section of the plans and specs and end up in conflict because they were required to perform work that was described in other sections.

RELATED INFORMATION IN OTHER CHAPTERS

Information provided in this chapter is supplemented by instructions provided elsewhere in this text. Before undertaking activities described in this chapter, read the related information in the following chapters:

- Legal requirements, Chapter 2
- Project pricing, Chapter 6
- Construction staking, Chapter 8

READING SPECIFICATIONS

Specifications provide written instructions about most aspects of project construction. To keep these instructions organized, specifications are typically divided into sections covering nontechnical aspects of the work, often called general conditions, and sections covering specific construction procedures, often called technical specifications.

General Conditions

General project information is located in the front of the spec manual and includes a wide range of documents. Based on the scope of the project, this information can be quite detailed and extensive. The information contained in these sections pertains to all work, and contractors should familiarize themselves with the details before beginning construction. Special requirements due to laws and regulations should be explained in this part of the specs. Typical documents found in the General Conditions section include:

1. Bidding information
   - Invitation to bidders
   - Instructions for bidders
   - Bid form
   - Noncollusion affidavit
   - Bid bond form
   - Receipt of addenda acknowledgment
2. Contract and other forms to be submitted
3. Forms used during construction
   - Notice to proceed
   - Change order form
   - Certificate of payment
   - Certificate of substantial completion
   - Lien releases

4. General conditions
   - Definitions of roles and responsibilities for owner, design professional, contractor, and subcontractors
   - Clauses addressing contract time, payment, completion, retention of fees, changes in work, correction of deficient work, bonding, insurance, royalty payments, related contracts, protection of the work, resolution of disputes, termination of the contract, and miscellaneous provisions
5. Supplemental conditions
   - Government requirements that will govern work, such as anti-kickback regulations, prevailing wage rates, adherence to civil rights, nondiscrimination, and labor legislation
   - Warranties and guarantees
   - Subcontractor approvals
   - Temporary fixtures and utilities
   - Request for material submissions and approval of substitution requirements
   - Other requirements relating specifically to the project

Technical Specifications

Technical specifications include descriptions of the methods, materials, and performance of components of the project. A variation of technical specifications, typically more common in nonlandscape trade areas, may specify a particular product name or specify how the installation should perform after completion.

To keep the instructions manageable, technical specs are typically grouped according to trade. In large projects, there could be several sections that contain building specifications that have limited impact on the landscaping work. Each professional will use an organizational style that he or she is familiar with, but the standard is a numbering system known as Masterspec, created by the Construction Specification Institute (CSI). Under this system, the work for a project is divided into sixteen sections. Landscaping and grading can usually be found under Section 2 (sometimes labeled 200 or 2000), concrete paving under Section 3, and lighting under Section 16.

Two basic types of technical specifications are typical to the construction industry. Material and workmanship specifications and performance specifications vary widely in their approach to directing the type of construction used on a project. The wording and manner of the technical specification influence the choices a contractor must make.

Material and Workmanship Specifications. Material and workmanship specifications detail the materials and construction procedures to be used to complete each component of a project. This type of specification restricts the contractor to use of only those elements chosen by the designer. A typical format for material and workmanship specification begins with a section covering aspects such as warranties, material delivery, and reference standards. A second part indicates what materials are to be used for the project. This description can be written as a closed or an open specification.

- Closed Specifications. Closed specifications limit the contractor's choice of methods and materials. Choices may be limited to a single product brand/model (proprietary specification), or a specific procedure to be used. When closed specifications are used, substitutions are uncommon.
- Open Specifications. Specifications written as open provide a method for the contractor to choose materials and the method to be used. Open specifications may list two to three alternative brands/models, or list a brand/model and the clause "or equal." The "or equal" clause allows the contractor to propose brands/models to the owner's agent for approval.

The last part of this type of specification describes the detailed installation methods to be used. An example of a material and workmanship specification is presented in Appendix A.

Performance Specification. Performance specifications prescribe the results that a component of construction must obtain without providing specific materials or methods. This type of specification limits the contractor only in producing a product that looks and performs as the designer intended. An example of a performance specification in landscaping is presented in Appendix A.

Reference Standards. Throughout specifications and construction material literature, references can be found to standards published by testing agencies, associations, and organizations. Design professionals often refer to these specifications and incorporate their parameters as part of legally binding contracts. These organizations spend considerable time preparing standards for the performance and installation of materials, many of which are related to the landscaping industry. A common reference is the American Society for Testing Materials, which prepares standards used throughout all industries. References relating to landscape trade areas include American Association of Nurserymen, American Standards for Nursery Stock (ANSI), Interlocking Concrete Paving Institute

(ICPI), Brick Institute of America, Portland Cement Association, National Electrical Manufacturers Association (NEMA), Asphalt Institute, American Concrete Institute (ACI), California Redwood Association (CWA), and American Institute of Timber Construction. When reference is made to one of these standards, contractors are responsible for obtaining the relevant information and verifying that their work conforms to the standard.

Additional Specification Documents

In addition to the written instructions provided by the specifications, other written instructions may be provided during the construction process. Two key documents that affect the process are Addenda and Change Orders. Addenda are official notifications of changes in the project, typically revisions of the construction plans or specifications, which occur before the time of bid opening. Addenda modify the project, and these modifications should be reflected in any bid submitted. Change Orders are revisions of the project that occur after the contracts have been signed. Change Orders are initiated by the owner's agent and require submission of price and/or project scope changes by the contractor.

READING CONSTRUCTION DRAWINGS

Plans are the graphic communication tools that provide information regarding location, grade, and type of improvements to be placed on the site. To interpret construction plans, a basic understanding of the elements of engineering drafting is necessary. Following are descriptions for reading drawing scales, dimensions, abbreviations, symbols, and textures. This presentation is an overview and should not be interpreted as a comprehensive explanation of all aspects of engineering graphics. Interpretation of specific plans prepared for landscape construction follows the information regarding reading drawings.

Drawing Types

Three types of drawings are typically used to show improvements for a project—plan view, cross sections, and elevations. Plan view is the primary drawing type used for landscape improvements. Plans show a view looking straight down on the top of a project. Only two dimensions of site elements are visible in a plan view—the length and the width. Cross sections show a side view that cuts vertically through an object, while an elevation is a side view that shows the vertical surface of the object. Height and either length or width are the only two dimensions of site elements visible in cross sections and elevations. Cross sections

and elevations used for landscape elements are typically used to show details of installations.

Drawing Scale

To assist in the interpretation of construction plans, a discussion of the concept of **scale** should precede plan reading. Since most drawings cannot be drawn life size, the site must be reduced to fit on a sheet of paper that is convenient to carry and read. To reduce the size in a manner that is consistent and measurable for all drawings, the concept of scale is used. Simply put, scale means that a certain measurement on a drawing is equal to a distance on the actual site. Whenever that scaled distance is measured on a plan, the same actual distance should be obtainable on the site.

Scale in landscape construction drawings is typically a reference of 1 inch on the plan equaling a certain number of feet in real life. Choice of scale depends on the size of the object/site being measured or the stated scale on a drawing. An example would be a scale stated as 1 inch = 20 feet, where 1 inch on a construction drawing equals 20 feet on the actual site. Two types of scales are used for construction drawings—an architect's scale using fractional divisions, and an engineer's scale using multiples of ten (Figure 3–1).

Special instruments, conveniently termed *scales*, are available for the measuring of plans. Rather than having to constantly measure the number of inches and multiply by a conversion number, scales indicate the number of actual feet on the instrument. While the reading of scales, particularly the architect's scales, takes some training and practice, they are useful tools for interpreting construction drawings. Landscape drawings and details relating to landscape work can be drawn in either type of scale. On each drawing, the reader should be able to locate a written scale and a graphic, or bar, scale.

Measuring lines with the architect's scale first requires finding the scale that matches the drawing being interpreted. After locating the scale, identify the line to be measured. Place the scale on the line, with one end of the line in the inch area of the scale (this is the finely divided area behind the zero). Adjust the scale left or right until one end of the line remains in the inch area and the other end rests directly on a foot mark. Read the foot reading and inch reading to obtain the measurement. Use caution when reading both the foot and inch readings on the architect's scale. Architect's scales have two scales running in opposite directions along the same edge of the scale—one reading the top set of numbers and the other reading the bottom set of numbers. Inch markings on the architect's scale may represent inches or

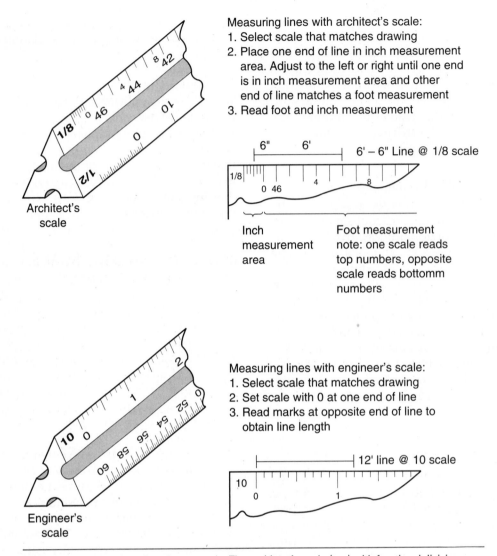

Measuring lines with architect's scale:
1. Select scale that matches drawing
2. Place one end of line in inch measurement area. Adjust to the left or right until one end is in inch measurement area and other end of line matches a foot measurement
3. Read foot and inch measurement

6' – 6" Line @ 1/8 scale

Inch measurement area

Foot measurement note: one scale reads top numbers, opposite scale reads bottomm numbers

Measuring lines with engineer's scale:
1. Select scale that matches drawing
2. Set scale with 0 at one end of line
3. Read marks at opposite end of line to obtain line length

12' line @ 10 scale

Architect's scale

Engineer's scale

Figure 3–1 Architect's and engineers scale. The architect's scale (top) with fractional divisions, and an engineer's scale (bottom) with divisions in multiples of 10.

fractions of inches, depending on the accuracy of the scale.

Measuring lines with engineer's scales also requires determining drawing scale before measurement. Locate the line to be measured and place the zero mark from the correct scale at one end of the line. Read the markings along the scale to determine line length. Each mark on an engineer's scale represents 1 foot. Helpful numbers are placed to identify increments along the scale.

A warning is in order when utilizing scales to obtain measurements from construction plans. Because the paper on which a drawing is printed can stretch and shrink, use of a scale to determine the dimension of objects shown on a construction plan is not recommended when dimensions are provided. Look for a labeled dimension, since labeled dimensions will always take precedence over scaled measurements. Plans that show plant locations may rely on field staking by the design professional or interpretation of the dimensions by the contractor to accurately locate plants. In addition, plant spacing may be noted elsewhere on the plan. In the absence of labeled dimensions, call the designer for a written verification of missing dimensions or for instructions on interpreting site element locations.

Line Types

Construction drawings utilize a hierarchy of line weights to identify the importance of drawing elements. Lines that are the darkest on the plans typically indicate objects that are to be constructed. A second

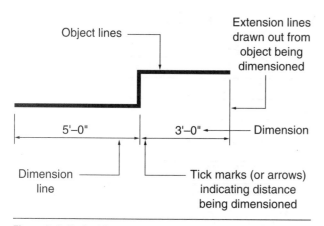

Figure 3–2 Typical line types on construction graphics.

tier of line weights is reserved for dimensioning objects (Figure 3–2). The variation in line weight is to improve readability of the plan and to underscore the importance of drawn elements. Dashed lines are often used to indicate existing contours, subsurface elements, hidden elements, or overhangs.

Dimensioning

Items shown on plans for which location is critical will have dimensions shown using special lines and labels (Figure 3–2). A typical dimension is located on a line that is slightly separated from and parallel to the object being dimensioned. The object being dimensioned and the dimension line are connected by extension lines that are drawn perpendicular to the object being dimensioned. The extension lines should line up with each end of the object being measured. In

instances where several short measurements are provided for an object, there may be an overall dimension placed outside the short dimensions.

Abbreviations, Symbols, and Textures Used on Construction Plans

Adding to the difficulty of interpreting construction plans is the variety of abbreviations (Table 3–1), symbols (Figure 3–3), and textures (Figure 3–4) used to represent construction measures and materials. Many of these items are standardized throughout all construction industries.

INTERPRETING LANDSCAPE CONSTRUCTION PLANS

Typical landscape construction plans are based on divisions of the work. Without experience in working with construction plans, there is the possibility of misinterpretation of the information presented by the documents. Included with each description are the types of data included on each plan, an explanation of how the data is graphically presented, and recommendations for interpreting plan information.

Site Preparation (Sometimes Identified as Site Demolition)

Site preparation plans show which parts of the existing site are to be removed and which are to be protected. Items that are to be protected or removed are indicated by notations and textures, and boundaries and work areas are identified by symbols and lines. A majority of the work on this plan may be done by

Table 3–1 Typical Construction Drawing Abbreviations

B&B	Balled and burlap	FG	Finished gradee	Nom	Nominal dimension
BC	Bottom of curb	FH	Fire hydrant	OC	On center
BM	Benchmark	FL	Flow line	OD	Outside diameter
BS	Bottom of slope	Ftg	Footing	PCP	Porous concrete pipe
Cal	Caliper	FT	Foot or feet	ROW	Right-of-way
CB	Catch basin	Ga	Gauge	SI	Storm inlet
CL	Center line	Gal	Gallon	SF	Square feet
CLF	Chain link fence	HB	Hose bibb	SY	Square yards
CMP	Corrugated metal pipe	HP	High point	TC	Top of curb
Conc	Concrete	ID	Inside diameter	TCP	Terra-cotta pipe
CF	Cubic feet	Inv	Invert	TW	Top of wall
CY	Cubic yards	LP	Low point	TS	Top of slope
Dim	Dimension	Lin	Linear	Typ	Typical
Dia	Diameter	LF	Linear feet	Var	Varies or variable
El	Elevation	Max	Maximum	VCP	Vitrified clay pipe
Exist	Existing	MH	Manhole	WWM	Welded wire mesh
Exp	Expansion	NIC	Not in contract	YD	Yard drain
FFE	Finish floor elevation	NTS	Not to scale		

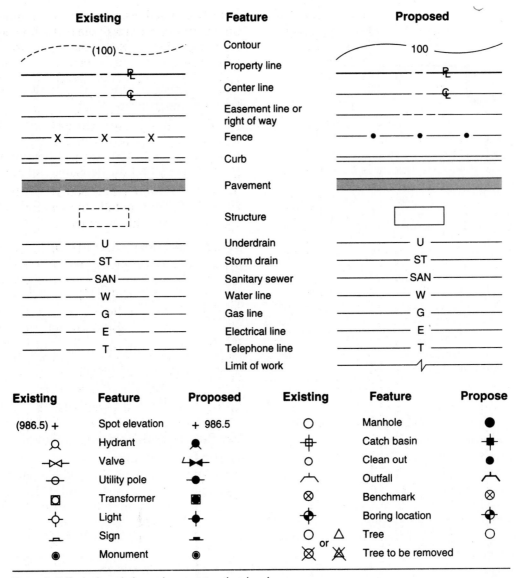

Existing	Feature	Proposed
(100)	Contour	100
	Property line	
	Center line	
	Easement line or right of way	
—X—X—X—	Fence	—•—•—•—
	Curb	
	Pavement	
	Structure	
—U—	Underdrain	—U—
—ST—	Storm drain	—ST—
—SAN—	Sanitary sewer	—SAN—
—W—	Water line	—W—
—G—	Gas line	—G—
—E—	Electrical line	—E—
—T—	Telephone line	—T—
	Limit of work	

Existing	Feature	Proposed	Existing	Feature	Propose
(986.5) +	Spot elevation	+ 986.5	○	Manhole	●
	Hydrant			Catch basin	
	Valve		○	Clean out	●
	Utility pole			Outfall	
	Transformer		⊗	Benchmark	⊗
	Light			Boring location	
	Sign		○ or △	Tree	○
	Monument		⊗ or ⊠	Tree to be removed	

Figure 3–3 Typical symbols used on construction drawings.

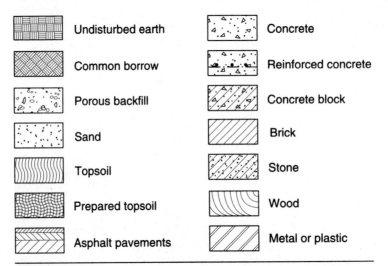

	Undisturbed earth			Concrete
	Common borrow			Reinforced concrete
	Porous backfill			Concrete block
	Sand			Brick
	Topsoil			Stone
	Prepared topsoil			Wood
	Asphalt pavements			Metal or plastic

Figure 3–4 Typical plan and section/elevation textures used on construction drawings.

other contractors, but occasionally the protection of existing site elements, particularly plant material, may be the responsibility of the landscape contractor.

To interpret the information on a site preparation plan, follow these steps:

- Obtain and read all plans and specifications related to site preparation for the project.
- Review the symbols, textures, and abbreviation used on the site preparation plan and locate and read any special notes placed on the plan.
- Become familiar with the drawing layout and locate major improvements such as buildings, roadways, parking lots, and natural features.
- Identify project boundaries and limits of work.
- Identify existing site elements to be preserved.
- Read notations regarding site elements to be removed.

Site Layout (Construction Staking)

Site construction staking plans show the horizontal location of all proposed improvements. Dimensions of walks, patios, drives, and any other landscape elements for which location is critical, are indicated on this plan. Design professionals use several formats for locating objects, including grids, **baselines,** and object dimensions. Grid format uses dimensions referenced

from two baselines. Measurements are made down both baselines from a single beginning point to locate an object (Figure 3–5). Baseline format measures down a single reference line to provide one dimension for a point, and makes a measurement at a right angle to that reference line to obtain the second dimension for a point. Using this format of two measurements, any point on a project can be located (Figure 3–6). Object dimensioning format uses measurements along an existing object to locate new points. Buildings, roadways, or any other existing objects can be used to dimension to new points (Figure 3–7). Dimensions should be carefully marked on a plan to avoid confusion with other numbers and lines.

Detail Keys. Details are identified on construction plans using a symbol termed a *detail key.* Although typically located on the construction staking plan, detail keys can possibly be found on any of the plans. The detail key indicates the sheet number on which the detail is located and the detail number (Figure 3–8). Two numbers are placed inside a circle, with the detail number located on top and the number of the sheet where the detail is located on the bottom. A line that divides the circle will have an arrow, indicating which way the detail is facing.

To interpret the information on a construction staking plan, follow these steps:

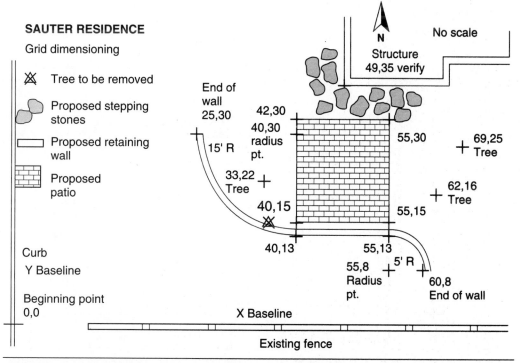

Figure 3–5 Layout plan using grid dimensioning. All points are located using X and Y coordinates laid out in reference to the beginning point.

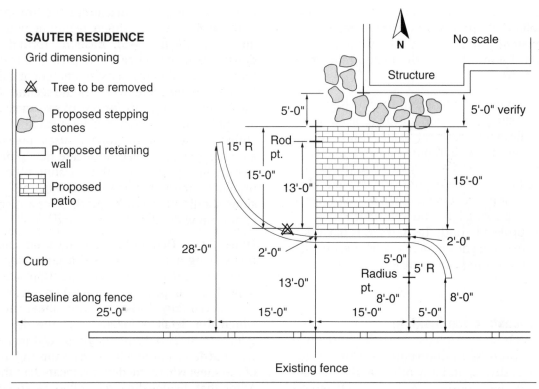

SAUTER RESIDENCE

Grid dimensioning

⊠ Tree to be removed

Proposed stepping stones

Proposed retaining wall

Proposed patio

Curb

Baseline along fence

Figure 3–6 Layout plan using baseline dimensions. Points are located measuring off a project baseline.

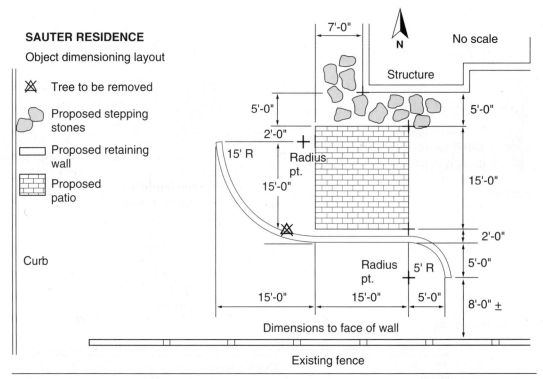

SAUTER RESIDENCE

Object dimensioning layout

⊠ Tree to be removed

Proposed stepping stones

Proposed retaining wall

Proposed patio

Curb

Dimensions to face of wall

Figure 3–7 Layout plan using object dimensioning. Project elements are located by measuring off other nearby objects which have already been located.

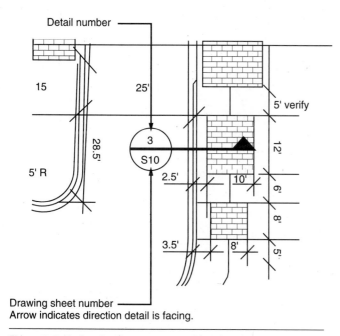

Figure 3–8 Keying details on a construction drawing.

- Obtain and read all plans and specifications related to construction staking for the project.
- Review the symbols, textures, and abbreviations used on the construction staking plan and locate and read any special notes placed on the plan.
- Become familiar with the drawing layout and locate major improvements such as buildings, roadways, parking lots, and natural features.
- Determine the layout method used.
- Find the reference point or beginning point used for layout.
- Read and mark overall dimensions for improvements.
- Locate specific dimensions for improvement.
- Identify which improvements have details, and use the keys to locate the details on the appropriate sheet.

Grading

Grading plans show the vertical location of all proposed improvements. Grading plans may also include information about water runoff, storm sewer construction, and storm water retention facilities. A key element of a grading plan is the **benchmark** for the site. The benchmark is a set elevation that serves as a reference for all existing and proposed grades for the site. Benchmarks should be located at a permanent location and should be easily accessible from the site. Using **contour lines** and **spot elevations,** the

grading plan indicates the desired height of site elements and direction of slope. All contours and spot elevations for a grading plan should be set in reference to the benchmark.

Contours are lines that connect points of similar elevation. Contours are useful for indicating general slope direction and steepness, but are not sufficient to install improvements at specific grades. Contours are read by locating the elevation number printed in or directly on top of the contour line. Larger numbers indicate elevations that are higher than contours with lower numbers. Grading plans should include existing contours shown as dashed lines and proposed contours shown as solid lines. The closer contour lines are spaced, the steeper the slope. Contours spaced farther apart indicate a shallow, or more level, slope. Landforms used in landscaping, such as **berms** and drainage swales, have a distinct contour signature. When grading a site, the finish grade must match the direction, slope, and elevation indicated by the contours (Figure 3–9).

Spot elevations indicate elevations of specific points in reference to the benchmark. Permanent improvements such as structures, footings, paving, and walls require specific elevations to build the improvement. The spot elevations provide that data (Figure 3–9). Spot elevations typically include a notation that indicates if the elevation is for the top, bottom, or some other part of the object being measured. Indications of direction and degree of slope can be obtained by calculating the difference between spot elevations.

To interpret the information on a grading plan, (Figure 3–10), follow these steps:

- Obtain and read all plans and specifications related to grading plan for the project.
- Review the symbols, textures, and abbreviations used on the site grading plan and locate and read any special notes placed on the plan.
- Become familiar with the drawing layout and locate major improvements such as buildings, roadways, parking lots, and natural features.
- Locate benchmarks and the range of elevations (high and low points) of the site.
- Identify major (multiples of five) existing contours and analyze the current "lay of the land."
- Identify major proposed contours to determine the proposed changes and areas where significant amounts of cutting or removal of earth, and filling or adding of earth, will be required (color-coding cut and fill areas can be helpful).

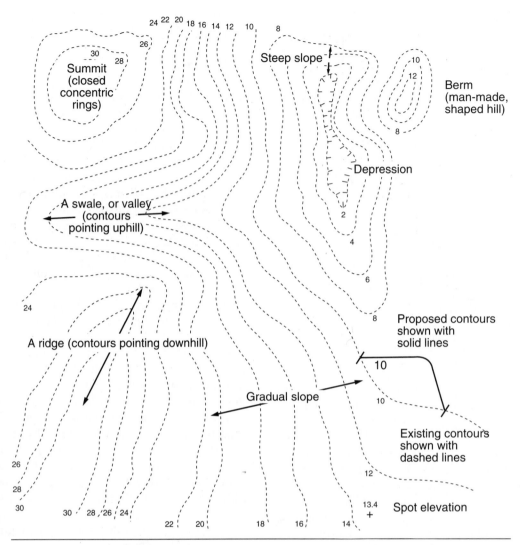

Figure 3–9 Typical topographic landforms and notations.

- Identify major proposed landforms (berms and drainage swales) by their contour signature.
- Locate proposed spot elevations of major structures and compare to existing grades to determine amounts of cut or fill required.
- Locate proposed spot elevations of other site improvements to determine if cut or fill is required.
- Look for drainage arrows and symbols indicating which direction walks, paved areas, and open areas are to be sloped.

Utility

Utility plans show location, type, size, and depth of proposed utilities. Many of these utilities will be installed by other contractors, but irrigation work may be included in landscape contracts. A separate sched-

ule on this sheet may contain relevant data regarding fixtures, such as pipe sizes and irrigation head model numbers. Irrigation installations are typically indicated by lines on the drawings with notations regarding the type, depth, and size of the utility.

To interpret the information on a utility plan follow these steps:

- Obtain and read all plans and specifications related to a utility for the project.
- Review the symbols, textures, and abbreviations used on the utility plan and locate and read any special notes placed on the plan.
- Become familiar with the drawing layout and locate major improvements such as buildings, roadways, parking lots, and natural features.
- Find the type, location, and size of utility lines.

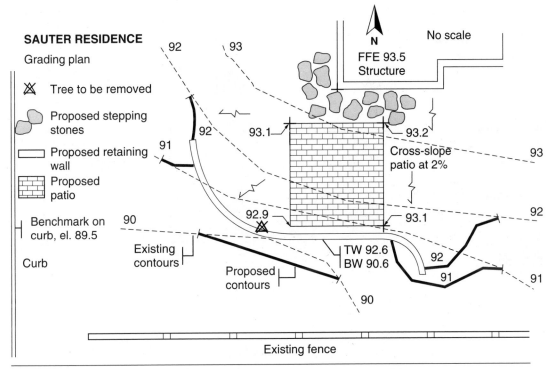

Figure 3–10 Grading plan showing spot elevations and existing and proposed contours.

- Determine depth of utility lines.
- Locate major utility structures, such as drainage inlets, manholes, utility poles, light footings, and note their grade (cross-reference the grading plan if necessary).

Planting

Planting plans indicate the location and type of plant material to be installed. Individual trees and shrubs are shown with circles indicating the approximate spread of the plant, and location of the center of each plant is indicated by an *X* or similar type of mark. The centers of identical plant materials are connected by a line, or leader, which will include a label of the type and number of plants in that "string"(Figure 3–11). Labels may include a key that refers to the **plant schedule.** Some labels may utilize multiple naming schemes, such as the plant's common name, or the plant's botanical name. Occasionally, the designer may include on the leader planting information that indicates spacing, planting elevation, or other pertinent data. It is common for a planting plan to include several of these leaders if there are several plants to label. Alternative methods used to label plants are to place a letter next to the planting location for each plant or to utilize a different symbol for each different type of plant. Both the letter and the symbol should

refer to a plant schedule for plant type, size, and details.

Masses of smaller plants, such as ground covers, may be shown using line or dot textures covering the area where plants are to be installed. Leaders for masses of such plants typically include the plant name, quantity of plants in the area labeled, and the spacing of individual plants.

Plant Schedule. Included with the planting plan is a plant schedule, which lists all proposed plants with their quantity, size, condition, and special instructions for planting (Figure 3–12). The detail from this schedule should be complete enough to allow ordering of all plant material to be used on the project. Notations may also be included on the planting plan that indicate the amount and type of mulch to be used, spacing of plants, staking or wrapping requirements, and other plant-related instructions. Written technical specifications typically provide other details regarding planting.

To interpret the information on a planting plan, follow these steps:

- Obtain and read all plans and specifications related to planting plan for the project.
- Review the symbols, textures, and abbreviations used on the planting plan and

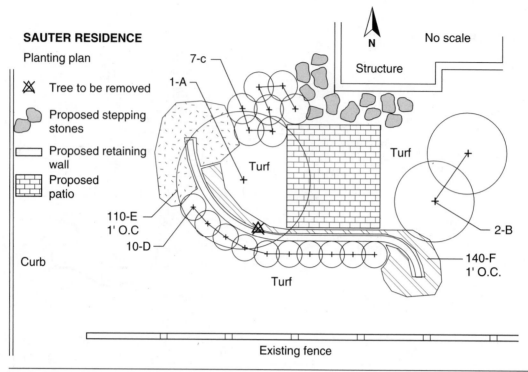

Figure 3–11 Planting plan.

SAUTER RESIDENCE					
PLANT SCHEDULE					
KEY	COMMON NAME	BOTANICAL NAME	QTY.	SIZE/COND.	REMARKS
A	Burr Oak	*Quercus macrocarpa*	1	2″ B&B	specimen quality
B	Shadblow Serviceberery	*Amalenchier canadensis*	2	8′ cont.	
C	Dwarf Compact Yews	*Taxus cuspidata "nana"*	7	7 gal.	
D	Cranberry Cotoneaster	*Cotoneaster apiculata*	10	5 gal.	
E	Purpleleaf Wintercreeper	*Euonymus fortuneii coloratus*	110	flats	space 1′ on center
F	Daylillies	*Hemerocallus sp.*	140`	1 gal.	space 1′ on center
					species selected by owner

Figure 3–12 Plant schedule.

locate and read any special notes placed on the plan.

- Become familiar with the drawing layout and locate major improvements such as buildings, roadways, parking lots, and natural features.
- Review the plant schedule for special symbols and keys.
- Locate each type of plant material in the order indicated on the plant schedule.
- Identify all plant locations, review the species, quantity, and spacing (if plants are grouped together).

- Note for each plant location the elevation, nearby utilities, and other site improvements in the vicinity.

Construction Details

Construction details provide information about specialized elements that need to be built for the project. Items that are usually detailed include bridges, specialty paving, amenities, stairs, walls, lighting, and other construction items that benefit from being shown in greater detail. Details may show cross sections, elevations, or large-scale plans that illustrate

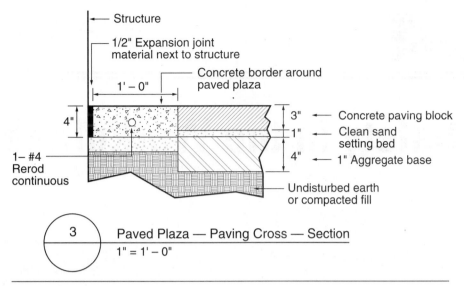

Structure

1/2" Expansion joint
material next to structure

Concrete border around
paved plaza

1' – 0"

4"

3" ← Concrete paving block

1" ← Clean sand
setting bed

1– #4
Rerod
continuous

4" ← 1" Aggregate base

Undisturbed earth
or compacted fill

3   Paved Plaza — Paving Cross — Section
1" = 1' – 0"

Figure 3–13 Construction detail.

more detail of specialty construction (Figure 3–13). Materials, dimensions, and notes are included to aid in proper construction. Scale for each individual detail is noted below the detail. Detail references, or keys, are typically placed on site construction staking plans.

To interpret the information on a site detail, follow these steps:

- Obtain and read all plans and specifications related to the area of work for the detail.
- Locate the detail on the appropriate detail sheet.
- Identify the type of drawing used for the detail (section, elevation, plan).
- Identify the scale of the drawing.
- Analyze the detail to determine the dimension of the materials used in the improvement.

Chapter 4

Safety in the Workplace

Like most construction occupations, landscape contracting carries with it the risk of disabling injury, serious injury, or death if the worker is unaware of the hazards and proper techniques for reducing potential accidents. The many benefits and rewards of landscaping can be quickly erased by a serious or debilitating injury. Fortunately, most of the hazards of the landscape workplace are identifiable. In this chapter, many common landscape workplace hazards and strategies for reducing risk are identified.

> **CAUTION**
>
> This textbook is intended to supplement supervised classroom instruction. Practice of the profession of landscape construction is filled with safety and health hazards, and use of this book without proper experience or supervision will place the reader at risk. *The reader assumes all risk when engaging in unsupervised activities.*

ACCIDENT PREVENTION

Prevention of accidents can begin by providing required information to workers regarding hazards in the workplace. Implementation of a safety training program will assist in improving awareness of work hazards for new and existing employees. Employees should be properly trained before operating equipment or engaging in new construction activities. Signage and postings that warn of health hazards should be present in all areas that workers frequent, including vehicles and structures. Be certain that warnings and instructions utilized for safety are placed in a highly visible space and are presented in language(s) understandable to all employees. Material Safety and Data Sheet (MSDS) sheets should be read when first received and periodically reviewed and updated.

Material Safety and Data Sheets (MSDS)

Businesses are required to keep on file in the workplace a Material Safety and Data Sheet for every material that poses a potential risk to employees and others who may come in contact with such materials. These sheets should be kept in a file in a vehicle at each job site, and in each shop and structure utilized for the business.

First Aid

Employees who are required to work in a potentially hazardous environment benefit from basic first aid and CPR training. Placement of first aid kits in vehicles and structures used in the business, and providing means to communicate problems with appropriate officials, will aid in addressing emergencies. In addition, knowledge of treatment facility locations by all employees, and identifying directions to these facilities in vehicles and structures, are beneficial.

Medical Conditions

Employers and employees should be familiar with chronic health problems that may become inflamed by work conditions. Allergies and asthma attacks may be triggered by dust or pollen from work sites, and treatment steps for such situations should be reviewed prior to beginning work. Current boosters for tetanus should be obtained prior to performing landscape construction activities.

PERSONAL INJURY REDUCTION

The most important tool in the contractor's repertoire is the human body. Protecting the body from injury is necessary to assure the worker can remain productive throughout his or her career. Following are several potential challenges facing landscape contractors in completing the tasks they are asked to undertake, and recommended methods to counter these threats to safety.

Use of Proper Clothing and Safety Equipment

Protection of the body is a logical starting place when attempting to reduce personal injury. Essential equipment includes proper work boots with steel toe protection, heavy-duty work gloves, and eye protection when using equipment or engaged in activities that create hazardous conditions. In addition, contractors should consider use of hard hats for overhead work, appropriate back support for lifting activities, and skin protection when exposure to the sun is long-term. Dust masks and ear protection should also be considered when performing landscape construction tasks that involve constant noise and/or dust. Kneepads provide protection when performing tasks that require long periods of kneeling.

Heat and Cold Injury

Since landscaping is typically considered fair-weather work, little consideration is given to protection from heat and cold injury. While the likelihood of cold injury is lower, proper clothing and protection are required to prevent hypothermia and frostbite when working in extreme temperatures. Threats from heat injury should be more obvious to the landscape contractor. In addition to protection of skin from burn and excessive exposure, caution should be used to protect from heat exhaustion and heat stroke when activities are planned for very hot days. Plan appropriate work hours and ample breaks, and make sure that water is available when temperatures climb.

Awareness of Wildlife

Working in the outdoor environment places the worker at potential risk from wildlife. Common problems from wildlife include threats from smaller animals such as bees, wasps, snakes, and scorpions, and possible threats from hostile, larger animals. Use cau-

tion when performing site preparation tasks that require the removal of potential animal habitat. Old structures, brush piles, dead trees, shrubs, wells, ravines, and other secluded areas can create harbors for potential deadly animals. Review the site carefully before beginning work in an attempt to locate animal habitats. Carry insect spray to ward off any insect attacks that are provoked. Examine equipment, tools, and clothing left at the site to determine that wildlife has not adopted the items for housing while left unattended.

If wildlife habitat is discovered, threatening or not, consider contacting local animal shelters or protection centers. Some organizations are willing to relocate animals to new habitat before problems arise.

Hazardous Plants

Workers should be aware of any plants may trigger dermatitis when contacted. Poisonous plants should be removed or avoided, or proper protective clothing should be worn to avoid direct contact with skin. With certain plants, contact should be avoided with all parts of the plant. Other plants, such as poison ivy, are caustic when burned. Obtain positive identification of all suspected plants and take appropriate precautions.

Back Injury

The work of landscaping has unavoidable risk built into the actions and motions required to complete every task. Improper lifting of heavy objects produces strain on muscle groups that were not intended to lift such weights. Repetitive improper lifting of light objects creates strain that can accumulate, even in muscle groups intended to lift such weights. Too often, workers are tempted to bend over and pick up materials and tools that they need for work. After performing such a motion numerous times without injury, there is little thought about potential risk. Placing undue strain on back muscles over time can eventually lead to injury. Learning how to lift and following simple rules will assist in reducing back injury from landscape construction activities. Performing simple stretching exercises before beginning work and following long periods of repetitive motion is also beneficial (Figure 4–1).

Proper lifting technique is designed to direct the strain of lifting to the proper muscle groups:

- Lift by bending at the knees rather than bending at the waist. This transfers the stress to the thigh muscles, which are better adapted to lifting such weight (Figure 4–2).

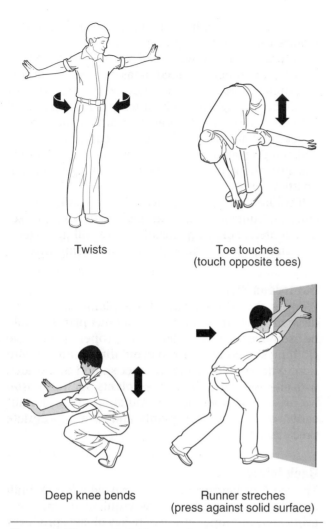

Twists

Toe touches
(touch opposite toes)

Deep knee bends

Runner streches
(press against solid surface)

Figure 4–1 Warm-up stretching exercises.

Figure 4–2 Proper lifting by bending at the knees.

- Avoid twisting motions while lifting. Move the feet rather than twist.
- Hold the load close to the body when lifting. Holding the load at arms' length greatly increases the stress on the back.

Other techniques that will help reduce back strain are:

- Get help when lifting heavy objects.
- Wear back support braces approved by the Occupational Safety and Health Administration (OSHA) and appropriate health organizations.
- Warm up properly and rest when engaging in **repetitive motions** such as laying brick or planting ground cover.
- Use labor-saving devices to lift when available.

Extended sitting and kneeling can also create potential for muscle injury. Interrupt periods of sitting

that are longer than 20 minutes with a stretching period. Utilize proper seating and cushioning for back and knees when performing tasks that require long-term sitting and kneeling. Additional suggestions on preventing back injuries are available from injury prevention and rehabilitation organizations.

WORKING AROUND UTILITIES

Cutting through or running into utility lines that serve a site introduces several potential risks to the landscape contractor. Personal injury or death can occur from electrocution, explosion, or fire if utility lines are disrupted. Financial risk is present in the cost of repairing damaged utilities. In addition, the project may suffer immediately or in the future from utility damage and activities required to repair damage. Particular care should be taken when working with electrical, gas, and fiber optic utilities. These pose the greatest risk to the contractor if damaged during construction.

Public Utilities

Risk can be substantially reduced by calling for utility locates prior to beginning your work. Any time plans require disturbing a site, utility companies should be contacted prior to construction to locate their utility lines. Most utilities require 48 hours advance notice for locates, and do not charge for locating services. To coordinate the task of locating utilities, many areas of the country have instituted the ONE-CALL service. ONE-CALL is a number that provides a centralized locate service for a portion of the utilities located on a site. ONE-CALL will *not* contact all utilities that are potentially located on a site. You will still be responsible for verifying which utilities are not contacted by ONE-CALL. To determine which ONE-CALL number is used in the area where you are working, call the nationwide directory at 1-888-258-0808. Suggestions for using this service include:

- Obtain the ONE-CALL number for the area in which work is planned.
- Call 48 hours prior to beginning construction.
- Have the address and specific location of the project ready for the service.
- Identify the type and extent of work.
- Indicate who is calling and provide an address.
- Provide construction starting dates and times.
- Provide a phone number the directory service can call with responses and questions.

In certain situations, the utility company may choose to have a representative present while work around its lines is in progress.

Landscape contractors involved in large planting projects or projects that have footings or new utility installation can also mark out their work prior to a utility locate. As long as markings are complete and accurate, this action will aid the utility company in safely identifying its lines. To aid in identification, it is suggested that white flags or paint be used to locate landscape improvements to avoid confusion with utility markings. Standard colors have been adopted by utility companies in identifying various types of utilities. This color coding systems is as follows:

- Red: electrical
- Yellow: gas
- Orange: phone, communication, possibly **fiber optics**
- Blue: water
- Green: sewer

Once utilities have been located, carefully hand-dig for 2 feet on either side of the utility marking. It is still possible that the utility is slightly mismarked, and hand-digging reduces the chance of disrupting the line. Certain utilities place a locator tape or cable with or slightly above the actual utility line to aid in locating and to warn that the utility is close. Trenching perpendicular to the marking and gradually working deeper aids in locating lines that are not directly below the mark. Work the soil gently with a shovel and in small amounts. Avoid chopping into compacted soil or using the shovel as a lever to remove large sections of soil.

Private Utilities

Before beginning a project, the site should also be reviewed for utilities that may not have been marked. Lines privately installed do not show up on utility company records and *are not marked* by locators. Look for outbuildings with power, yard lights, wells, gas tanks, septic tanks, or other site facilities that would obviously have some sort of connection to a utility line. Inspect inside buildings, particularly in the basement, for utility lines leaving a structure. Check the buildings on site for electrical breaker panels. Shutting off service at any unknown or unmarked breaker will stop service to areas served by those controls, but may also reduce the chance of accidental shock. Utility-locating equipment can be rented to help find private lines. Instructions for operating this equipment are critical, since connections are often made to the utility line to trace its location. If utilities cannot be located, no choice may be available but to unearth the entire utility line from source to end to find the location.

WORKING BELOW GRADE

Trenches pose a risk of collapse where the soil is unstable or the trench is deep. Workers should be aware that any trench poses a hazard when the worker needs to bend down to perform work tasks near the bottom. Special precautions should be taken any time the top of the trench is above the chest of the worker.

Reducing the risk when trenching involves placement of shoring when the trench exceeds a safe working depth. Sheets of plywood separated by 2 × 4s can provide sidewall protection for shallow trenches in unstable soils. Larger trenches will require the use of metal trench shoring. Avoiding the operation of equipment near the edges of open trenches also reduces the hazards of trench collapse. Always work in pairs when performing trench work. Nothing can be done by the person who is caught in a trench collapse.

OPERATING POWER EQUIPMENT

Personal injury, disability, and death are common risks from the improper operation and use of power equipment. See Chapter 7 for instructions on performing basic construction activities, including operation of many pieces of equipment used by the landscape contractor. Tractors, skid-steers, chain saws, and power hand-tools all pose a risk of serious injury if the rules of operation are not followed. Never modify equipment or remove safety guards. Read all manufacturer's safety instructions for each tool or piece of equipment that is used. In addition, safety risk labels are found on nearly every piece of equipment used. Reading these instructions and heeding these labels are the foundation of workplace accident avoidance.

Care should be exercised when using electrical equipment in landscape applications. With cords strung around a construction site, it is easy to accidentally sever a cord, creating a potential for shock. Water is also a potential hazard for use of electrical equipment outdoors. Verify that electrical circuits being used have ground fault protection against shock and electrocution.

Operation of tractors, loaders, and skid-steers creates additional hazards for the worksite. Use care when backing such equipment, since the operator is often screened from clear vision behind the vehicle. Warning bells for backing should be kept in working order, and all workers in the area of such equipment should be alert when backing occurs. Potential for tip-over increases for equipment operated on a slope or when loads become top-heavy. The concept of center of gravity is what keeps most equipment operating on all four wheels (or two tracks). When this center of mass approaches the outside edge of a piece of equipment's operating base (wheels or tracks), instability results. If the center of mass moves outside the operating base, a tip-over occurs. Most equipment is designed with a low mass, but when carrying a load in a bucket or operating on a slope, that center will change (Figure 4–3). Good rules to follow include working slopes steeper than 3.5:1 (3 1/2 feet of horizontal distance for each foot of vertical fall) from bottom to top rather than from side to side, and always carrying loads low.

Caution should also be exercised when using power washers and sprayers. Never aim the spray at exposed skin or body parts. Serious damage to eyes and skin

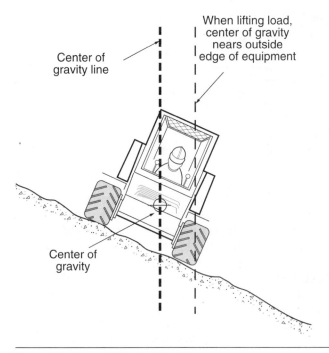

Figure 4–3 Center of gravity for equipment.

can result from a direct spray at high pressure, including contusion and cutting of skin or the penetration of materials into skin.

WORKING WITH CHEMICALS

Serious injury, particularly burns, can result from misuse of chemicals associated with landscape construction. Painting, staining, and cleaning masonry work are primary activities for practicing safety techniques aimed at reducing chemical injury.

Acids and Volatiles

Acids and volatile chemicals found in exterior finishes can present a risk to the landscape contractor. Following a few simple rules when working with muriatic acid, wood preservatives, and stains/paints can reduce the potential risk. Safety rules are include:

- Wear proper protective clothing when mixing and applying volatile chemicals to reduce the chance of injury. Eye and face protection, rubber gloves, and complete covering of body and limbs will afford protection when working with chemicals.
- Mix and use chemicals in open areas with ample ventilation to avoid being overcome by high concentrations of fumes.
- Follow the procedures recommended by the manufacturer when applying any chemicals.

Chapter 5

Construction Math

Whether ordering materials, establishing grades, laying out a walkway, or performing any one of the many activities of landscape construction, the need to perform math operations is basic. In this chapter are the building blocks of math operations for landscape construction. Included as basic calculations are the concepts of item counts; linear, perimeter, area, and volume measurements; and weight conversions. Formulas that utilize these basic measurements to obtain specialty calculations are also included in this chapter.

Math and project measurements can be made easier if the decimal system is used, rather than measuring in inches and fractions. Most math functions benefit from converting numbers to feet and tenths.

CALCULATING AVERAGES

In construction, calculating the average of several numbers may be required. The formula for calculating an average is:

Sum of all numbers/N, where N equals the number of numbers added together (Figure 5–1).

$$\frac{A + B + C + D \ldots X}{N} = \text{AVERAGE}$$

N = QUANITIY OF NUMBERS ADDED TOGETHER

EXAMPLE:

$$\frac{5 + 3 + 12 + 17 + 10 + 8}{6} = 9.16$$

Figure 5–1 Calculating averages.

ITEM COUNT

Items such as benches, lights, plant material, and other amenities are measured and ordered by the quantity of each item used. An item count requires only a "head" count of each separate item being used.

LINEAR MEASUREMENTS

Linear measurements are used to calculate quantities for items that are purchased by length, such as edger and fence. Expressed as linear feet (LF), the linear measurement is also the building block for area and volume calculations. Linear measurements are performed by measuring directly from one point to a second point.

Three methods for performing linear measurements are common in the landscaping field. Choice of a method for making linear measurements is determined by the level of accuracy that is needed. Direct measurement should be selected if an accuracy of 2% is required. **Pacing** can be used if the margin of error can range up to 5%. Estimating is acceptable if the level of error can be as high as 10%. Actual percentage of error for each measurement technique will vary depending on how the user applies the technique. Techniques for each method are identified in the following paragraphs.

Direct Measurement

Direct measurement involves the use of a tape measure or measuring wheel to determine the distance between two points. Accuracy of this method can be affected by obstacles or slopes between the two points. If obstacles prevent a direct measurement, take the measurement along a route parallel to the

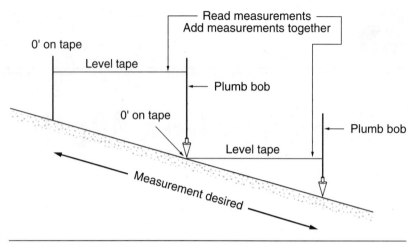

Figure 5–2 Making accurate measurements on a hillside.

one that requires measuring. Use of a screwdriver to anchor the zero end of the tape allows one person to take measurements that normally require two people. A measuring wheel is easy and convenient to use, but most wheels have an accuracy of 1 foot and provide no increments between feet (inches or tenths).

The accuracy of linear measurements is also affected by slope. Horizontal distances are less when measured on a flat surface than when measured on a sloped surface. To accurately find a location on a sloped surface, begin by extending a tape measure from the point where the measurement begins. Hold the tape level and along the alignment for the measurement (Figure 5–2). At the correct measurement on the tape, lower a **plumb bob** to obtain the desired location on the ground. Measurements that are longer than one length of the tape can be made by repeating these steps, extending the level tape from each new location found with the plumb bob. Very steep slopes may require several short measurements to maintain accuracy. To accurately measure a point on a slope, hold the measuring tape level and hold the plumb bob directly over the point being measured. Move the tape adjacent to the plumb bob and read the dimension where the plumb bob line intersects the tape.

Pacing

Pacing can be useful because it requires no special tools and can be adapted to nearly every site or terrain. More than walking, pacing utilizes the consistency in the normal pace of a person to derive a linear measurement for a site. To determine average pace, set up a course that is 100 feet long, and walk the course several times, counting the number of steps each time. Divide the distance by the average number

1. Set up and pace course three times:

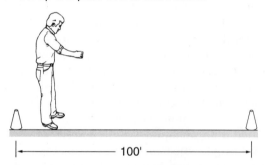

(note: course should be level and free of obstacles)

2. Average the number of paces:

   1st pass: _____

   2nd pass: _____

   3rd pass: _____

   Total paces $\div\,3 = \dfrac{1+2+3}{3}$ = average number of paces

3. Divide average number of paces into 100' to obtain typical pace:

   $$\dfrac{100'}{\text{Average number of paces}} = \text{average pace}$$

Figure 5–3 Calculating your pace.

of steps to calculate the average pace (Figure 5–3). It is important, both when calculating average pace and when performing measurements with this method, that a normal step is maintained. Attempting to step an even distance or altering the normal step will result in inconsistent results.

Estimation

When accuracy is not critical, estimating linear dimensions is a viable option. Estimating requires determining a distance based on visual observation or comparison with a known dimension. Visual determination is developed by experience in working with outdoor spaces, while comparison requires visually comparing the distance to be measured with an object of known measurements. Items such as structures, door openings (usually 7 feet high), and sidewalk squares can be used to make comparative measurements.

PERIMETER CALCULATIONS

Installation of edging or materials placed around a shape requires calculation of the perimeter of an object. To perform perimeter calculations, identify the shape being measured, and then make linear measurements of portions of that shape. If working with an irregular shape, the fastest method for measuring a perimeter may be to pace or take a direct measurement. If measuring standard geometric shapes, utilize the formulas identified in Figure 5–4 to obtain perimeter measurements. Figure 5–5 provides sample calculations. Common shapes include:

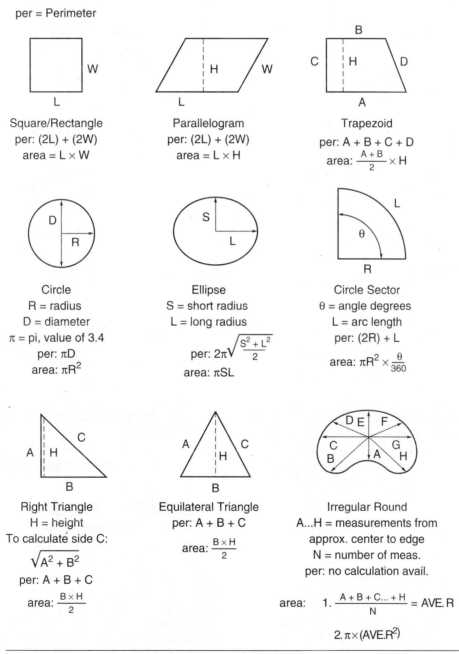

Figure 5–4 Perimeter and area formulas for common shapes.

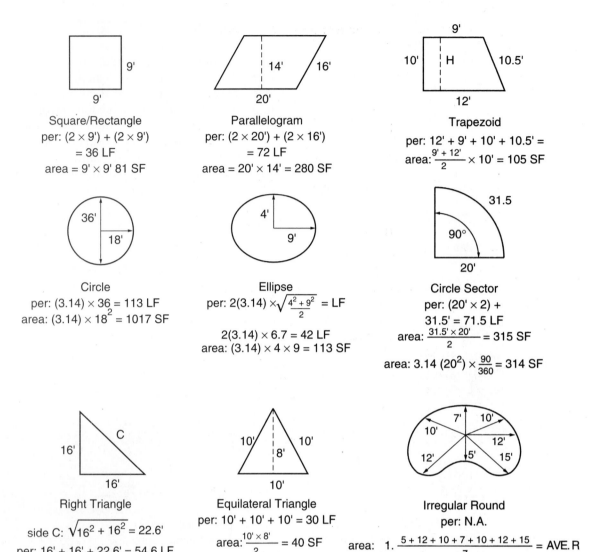

Figure 5–5 Calculating perimeter and area for common shapes.

- Square/Rectangle/Parallelogram: Shapes with two sets of parallel sides. Squares and rectangles have right-angled corners and parallelograms have no right-angled corners.
- Trapezoid: Shape with one set of parallel sides and two right-angled corners, one adjacent to each parallel side.
- Circle: Round shape with edges equally distant from the center.
- Ellipse: Rounded egg shape with edges at variable distances from center.
- Circle Sectors: Partial segments of a circle. Defined by length of the outside arc or the enclosed angle of circle measured in degrees.
- Triangle (right or equilateral): Shape with three sides. Right triangle has one right angle;

equilateral triangle has three sides of equal lengths.
- Irregular Round: Kidney or irregular shaped circles for which a center radius can be located.

AREA CALCULATIONS

Materials such as brick, seed, and sod are purchased based on the area of the space they cover. Quantities of materials necessary to cover flat surfaces require area calculations. Expressed as square feet (SF) or square yards (SY), measurement of area requires recognizing the shape (or collection of shapes) of the space to be covered and calculating key dimensions for that shape using linear measurements. Formulas for calculating the area of standard shapes

1. Identify shapes and measure dimensions of each

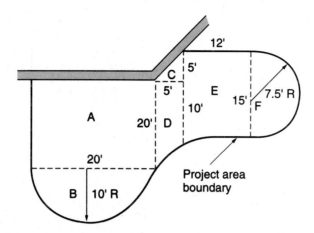

2. Add together areas for each of the shapes identified within the project area boundary (refer to Figure 5–4 for formulas for each shape)

A Square		= 400	SF
B (Half) Circle	314/2	= 157	
C Triangle		= 12.5	
D Trapezoid		= 75	
E Rectangle		= 180	
F (Half) Circle	176/2	= 88	
	±	912.5	SF

Figure 5–6 Calculating the area of an irregular shape.

are shown in Figure 5–4. Figure 5–5 provides sample calculations.

Many methods are available for obtaining area calculations. New computer design programs automatically calculate the area of shapes. **Planimeters** are an engineering tool used to measure the area of random shapes. Lacking these tools, any shape can be broken down into a collection of measurable geometric spaces. Once the dimensions of these shapes are determined, the areas can be calculated and added together to obtain the area of irregular shapes.

To calculate square footages by using the sum of shapes, begin by analyzing the space for which an area measurement is required. Observe if the space is an easily measurable single shape, or if one or more of the geometric shapes previously mentioned is recognizable (square/rectangle, parallelogram, circle, or triangle) within the boundaries of the shape. These shapes should closely cover the majority of the area being measured (Figure 5–6). Using the formulas in Figure 5–4, make the measurements and perform the calculations required for each shape, then total the answers to obtain the area for the shape. Be certain all calculations are correct, and if only a portion of the shape is utilized (such as one-quarter of a circle), reduce the total appropriately. Be forewarned that this method will have a substantial margin of error, so if material orders are critical, a more precise method of measurement should be selected.

Some conversions applicable to area measurements follow. Appendix B includes additional conversions.

- If the measurement required must be expressed as square yards, divide the square foot total by 9 to obtain square yards.
- To obtain the number of acres, divide the square foot total by 43,560.
- To obtain squares for sod, divide the square footage by 100.
- To obtain rolls of sod (each sod roll being 1 SY), divide square footage by 9.

VOLUME CALCULATIONS

For materials that are purchased in bulk, a volume measurement is typically required. Bulk measurements are expressed as cubic feet (CF) or cubic yards (CY), and require an area measurement and knowledge of the depth of the material to be calculated.

To calculate volumes for landscaping, begin with the square footage of the area that is to be filled. Determine the depth of the layer to be filled and

Volume formula:

$$\frac{\text{area (in SF)} \times \text{depth (in inches)}}{12} = \text{CF (cubic feet)}$$

Examples from Figure 5–5:

1. 6" layer of topsoil over square shape of 81 SF

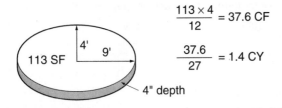

$$\frac{81 \times 6}{12} = 40.5 \text{ CF}$$

$$\frac{40.5}{27} = 1.5 \text{ CY (cubic yard)}$$

2. 4" layer of mulch over ellipse shape of 113 SF

$$\frac{113 \times 4}{12} = 37.6 \text{ CF}$$

$$\frac{37.6}{27} = 1.4 \text{ CY}$$

Figure 5–7 Calculating volumes for materials placed in layers.

place the numbers in the following formula (Figure 5–7):

$$\frac{\text{Area (in SF)} \times \text{Depth (in inches)}}{12} = \text{CF}$$

Note that the depth is not converted to feet, but is left in inches.

Determining the volume of plant root balls requires a formula used to calculate the volume of a sphere. Once the volume has been calculated, use a weight conversion to convert to pounds. The formula for calculating the volume of a sphere is as follows (Figure 5–8):

$$4.18 \times R \times R \times R = \text{CI (cubic inches)}$$
when R is expressed as inches.

Expressing R as feet (or decimal portions of feet) will provide an answer in CF (cubic feet)

To calculate the volume of a cylinder, use the following formula (Figure 5–8):

$$\text{Area} \times \text{Height} = \text{CF (both area and height must be expressed as feet or decimal portions of feet)}$$

Some conversions applicable to volume measurements include:

Volume of Shpere:

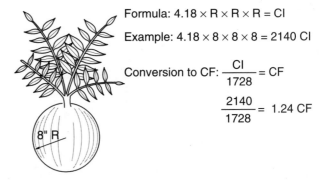

Formula: $4.18 \times R \times R \times R = \text{CI}$

Example: $4.18 \times 8 \times 8 \times 8 = 2140 \text{ CI}$

Conversion to CF: $\frac{\text{CI}}{1728} = \text{CF}$

$$\frac{2140}{1728} = 1.24 \text{ CF}$$

Volume of Cylinder:

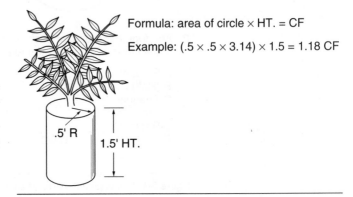

Formula: area of circle × HT. = CF

Example: $(.5 \times .5 \times 3.14) \times 1.5 = 1.18 \text{ CF}$

Figure 5–8 Calculating volumes for spheres and cylinders.

- If the measurement required is in cubic inches and must be converted to cubic feet, divide the cubic inches total by 1,728 to obtain cubic feet.
- If the measurement required is in cubic feet and must be expressed as cubic yards, divide the cubic foot total by 27 to obtain cubic yards.

WEIGHT CONVERSIONS

Materials purchased in bulk may require a weight rather than a volume measurement. In this case, use the following chart to convert volume to weight for each of the following types of materials by multiplying the CY or CF by the conversions:

- 1 CF soil, dry and loose = 90 pounds
- 1 CF soil, moist = 75 to 100 pounds
- 1 CF sand, dry = 100 pounds
- 1 CF limestone, uncrushed = 160 pounds
- 1 CF concrete = 140 pounds
- 1 CY fill dirt, dry and loose = 1.2 tons (2,400 pounds)
- 1 CY concrete sand, dry = 1.5 tons (3,000 pounds)
- 1 CY aggregate (class 5 aggregate, 1 inch roadstone, or equal) = 1.25 tons

SPECIALTY CALCULATIONS

Certain work areas within landscape construction require adaptations of the basic math operations to perform required calculations. These calculations are presented in the sections that follow.

CALCULATING RIP-RAP QUANTITIES

Calculations for rip-rap (large stone used for erosion control) require measurement of the area to be covered. Once the area is determined, use a layer thickness of 6 inches in the volume calculation formula to determine the CF required. Convert the CF to CY, and multiply by 1.5 to determine tonnage.

CALCULATING LENGTH OF FRENCH DRAINS

French drains are sized according to the length of trench required to store water now ponding on the surface. In areas where drainage problems are serious or surface storm water storage is required, a design professional who is familiar with calculating water storage based on storm intensity, rainfall amounts, runoff coefficients, and allowable outflow should be consulted to address the problem. If the drainage problem does not create a hazard to life or property, french drain size can be approximated by the contractor. Before performing this calculation, observe the drainage area during a rain to estimate the area and depth of the pond. The following steps can be used to determine the length of the french drain (Figure 5–9):

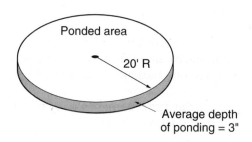

Area calculation: $3.14 (20^2) = 1,256$ SF

Volume calculation: $\dfrac{1,256 \times 3}{12} = 314$ CF

French drain length required: $\dfrac{314}{1.2} =$ 262 LF of french drain required (drain dimensions of 12" wide × 42" deep, with 6" of soil cover)

Figure 5–9 Calculating length for french drains.

- Estimate the CF of water that requires storage (volume calculation for ponded area).
- Divide the CF of storage by 1.2 (one and two-tenths).

The answer is the length of trench required to store water. The division factor used in the second step provides an answer that is based on a trench that is 12 inches wide, 42 inches deep, and has 6 inches of soil cover. Trenches that have different dimensions will require a different division factor. As trench dimensions diminish, the factor value decreases; as trench dimensions increase, the factor value increases.

SLOPE CALCULATIONS

When establishing grades or installing storm water drainage systems, it is critical to maintain downward slope. While many contractors accomplish this task using visual estimation or a level, the ability to calculate slope is a valuable math skill. Slope calculations are presented for two applications: calculating the existing slope between two points and determining the amount of vertical change required to obtain a certain slope over a given distance.

Calculating the Slope Between Two Points

To determine the slope of an uniform grade between two points, use the following steps. Measurements must be in the same units (feet and tenths are recommended) (Figure 5–10).

- Measure the vertical difference between the two points (V). Use a carpenter's level resting on a straight 2 × 4 or a survey instrument to accomplish this.
- Measure the horizontal distance between the two points (H).
- Insert the measurements into the following formula:

$$(V/H) \times 100 = \% \text{ of slope.}$$

- The answer is the slope percentage.

Calculating the Vertical Change Required to Obtain a Given Slope Between Points

When staking improvements, it may be necessary to construct a slope at a certain percentage from a first point to a second point. To calculate the amount of elevation change, use the following formula. Measurements must be in the same units (feet and tenths are recommended) (Figure 5–10).

Calculating slope between two points.

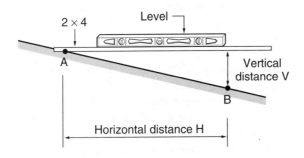

Example:
calculate slope between points A & B
V = 1.5' H = 13'

$$\frac{1.5}{13} = .115 \times 100 = 11.5\% \text{ slope}$$

Calculating the vertical change to obtain a
required slope between two points.

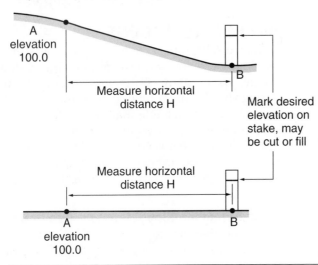

Example:
calculate a 2% downslope
between points A and B,
H = 13'% slope = – 2% elevation A = 100.0

$13 \times – .02 = – .26$ elevation change for point B

$100 – .26 = 99.74$ elevation of point B

Example:
calculate a 3% upslope
between points A and B,
H = 13'% slope = + 3% elevation A = 100.0

$13 \times .03 = .39$ elevation change for point B

$100 + .39 = 100.39$ elevation of point B

Figure 5–10 Calculating slopes.

- Measure or set the elevation of the first point (use 100.0 if the actual elevation is not known) (X).
- Measure the horizontal distance between the first point and the second point (H).
- To determine the change in elevation (Change), multiply the horizontal distance times the slope required between points: H × .PS = Change (PS is the percent slope expressed as a decimal, such as a 2% slope shown as .02).
- If the slope is to go down toward the second point, *subtract* the Change from A.

     A – Change = Elevation of second point.

- If the slope is to go up towards the second point, *add* the Change to A.

     A + Change = Elevation of second point.

WALL MATERIAL CALCULATIONS

Determining wall material needs requires calculating the surface area of the wall to be built and information from the manufacturer regarding how much surface area is covered by a single unit of wall material (Figure 5–11).

To determine the surface area of a wall that has a consistent height (Figure 5–11A):

- Perform an area calculation by measuring the height and the length of the wall.
- Add 1 foot to the height measurement to accommodate a buried base, and multiply by the length to obtain square footage of the wall.

If the wall is variable in height over its entire length (Figure 5–11B):

- Obtain the average height by making multiple measurements at points where the wall is tallest and shortest, adding them together, and

Formulas: **Examples:**

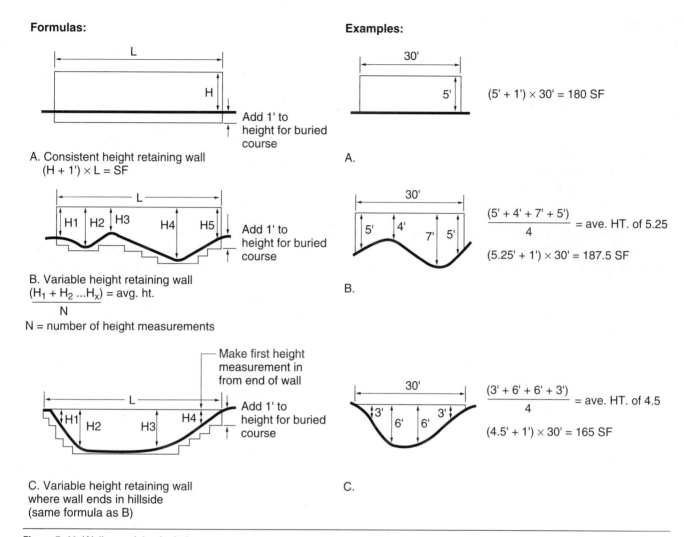

A. Consistent height retaining wall
$(H + 1') \times L = SF$

B. Variable height retaining wall
$\dfrac{(H_1 + H_2 \ldots H_x)}{N}$ = avg. ht.

N = number of height measurements

C. Variable height retaining wall
where wall ends in hillside
(same formula as B)

Figure 5–11 Wall material calculations.

dividing by the number of measurements taken.

- Add 1 foot to the height measurement to accommodate a buried base, and multiply by the length to obtain square footage of the wall.

If the wall is consistent in height except for a few small sections, such as return or cheek walls (Figure 5–11C):

- Measure the small sections 5 feet from the ends and several times along the consistent height portion, then calculate as a variable height wall.
- Add 1 foot to the average height measurement.
- Multiply by the length to obtain the square footage of the wall.

If the project contains multiple walls, the square footage of each wall must be measured and added to obtain a total.

Segmental Unit (Precast) Calculations

Add the area of all wall sections together when square footage calculations are complete. To determine how many segmental units are required, the square footage of a single unit will need to be obtained. Suppliers typically can provide this information or it can be calculated by performing an area measurement on the surface of one unit (typical areas range from .5 SF to 1.25 SF). Divide the wall square footage by the surface area of one unit to obtain the number of total units required for the wall.

Segmental Unit Capstone Calculations

Capstones are usually ordered by the number of linear feet they will cover. Use a direct measurement of the total linear feet of the wall and multiply by the conversion number (number of capstones per linear foot) supplied by the manufacturer.

Stone Calculations

To calculate the tonnage of stone required for a wall, divide the square footage of the wall by the number of SF 1 ton of stone will build. This conversion number can be provided by the quarry or supplier (typically around 16–25 SF of wall per ton of stone).

Tie/Timber Calculations

Calculations for ties and timbers require the same area calculations just performed, but the area is divided by conversion numbers to determine the number of ties and timbers to order. Find the material being used in the following list, and then divide the wall SF by the conversion number (Note: Height is the vertical dimension when looking at the side of the material.):

- Railroad tie, 8 feet long by 8 inches in height: divide by 5.0
- Railroad tie, 8 feet long by 7 inches in height: divide by 4.0
- Railroad tie, 8 feet long by 6 inches in height: divide by 3.5
- Timber, 8 feet long by 6 inches in height: divide by 3.7
- Timber 6 feet long by 6 inches height: divide by 2.8

To calculate the number of ties/timbers required for deadmen or verticals, divide the total wall length by 5. Add the wall ties/timbers numbers to the deadmen or vertical numbers to determine the number of ties/timbers to order.

STAIR CALCULATIONS

Calculation of stair dimensions requires an understanding of both slope and the stair components. There is a relationship between the dimension of the vertical portion of the stair, or the **riser,** and the horizontal portion of the stair, or the **tread** (Figure 5–12). The combination of these two dimensions will determine if the stairs are comfortable to use and fit the space reserved for stair construction. The math presented here applies to exterior stairs constructed of wall materials, concrete, and wood.

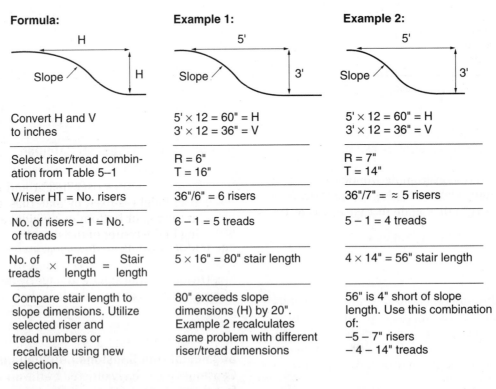

Formula:	Example 1:	Example 2:
Convert H and V to inches	5' × 12 = 60" = H 3' × 12 = 36" = V	5' × 12 = 60" = H 3' × 12 = 36" = V
Select riser/tread combination from Table 5–1	R = 6" T = 16"	R = 7" T = 14"
V/riser HT = No. risers	36"/6" = 6 risers	36"/7" = ≈ 5 risers
No. of risers – 1 = No. of treads	6 – 1 = 5 treads	5 – 1 = 4 treads
No. of treads × Tread length = Stair length	5 × 16" = 80" stair length	4 × 14" = 56" stair length
Compare stair length to slope dimensions. Utilize selected riser and tread numbers or recalculate using new selection.	80" exceeds slope dimensions (H) by 20". Example 2 recalculates same problem with different riser/tread dimensions	56" is 4" short of slope length. Use this combination of: –5 – 7" risers – 4 – 14" treads

*Stairs that do not match slope dimensions with riser/tread combinations in Table 5–1 will require special design considerations. Consult a design professional.

Figure 5–12 Stair calculations.

Riser/Tread Relationship

This relationship, which can be expressed in a formula, maintains dimensions that produce stairs matching a human's normal stride. If one of these two dimensions is beyond the normal standards, the user may have difficulty navigating the stairs. If the sum of these two dimensions is beyond the normal standards, the user may have a difficult time maintaining a normal pace up the stairs. Considering the limitations of many of the materials from which steps are constructed, stairs in landscaping should follow this relationship as closely as possible.

A formula for riser/tread relationships and a chart illustrating standard riser heights and tread derived from the formula are shown in Table 5–1.

Having no set standard for the maximum width of a tread provides flexibility, and problems, for stairs constructed in the landscape. It should be noted that using a tread width longer than those determined by the formulas shown in Table 5–1 will create an uneven, perhaps even dangerous, stepping pattern. Stair dimensions in the landscape are complicated further by the fact that many materials used to construct stairs are materials with dimensions that cannot be adjusted. Risers heights may be limited to a wall material thickness, and the corresponding tread length set by the riser dimension.

In addition to these considerations, stairs that serve the public must follow local, state, and national regulations. Specific requirements for access by persons with differing abilities may determine the actual stair dimensions. Stairs accessed by the public should be designed by a design professional.

Process for Calculating Risers/Treads

Calculating the number of risers and treads required for a project requires measurements for the thickness of materials from which the stairs will be constructed (timbers, wall units, wall units plus caps, stone, etc.), the horizontal length of the slope that must be tra-

versed (measured from **top of slope** to **toe of slope**), and the vertical distance that must be covered. To obtain the horizontal and vertical distances, consider using the measuring techniques described in the construction staking in Chapter 8. These measurements can be used to calculate stair requirements.

Using the following steps and the riser/tread formula, determine how many risers and treads will be required for the project (Figure 5–12):

1. Convert the vertical and horizontal distances into inches.
2. Divide the vertical distance by a selected riser height. This answer will indicate how many risers are required to traverse the slope. Numbers over .5 can be rounded up; numbers under .5 can be rounded down. As mentioned earlier, the thickness of the construction material may dictate the riser height of the stair. If material thickness is not an issue, speed the process by selecting a high riser value if the slope is steep, or a low riser value if the slope is shallow.
3. Subtract one from the number of risers.
4. Multiply the answer from the previous step by the tread width required for the selected riser height. Use Table 5–1 to find the corresponding tread width for the selected riser height. This answer will provide the horizontal distance covered by the stairs.
5. Compare the horizontal distance required for the stairs and the measured distance.
   - If the distances are equal or very close, use that riser/tread combination.
   - If the difference between the measured and required distances is less than the width of one tread, add the distance to either the top or bottom to create a landing.
   - If the difference exceeds the length of one width, choose another riser height and recalculate. If riser height is fixed by materials, add a landing at the top or bottom or consider adding a landing in the center of the stairs.
6. If horizontal dimensions cannot be matched with adjustments, the stairs will require excavation farther back from the top of the slope to accommodate the longer horizontal distance required.

EDGE RESTRAINT CALCULATIONS

The amount of edge restraint required for a project is determined by calculating the perimeter of the area being edged. Most edging is purchased by linear foot, and the perimeter measurement will be adequate. Stone edging materials are ordered by the tonnage, similar to wall material. To determine the tonnage

Table 5–1 Relationship Between Riser Height and Tread Length in Stairs

2r + t = 28″ where: r = riser; t = trend

if r is . . .	then t is . . .
4 inches	20 inches
5 inches	18 inches
6 inches	16 inches
7 inches	14 inches
8 inches	12 inches
9 inches	11 inches

required, calculate the square feet of edging and divide by the number of square feet per ton for stone.

SETTING BED AND BASE COURSE CALCULATIONS

Setting Bed

Setting bed material should be placed in a layer over compacted base. This material should be ordered by either the cubic yard or ton. To determine the quantity needed, calculate the area to be paved and insert that number in the following equation:

- Area paved (SF) × depth in inches/12 = CF of material required
- CF/27 = CY of material required
- CY × 1.5 = tons of material required

Base Course

Base course material should be placed in a layer over the compacted subgrade at a depth specified by the designer. This material should be ordered by either the cubic yard or ton. To determine the quantity required, calculate the area to be paved and insert that number in the following equation:

- Area paved (SF) × depth in inches (X)/12 = CF of material required
- CF/27 = CY of material required
- CY ×1.25 = tons of material required

PAVING MATERIAL CALCULATIONS

Determining the quantity of paving material required for a project will utilize the area formulas for a unit paving material such as brick or limestone, or the volume formulas for concrete of granular paving materials. It is advisable that extra unit pavers be ordered to allow for cutting and variation in pallet quantities. Orders increased by 10%–15% typically provide an adequate surplus.

Unit Paving

To calculate unit paver quantities, determine the area of the space to be paved. Obtain from the manufacturer the number of unit pavers per square foot for the style of paving to be used. Most pavers range from 3.5 to 5.0 units per square foot. Standard 4 inch × 8 inch brick requires 4.5 pavers per square foot. Use the following formula to determine material requirements:

- SF of paved area × units per square foot = number of pavers to order

Stone

To calculate stone quantities calculate the area of the space to be paved in square feet. From the supplier, determine the amount of square footage covered by 1 ton of stone material (standards for stone are typically 40–80 square feet of coverage per ton). Use the following formula to determine material requirements:

- SF of paved area/SF coverage per ton = tons of flagstone to order

Concrete

Concrete orders require measurements in cubic yards for the area paved. Concrete is ordered by the CY, so once a volume calculation is complete, no further calculations are needed. The following formula can be used for this calculation:

- Area paved (SF) × depth in inches (X)/12 = CF of material required
- CF/27 = CY of material required

Granular

Granular paving projects do not require a leveling course, but do require calculations for volume of wearing course material and linear feet of edge stabilization if it is used. To perform volume calculations, use the following formula:

- Area paved (SF) × depth in inches (X)/12 = CF of material required
- CF /27 = CY of material required
- CY × 1.25 = tons of material required

CALCULATING CONTRACTION JOINT SPACING IN CONCRETE

When tooling or sawing contraction joints in concrete, the spacing is important to control cracking in the surface. To determine an approximate horizontal distance between contraction joints, use the following formula:

- Multiply the thickness of the concrete times 12.
- The answer is a typical spacing (in inches) between joints.

This formula produces an approximate spacing. Other considerations for locating joints include width, shape, and configuration of paved areas.

WOOD MATERIAL CALCULATIONS

Ordering wood materials requires an item count for each size of structural member used. The most efficient method for performing this count is to prepare a plan for the project and make a list of all types of lumber pieces, hardware, and connectors required. Count the number required, paying special attention to those members that will require specific lengths.

The concept of board foot is sometimes used when referring to lumber. One board foot is the equivalent of a 1 foot × 1 foot × 1 inch piece of wood and is a calculation used by lumber suppliers when purchasing and pricing wood. Materials in landscape applications seldom use this method for ordering, but the term may appear on billing statements. To calculate board feet, multiply the width of the board, in inches, by the thickness of the board, in inches, by the length of board, in feet, then divide the answer by 12. Dimensions less than 1 inch are considered as 1 inch.

Chapter 6

Project Pricing

Determining the price of a project requires a combination of calculating the costs of the various aspects, and formatting the costs in one of several ways for presentation to a client. Preparation of costs requires application of many of the formulas and concepts presented in the chapter on Construction Math (Chapter 5), but the formulas are not the only variable important to project pricing. Effective preparation of estimates and bids requires input from historical records of a firm's activities, data obtained from estimating textbooks, and judgments from the contractor regarding profit and contingency. Without this external input, the methods presented here create only a framework for calculating project pricing.

ESTIMATES AND BIDDING

Prior to engaging in the preparation of cost estimating, the contractor should determine if the client desires an estimate or a bid. The level of accuracy in calculating project costs is affected by the way the numbers are used. There is little need to spend hours calculating detailed project costs if the only question the client needs answered is whether the project is within his or her general price range. On the other hand, a rough estimate is risky if the client is expecting to receive a bid to compare with bids from other contractors. Knowing what the stage the project is at and preparing the proper type of cost estimate or bid are essential to an efficient operation.

Estimates

Estimates should be considered exactly what the term implies. When a potential customer wants to know how much a project will cost so that he or she can decide whether to proceed with more detailed work,

an estimate may be in order. Preparation of an estimate may be as simple as quoting a range of prices based on similar types of projects. Estimates may also be prepared based on unit costs for general types of materials the customer inquires about, such as deck (SF), plantings (cost per tree), or fill dirt (CY).

Be cautious when asked to prepare an estimate for a client. Misunderstandings can develop when the client is not informed that the estimate is not a firm price. If an estimate is to be converted into a contract, it is important to qualify the proposal, stating that additional costs may be incurred if the project is changed or if unknown conditions are uncovered.

Bidding

Formal proposals to clients in the form of bids serve as a prelude to binding contracts. Prices prepared for bids should be carefully calculated and the contractor should be prepared to uphold the price if accepted by the owner. Bids are common for all types of projects, public and private. Calculation of a bid requires that a design exists that provides all potential bidders a common document(s) from which to prepare the bid.

The bidding process typically begins with the advertisement, or invitation, to submit a bid. This notice provides instructions that describe the project, client, method of bidding, form of bid, bonding requirements, and the time and place where bids must be submitted. The advertisement typically informs the contractor where contract documents may be obtained or reviewed and when bids will be opened and reviewed. After bids are opened, the design professional will canvass the bids to assure there are no irregularities or mistakes in the proposals. If all is in order, a contract is awarded, bonding and insurance documents are submitted to the client's representative, and a notice to proceed is issued.

CALCULATING PROJECT COSTS

Several components play key roles in calculating project costs. Accurate take-offs, or calculations, of materials required to build the project must be determined. Labor, or the time spent by employees installing projects, must be estimated. While materials can be calculated with accuracy, labor can vary greatly, based on the type of operation and project situation. **Overhead** or nonlabor or material costs of operating a business, can also range widely among firms of different size and management styles. Overhead can be separated into assignable overhead and unassignable overhead. Providing a percentage of project costs to cover contingency issues is another related project cost. Determining profit is also an essential piece of the pricing scheme. The totals from this list can be combined in many formats to arrive at a price presentation for the client.

Materials

Material costs are calculated using the design prepared for the site. Begin by performing a material take-off, or a calculation of the quantity of all materials to be used for the project. Organization of a material take-off can take whatever form provides the results required. Thoroughness and consistency from project to project are important when performing take-offs in order to develop a method that reduces the potential for missing aspects of the proposed work. Utilizing a method that interfaces easily with the system used by the accounting department to track work activities will provide information that is valuable for the current project, and also for future projects.

When performing a material take-off, consideration should be given to increasing the actual quantity required for certain materials. Landscaping projects typically utilize many materials that could settle during shipping, be blown away, eroded, or otherwise lose "bulk" from the time they are first loaded to the time they are placed on the site. To accommodate this loss, it is not uncommon to add 10%–15% to bulk material orders to assure that adequate supplies exist to complete the job. Materials that are typically treated in this manner include soil, sand, base material, and other bulk granular materials.

Bulk materials are not the only material that should be over-ordered to accommodate construction conditions. Unit pavers and wall materials typically experience waste as high as 5% when cutting is required by the design. Concrete is another material that should be over-ordered by 2%–3% to avoid short load charges (additional fees added to concrete prices when customers order less than a full load). Ground cover plants and sod are typically over-ordered by as much as 5% to accommodate plant mortality and waste in sod placement.

This concept of over-ordering may require that excess material be hauled from the site, but this expense compares favorably to the cost of additional deliveries. Excess hauling and material storage can add to the cost of operating a business, so careful review of the project is required to determine which materials should be considered for this approach to ordering. Maintaining stockpiles of materials that can be loaded and taken to the site with a work crew can reduce the quantity of over-orders required. Items calculated using an item count should not be over-ordered.

Labor

Calculating the time required to install a project is one of the most difficult tasks of the landscape contractor. Experienced contractors are able to calculate with a reasonable degree of accuracy the amount of time employees will require to complete most tasks, but they are sometimes challenged by unusual site conditions and unforeseen problems.

For purposes of calculating project costs, time is typically presented in a hours-per-unit format. This format presents the number of hours it will take a laborer to complete one unit of a task. Units are typically the measurable units of a work task, such as square feet of retaining wall or square of sod, and hours are typically one hour of a worker's time. A theoretical example might be a wall construction hour-per-unit rate of one hour per square foot of wall. If a project requires 100 square feet of wall, the take-off would calculate 100 hours for construction of the wall. Rates can also be calculated for work teams rather than for individual workers.

Hours-per-unit rates are typically determined based on reviews of project designs, the site, and skills and quantity of labor available. Maintaining records that track past projects provides a historical basis for calculating project time and material requirements. When calculating hours, it is important to distinguish among the different types of tasks being tracked. Granular paving and unit paving are both surfacing tasks, but they will have significantly different hours-per-unit rates. Businesses can divide the total hours required to complete a particular task for a project by the size of the task to obtain hourly rates-per-unit of work for various types of construction activities. Tracking several projects under different conditions improves the accuracy of the system.

Assistance in determining hourly rates is also available in the form of time standards developed by firms such as Kerr and Associates. Time standards provide typical hourly rates-per-unit of construction for most types of landscape work. When using such standards, be certain to compare to actual costs, and adjust based on the region and construction methods used.

Overhead

Another cost that can be directly or indirectly associated with the work being done is overhead. For purposes of calculating project costs, overhead should be split into two categories—assignable overhead and unassignable overhead.

Assignable Overhead. Assignable overhead includes nonlabor and nonmaterial costs that can be directly attributed to a project. Included in this category are special permits, equipment rental, special insurance or bonding requirements, temporary utility costs, and related project activities that do not benefit directly other projects. These costs should appear as a line item on the take-off for identification purposes. For presentation purposes, they may be presented as such or folded into other project costs as required by the proposal format.

Unassignable Overhead. Unassignable overhead includes the costs of operating a business that cannot be attributed directly to a specific project. Included in this category are the cost of equipment and vehicles, gas, tools, insurance, payroll expenses, debt service, mortgages, office expenses, maintenance, fees, and many other additional expenses. Approaches to recovering these expenses include calculating costs and adding them on a per project basis or establishing a hourly cost for overhead and adding this to hourly labor costs.

This addition to the hourly cost, often called a multiplier, is determined by calculating total overhead expenses for the year (or reporting period) and dividing by the number of hours billed to clients for work performed during the year (or reporting period). Because not all worker hours are consumed with working of projects, using billed hours rather than total hours worked by employees is essential to obtain an accurate markup. The multiplier used may be adjusted up or down based on anticipated changes in upcoming work.

Contingencies

No contractor can accurately determine the extent of weather and material delays, labor shortages, equipment breakdowns, and unknown underground obstructions. Due to the unpredictable nature of such conditions, a portion of the project cost should be reserved as a contingency fund. Typical contingency funds range from 5%–10% of the total material and labor costs. When costs are presented, this number is typically included in the overhead multiplier to avoid disputes regarding refunds of unused contingencies.

A method for developing contingency funds that minimizes the issue of unused balances is to have the client establish a contingency fund. A dollar amount equal to a fixed percentage of construction costs is available to the contractor when specifically defined conditions arise. The owner retains unused contingencies and the contractor has a method for recovering unforeseen expenses arising from circumstances beyond his or her control. Definition of this arrangement should be included in the contract with the client.

Profit

Determining what amount of profit is appropriate and knowing how to calculate that amount require insight regarding the market and competition, judgments about income requirements, anticipation of business conditions during the upcoming season(s), and advice from trusted financial advisors. Profit margins are far from standardized in the green industry, and the choice for determining profit remains a personal and business decision.

The methods for calculating this important cost range even wider than the margins expected. Profit is sometimes generalized as a percentage of the cost of materials and labor used in a business. Large projects may operate on a smaller margin, while services that are in demand may have a significant markup. Businesses working in the field have materials marked up as much as two to three times their cost to cover overhead and profit. Labor costs are sometimes doubled to provide adequate income for overhead and profit. Profit expressed as a percentage of total project costs ranges from lows of 3%–4% to highs of 25%.

Profit can be included with overhead when presented to the client as part of the hourly charge for employees, or, as in some presentation formats, presented as a separate, negotiable line item in a bid.

Material and Labor Take-Off Form

One format for calculating project costs that includes many of the aspects previously identified is illustrated in Figure 6–1. This form can be created and modified to suit individual business needs by using a computer spreadsheet program. In this illustration, the work tasks are listed as headings down the left-hand side of the chart, with specific materials utilized in the task listed below the headings. Quantities of materials required and overages (over-orders) from the next two columns are added and multiplied by the current unit

MATERIAL AND LABOR TAKE-OFF FORM

Job: Sauter residence					Job No.: 9969	Address: 2852 H. Ave.		Date:6/23/99			
MATERIALS							LABOR			NOTES/RELATED EXPENSES	
Description	Unit	Qty.	Over.	Total	Unit Cost	Mtl. Total	Hrs.	Rate	Labor Total	Description	Related Total
				0		$ —			$ —		
SITE PREPARATION				0		$ —			$ —		
sod removal	SF	700		700	$ —	$ —	4	$ 25.00	$ 100.00		
excavation	CY	10		10		$ —	2	$ 45.00	$ 90.00	skidsteer operator rate	
small tree removal	EA	1		1		$ —	4	$ 20.00	$ 80.00		
WALL INSTALLATION				0		$ —			$ —		
segmental wall material	SF	150	10	160	$ 5.25	$ 840.00	20	$ 25.00	$ 500.00		
4" diameter perf socked tile	LF	50		50	$ 1.50	$ 75.00	1	$ 25.00	$ 25.00		
aggregate backfill	TON	2		2	$ 20.00	$ 40.00	4	$ 25.00	$ 100.00		
clean topsoil backfill	TON	24	5	29	$ 15.00	$ 435.00	4	$ 45.00	$ 180.00	skidsteer operator rate	
PAVING INSTALLATION				0		$ —			$ —		
4" deep aggregate base	TON	3.5	0.5	4	$ 20.00	$ 80.00	6	$ 25.00	$ 150.00		
1" deep sand settling bed	TON	1.5	0.5	2	$ 20.00	$ 40.00	4	$ 25.00	$ 100.00		
interlocking concrete pavers	SF	225	20	245	$ 3.50	$ 857.50	20	$ 25.00	$ 500.00		
plastic paving edger	LF	60	10	70	$ 2.00	$ 140.00	1	$ 25.00	$ 25.00		
flagstone stepping stones	SF	75		75	$ 2.50	$ 187.50	4	$ 25.00	$ 100.00		
PLANT MATERIAL				0		$ —			$ —		
trees	EA	3		3	$ 150.00	$ 450.00	1.5	$ 25.00	$ 37.50		
shrubs	EA	17		17	$ 35.00	$ 595.00	3	$ 25.00	$ 75.00		
ground cover	EA	250	10	260	$ 2.50	$ 650.00	4	$ 25.00	$ 100.00		
mulch	CY	4		4	$ 25.00	$ 100.00	2	$ 25.00	$ 50.00		
staking materials	EA	3		3	$ 20.00	$ 60.00	1	$ 25.00	$ 25.00		
sod	SQ	8		8	$ 28,00	$ 224.00	5	$ 25.00	$ 125.00		
BILLABLE OVERHEAD									$ —	half-day rental of sod cutter	$60.00
ITEM TOTALS						$4,774.00			$2,362.50		$60.00
									PROJECT TOTAL	**$7,196.50**	

Figure 6–1 Example form used for material and labor take-offs for landscape projects.

cost from the column titled "Unit Cost". The total for materials is then placed in the column titled "Mtl Total". Hours for completing a specific task are calculated and placed on the same row as the task heading. Hours are multiplied by the current hourly rate to obtain the total labor cost. Related construction items, costs, and notes are placed in the far-right column. Rows at the bottom of the form provide space for adding additional billable costs described in previous paragraphs.

PRESENTATION FORMAT FOR PROJECT PRICES

Totals obtained from calculating project costs can be combined in a number of different ways for presentation to clients. The following sections identify methods to arrange the various cost aspects into bid or estimate formats. Selection of an appropriate method will be based on project requirements. Examples of how the material and labor take-off from Figure 6–1 can be presented in the following formats are provided in Figure 6–2.

Lump Sum

Lump sum is a bid format in which all projects costs are provided to the client in a single number. All components of the project are added together to obtain a lump sum bid. While this bid is easy to compare from one contractor to another, it is difficult to change because there are no itemized materials or labor costs. Bids may sometimes include an add or deduct request in which a specific amount is added or deducted from the lump sum total for the addition or deletion of a project component.

LUMP SUM

Total project costs: $ 7,196.50

ITEMIZED

Prices would be presented in the same format as **Figure 6–1.**

TIME AND MATERIALS

Material costs:
- walls $ 1,390.00
- paving   1,305.00
- plant material   2,079.00

Assignable overhead:
- equip, rental $ 60.00

Time (estimate only):
- 6 hrs. @ $45.00 $ 270.00   skidsteer operator
- 80.5 hrs. @ $25.00   2,012.50   laborer
- 4 hrs. @ $20.00   80.00   assistant

Total estimated project costs: $ 7,196.50

Actual costs will be based on hours worked multiplied by above rates.

COST PLUS FIXED FEE

Materials:
- walls $ 1,251.00
- paving   1,174.50
- plant material   1,871.10

Assigned overhead:
- equipment rental $ 60.00

Labor: $ 2,126.25

Fixed fee: $ 713.65

Total estimated project costs: $ 7,196.50

Hourly rates for work beyond estimated hours: $18/hour assistant, $22.50/hour laborer, and $40/hour skidsteer operator (hourly rates minus 10% fee).

UNIT COSTS

Site preparation: $ 270.00

Walls: 2,195.00

Paving: 2,180.00

Plant material: 2,491.50

Assignable overhead: 60.00

Total project costs: $ 7,196.50

Above costs include materials, labor, and overhead.

Figure 6–2 Typical presentation formats for landscape pricing, from information in Figure 6–1.

Itemized

An itemized bid or estimate provides a total project price, but also provides a detailed listing of project materials or work tasks. This form of bid is easy to compare and adjust if work is changed. Prices can be itemized by unit of work (see following paragraphs) or with lines for each item used in calculating costs. A detailed itemized cost presentation is similar to the format presented in Figure 6–1.

Time and Materials

When either the customer or contractor is wary about the amount of time required to complete a project, a time and materials bid may be submitted. This is a form of bid that provides for a set price for materials and assignable overhead, plus an hourly cost for labor. Labor costs on a time and material bid may be represented by an estimated number of hours multiplied by the hourly rate to obtain a general idea of labor costs. Materials and assignable overhead costs do not change significantly during the project, and compensation for labor is based on the rate times of the actual hours worked. Unassignable overhead, contingency, and profit are assigned to hourly rates for employees.

Cost Plus Fixed Fee

When there is uncertainty about the amount of time required to complete a project, the cost plus fixed fee form of bid may be useful. The cost plus fixed fee bid is a time and materials bid with the profit listed as a fixed line item, rather than being assigned to the employee hourly rate. The contractor is compensated for actual time and materials, but the owner is not responsible for additional profit due to delayed completion. Unassignable overhead and contingency costs remain assigned to the hourly rate for employees.

Unit Costs

Unit costing is primarily an estimating tool used to provide general project costs. Rather than itemizing the costs by traditional divisions such as materials, labor, etc., costs are presented according to units of work areas, such as square foot of paving.

Development of unit costs requires tracking previous jobs over a variety of installation situations. Calculate unit costs by dividing the total costs associated with a particular aspect of the job by the size of that portion of the job. For example, the total costs for all items associated with the decking for a project totaled $5,000, and the total area of the deck was 500 SF. The unit cost for that project would be decking for $10 per SF. These types of costs can be tracked over several projects to obtain a more accurate picture of actual costs.

Pricing to a General Contractor

On projects for which you function as a subcontractor for landscape portions of the work, a price will be required by the general contractor. The process used for this pricing should be formal, with a written proposal or letter of agreement submitted. Typical formats used for pricing to a general contractor include lump sum bids and itemized bids.

Chapter 7

Basic Construction Techniques and Equipment Operation

Included in this chapter are basic construction techniques and equipment operation instructions that will be utilized in several of the following chapters. Construction techniques that are specific to a particular aspect of landscaping are described in the appropriate chapter.

RELATED INFORMATION IN OTHER CHAPTERS

Information provided in this chapter is supplemented by instructions provided elsewhere in this text. Before undertaking activities described in this chapter, read the related information in the following chapter:

- Safety in the workplace, Chapter 4

EXCAVATION

Use of hand tools cannot be avoided in the field of landscape construction. While the work is labor inten-sive, proper use of these tools can make work easier and more productive. When the work with a particular hand tool on a job is extensive, consideration should be given to the economy of renting or purchasing the powered alternative. Several powered options can be substituted for much of the hand labor that is common on landscape jobs. The most common tools include power augers, sod cutters, edging machines, trenchers, backhoes, skid-steers, tractors, and dump trucks, and the variety of attachments for these pieces of equipment. Operation of much of the power equipment is not covered in this text and should be learned under the supervision of an experienced practitioner.

Digging

Nearly every task in the landscape construction field requires digging in some form. Consideration should be given to use of a skid-steer for excavations of more than 2–3 CY, but smaller excavations and portions of larger jobs will require hand excavation. Proper selection, use, and care of shovels greatly eases the burden of hand digging.

Selection begins with determining whether a spade or shovel is required (Figure 7–1). A spade has a flat surface with no edges and is designed for excavating, while a shovel has a curved surface with raised edges and is designed for scooping material. Spading shovels combine the benefits of both and are the most common tool used by the landscape contractor. Digging tools can also be classified by their shape. Scoop shovels have very wide blades (12–18 inches wide) designed for large-scale moving of mulch, rock, and other landscape materials. Round point shovels have a medium width blade (8–10 inches wide) with a rounded point and are an excellent general purpose

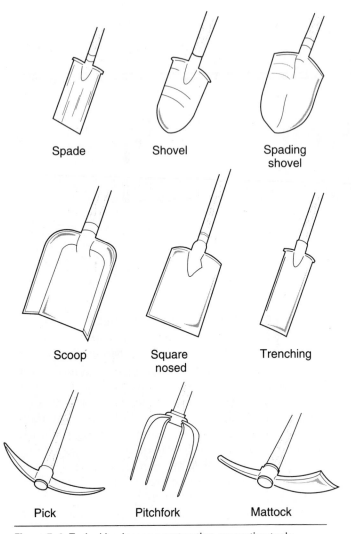

Figure 7–1 Typical landscape construction excavation tools.

Figure 7–2 Proper spading technique.

spading and scooping tool. Square-nosed shovels have a medium width blade with squared corners designed for cleaning out wide base or utility trenches and skimming soil. Trenching spades have long, narrow blades (6 inches wide) that are designed for excavating narrow utility trenches. Shovels and spades also come with long handles, which provide more leverage, and short handles, which work better in tight situations. Additional tools that are helpful in moving and removing materials are picks, mattocks, and pitchforks. Identify the type of digging tasks required and select the proper tool to accomplish the task. The experienced contractor carries a variety of tools to accommodate almost any digging chore.

Proper excavation technique is obtained by positioning the digging tool so that the blade is perpendicular with the ground (Figure 7–2). Use the foot to push the blade completely into the ground and the handle to leverage the material out. Alternate sides and dig-

ging positions to avoid overexertion of one muscle group. Excavate in reasonable amounts that fit onto the spade or shovel without spilling. Proper digging involves many repeated steps, rather than two or three giant shovelfulls. Keep equipment clean and sharp. Use of a file to maintain an edge on the blade of the shovel or spade will ease cutting through tough soils and roots. When not in use, clean digging tools completely, oil their surfaces, and store them in dry locations.

Excavating Postholes

Postholes can be excavated using clamshell diggers, augers, and powered augers. Augers work by positioning the tool over the hole location and twisting until the auger is full (Figure 7–3). Lift from the hole and tap on the ground away from the hole to remove the excavated material. Continue auguring until the proper depth is

Figure 7–3 Operating auger type posthole excavator.

Figure 7–4 Operating clamshell type posthole excavator.

reached. Clamshell diggers are operated by holding the handles together and driving the blades into the ground at the hole location by dropping or thrusting the tool downward with the arms (Figure 7–4). Avoid pinching fingers between the handles. Spread the handles to capture a load of material and lift out of the hole. Continue this process until the hole is excavated. Excavating is faster with either tool if depths are marked on the handle. If soil in the hole is dry and spills out of the digging tools while extracting from the hole, add a small amount of water into the hole to create a light mud that can be excavated. Clamshells are difficult to use on deep excavations, while augers can be difficult to maintain desired direction. Starting an excavation with a clamshell and finishing with an auger will more effectively excavate a hole. Keep a sharpened, long-handled trenching spade handy to cut any roots that disrupt hole excavation.

Hole excavation can be a laborious task, so if several holes are planned, consider renting a power auger to perform the work. Power augers have a gas engine that turns a spiral auger to excavate the hole. Two people are required to operate the power auger. To use a power auger, start the engine and position the auger over the hole. Accelerate the engine and push down on the auger. The auger will bite into the soil and pull the material out of the hole. Maintain the auger in a

plumb alignment for the entire depth of the hole. Work the auger in an up-and-down motion to periodically clear the blade, but avoid pulling the auger completely out of the hole.

Trenching

Use of the proper hand trenching method will allow production of more stable trenches in shorter amounts of time. Begin with the selection of the proper tool. A trenching shovel should be used to dig narrow slit trenches. Using the full length of the shovel blade allows trenches to reach maximum depth. When trenching, turn the spade at a 45 degree angle to the side of the trench rather than digging perpendicular to the edge. This will limit trench caving when the soil is levered out.

Compacting Fill Materials

Compaction of fill materials can be accomplished by hand or mechanical means. All filling activities are more effective if fill material is placed and compacted in layers 6 inches thick or less. Hand tamping requires pounding the fill area repeatedly using a hand tamping plate or the end of a 4×4. Mechanical pounding can be accomplished with the use of a **vibratory plate compactor.** The plate compactor operates much like a self-propelled lawnmower. Start the compactor and adjust the throttle and settings. The compactor moves forward without pushing and steers using the handle. To turn sharp corners, lift the handle vertically and spin the compactor.

STAKING AND FENCING

Driving Stakes
Driving, or forcing stakes into the ground, typically requires the use of a small (1–2 pound) or large (4–5 pound) sledgehammer. Position the stake in the proper location and lightly strike the top of the stake to start it into the ground. Stakes may shift location when driven, depending on the size of the stake, or the taper on the end being driven into the ground. If the stake shifts when being struck, use the foot to hold it in position. When the stake is standing on its own, remove your foot and continue striking while increasing the force of the blows until the stake is driven to the required depth. Maintain eye contact with the top of the stake while striking with the sledge.

Driving Metal Fence Posts
When installing chain link fence or other fence with metal posts, the posts will need to be driven into the ground. To drive metal posts, place the base of the post at the location where the post is to be set. Lean the post down until the post driver can be slipped over the top of the post (Figure 7–5). Check the post to make sure it is in the proper location and oriented properly (notches to the fabric side for fabric fences). Straighten the post and driver to a plumb position, lift the driver, and let it drop on the top of the post. If soil conditions are dry or if driving the post is difficult, apply downward force on the driver to sink the post to the desired depth.

Temporary Fence Installation
Construction fence requires that posts be driven and surfacing attached to control traffic at the project boundary. Fence surfacing is leaned against the fence and connected by passing plastic or wire ties through an opening in the fence, around the post, and back through another opening in the fence. The two ends are then fastened together by twisting or tying.

Figure 7–5 Preparing to drive a fence post by leaning the post at an angle and inserting post into driver. Return post and driver to plumb and being driving.

CUTTING AND JOINING PIPE

Measuring Pipe Before Cutting
Two methods work best for measuring pipe before cutting. The first is to measure the distance between fittings at either end of a length of pipe and add the length of pipe that will be placed inside the fittings. The second method is to hold the pipe section up to the opening where the pipe will fit and directly mark the location of each end in the fitting at either end of the pipe.

 CAUTION　Cutting pipe may create sharp edges. Wear gloves and use care when handling.

Cutting and Joining Corrugated Plastic Drainage Pipe
Corrugated plastic drainage pipes are manufactured in a variety of types and diameters and include a wide range of drainage fittings and accessories. Drainage pipe is typically manufactured in diameters of 3, 4, 6, 9, and 12 inches and in varying gauges. Lightweight tubing gauges are very flexible and are available in

perforated, nonperforated, and socked versions for use in various drainage situations. Heavier piping gauges are very rigid and can be used for subdrains and culverts. Fittings include connectors to downspouts, inlets, sump pump basins, and a variety of Ts, Ys, and elbows.

Corrugated pipe can be cut using a utility knife or hacksaw. Place the pipe on a stable surface, mark the cut locations, and make a perpendicular cut (Figure 7–6). Most fittings for drainage pipe are available with snap fittings. These fittings allow the pipe to be slipped into the fitting and held in place with flanges on the fitting (Figure 7–7). Wrap the joint with duct tape (Figure 7–8) to obtain a longer-lasting fit.

Cutting and Joining Poly Pipe

Poly pipe is a flexible plastic pipe commonly used for irrigation and potable water supply lines. Poly pipe can be cut with a hacksaw or a special ratchet cutting tool that slips over the pipe and slices it. Cutting with the ratchet cutter may require multiple squeezes on the handles to cut through the pipe. Connections are made using automotive type or specialty hose clamps and ridged plastic fittings.

To cut and join poly pipe, use cutters or a saw to make a square cut at the proper location (Figure 7–9).

Slide a clamp over the end and down the pipe a short distance. Fully insert the fitting into the open pipe and slide the clamp back down the pipe until it is resting completely over the fitting (Figure 7–10). Tighten the clamp securely (Figure 7–11).

Cutting and Joining PVC Pipe

Polyvinyl chloride, or PVC, pipe is a rigid pipe that comes in a variety of diameters. PVC can be thin-walled or thick-walled (Schedule 40). In addition to the many sizes of piping, PVC offers a full range of fittings, valves, and accessories for exterior applications. PVC is suitable for use as electrical conduit, irrigation water lines, and sewer lines. As part of the installation planning, verify the type of pipe required for the application with local plumbing codes.

PVC pipe can be cut with a pipe cutter or a hacksaw (Figure 7–12). Mark the cut location and make a square cut through the pipe. To use the pipe cutters, slip them over the pipe and squeeze the handle several times, until the pipe cuts. Use a reaming tool, wire brush, or sandpaper to clean burrs from the inside of the cut pipe.

The outside of the PVC pipe and the inside of all PVC connectors must be cleaned before the pipe can be joined. Wipe dirt and dust from the end of the pipe

Figure 7–6 Cutting corrugated plastic drain pipe with carpet knife.

Figure 7–7 Joining corrugated plastic drain pipe to a premanufactured elbow fitting. Pipe "snaps" into fitting.

Figure 7–8 Securing corrugated plastic joints with duct tape.

Figure 7–9 Cutting poly pipe using a tubing cutter.

Figure 7–10 Joining poly pipe with a plastic fitting.

Figure 7–11 Clamping poly pipe around the fitting. Each pipe connected to a fitting should be clamped.

Figure 7–12 Cutting PVC pipe with a hacksaw.

and liberally brush the primer over the parts to be joined (Figure 7–13). While the primer is still wet, apply PVC joint compound to both the pipe end and the connector, push them firmly together, and twist one-eighth to one-quarter turn (Figure 7–14). Verify that the alignment of the fitting is correct. The compound will set in approximately 10 seconds, and errors cannot be corrected without cutting new sections of pipe.

Cutting and Joining Copper Pipe

Copper pipe is typically used for supplying water and LP gas to service points within a site. Copper is available as rigid lengths of pipe or as flexible tubing. Each type has a special method for joining. A variety of pipe sizes, fittings, valves, and accessories are available for exterior uses.

Both rigid copper pipe and copper tubing are cut using a pipe cutter. The pipe cutter clamps onto the pipe at the cut location and is rotated around the pipe. With each successive turn, the clamp is tightened until the pipe is cut. A reamer should be used

to remove burrs from the inside of the pipe after cutting.

> **CAUTION**
>
> Use caution when lighting and operating a torch. The heat zone for a torch extends well beyond the visible flame. Always direct flames away from people and flammable objects. Let all heated items cool before handling.

Rigid copper pipe is joined using sweat fittings. Sweat fittings are heated joints into which solder is drawn to seal the joint. To sweat a fitting, begin by cleaning both the end of the pipe and the fitting with sandpaper or wire pipe reamer. Apply a substance termed *flux*, which aids in the chemical conversion between pipe and solder, to both the cleaned pipe and connector. Slide the fitting over the end of the pipe and position at the proper angle. Set the joint on a wood block or use a locking pliers to hold the pipe, then heat the fitting using a propane torch. When the

Figure 7–13 Cleaning PVC pipe with pipe-cleaning solution. Clean the outside of the pipe and the inside of the fitting. When dry, liberally apply glue in the same manner. Most cleaning and joining solutions provide a swab applicator in the can.

Figure 7–14 Joining PVC pipe after applying glue. Use a twisting motion to push pipe completely into fitting.

joint starts to turn a darker color, touch a piece of solder to the joints and hold it there while it melts and is drawn into the joint. When the solder starts to drip, the joint is filled. Let the joint cool before handling.

Copper tubing is joined using solderless compression fittings. The compression fitting has two parts that sandwich around the end of the tubing and are twisted together using a wrench to complete the joint. Slip the end of the fitting with the open center over the tubing. Insert any washers provided with the fitting. Ream the end of the tubing and insert the part of the fitting with the rounded end. Apply joint compound or joint compound tape over the threads of the second piece. Slide the first part up the tubing to the second part and twist together tightly.

Copper installations should be laid out beginning at the supply point, cutting and joining pipe as the installation moves toward the service point. If placing it in a trench, carefully lay the piping and pressurize the system, checking for leaks. Carefully backfill the trench.

Joining Galvanized and Black Pipe

Galvanized pipe and black pipe are rigid steel pipes. Galvanized pipe is used primarily for water distribution and black pipe is used for gas distribution. Both pipe types come in a range of sizes and have a variety of fittings and valves available. While either pipe can be used in exterior applications, rust resistant galvanized pipe is a better choice.

Galvanized and black piping are very difficult to trim to appropriate lengths in the field because the pipe needs to be rethreaded for joining after each cut. The best solutions are to have a variety of prepared lengths available, or to measure the piping requirements and take to a hardware store for cutting and rethreading. Both piping types are joined by applying a joint compound to the threads, threading fittings onto pipe, and then tightening.

CUTTING PAVING AND WALL MATERIALS

> **CAUTION**
>
> When cutting any wall material, wear proper clothing and safety equipment. Use cutting equipment according to manufacturer's instructions.

Marking Materials

The easiest way to mark for cutting is to hold the material in position and use a pencil to transcribe the cut location directly onto the material. Measuring can be done if a high degree of accuracy is maintained.

Marking with a magic marker provides better visibility of the mark when using a wet saw. Make the mark on the scrap side of the cut location so the mark will not be visible when the material is placed. Most paving and wall material must be cut to within 1/8 inch of required dimensions to avoid problems with fit. Irregular and angle cuts may take two pieces to get the correct shape. Cutting methods make curved cuts difficult, and notching is difficult to obtain with all cutting tools except a masonry wet saw.

Brick or Stone Set

Clay brick can be cut, or cleaved, with a wide chisel-like tool called a set (also called brick set or **stone set**) (Figure 7–15). Place the brick on a solid surface and align the brick set along the mark for the cut. Lightly tap the set with the hammer to create a weakened joint. Strike the end of the set with a hammer. The brick should cleave along that angle. Concrete paving block can be cut in this method with reasonable success, but precast concrete does not cleave well with a

Figure 7–15 Cutting paving materials with a brick set. Set the paving material on a flat surface. Place the blade of the set along the line where you want the stone to cleave, then strike firmly with a hammer.

set. Stone is cut in a manner similar to that used for clay brick, although there may be some irregularity in the edge of stone after being cleaved. Stone can also be scored along the cut line with a cutoff saw and then struck on the opposite side with a stone set to cleave.

Brick/Stone Hammer

Striking stone with the pointed end of a brick hammer can chip or cleave certain types of stone (Figure 7–16). The blunt end of the hammer can be used to knock off bumps or edges that prevent a tight fit. To reduce splitting of the stone, strike the offending piece from the side rather than from the top.

Hydraulic Block Splitter

The double blade **hydraulic splitter** is a piece of equipment that will accurately cleave stone, brick, or precast units. Wear gloves and safety glasses when operating a splitter. The material is set on a table with the cleave location directly below a striking bar. The

Figure 7–16 Shaping stone edge with a brick hammer.

bar is lowered to the stone manually and is tightened. An operator then repeatedly presses a foot pedal that applies pressure to the striking bar through a hydraulic cylinder. When the cylinder reaches the right pressure, the material is cleaved. While this is a fast way to cleave materials, waste can be excessive with a hydraulic cutter.

If the splitter is equipped with a tilting base plate, it will help improve the cleaving of interlocking concrete pavers. When the bar is lowered, angle paving material downward away from the cut location. This will provide a slightly undercut angled end that will improve the pavers' placement.

Cutoff Saw

A cutoff saw is a handheld cutting tool powered with a gas motor and a circular diamond or carborundum blade for cutting. Operated much like a chain saw, the cutoff saw creates a large amount of dust, but is portable and useful for remote sites without electricity. To cut with a cutoff saw, mark the cut location and anchor the material to avoid movement while cutting. If necessary, place a block underneath the material to avoid cutting into the surface below. Wear gloves, safety glasses, dust mask, and ear protection. Start the saw and position it over the material to be cut. Run the saw to full speed and slowly lower onto the paving material. Let the saw pass completely through the material before removing the blade from the cut.

Wet Masonry Saw

A wet masonry saw uses a moving table to pass materials through a fixed diamond blade. A fine spray of water helps remove debris and reduce dust. To cut with a masonry saw, mark the cut location and adjust the saw blade to the proper depth. Wear gloves, safety glasses, and ear protection. Start the saw and make sure the water is flowing. Set the paver on the moving cutting table and align the cut mark on the paver with the saw blade. Slowly pass the paver through the blade (Figure 7–17). If the motor begins to slow (indicated by a lowering of the pitch of the motor), back the material up and restart the pass through the blade at a slower pace.

Notching with a Wet Masonry Saw. When cutting unit pavers to fit around objects in the paved surface, notching may be required. This difficult task can be accomplished with varying degrees of success using a wet saw to make the cuts (Figure 7–18). Mark the cuts. Turn the paver over and cut along the long dimension one inch beyond the mark. This extra cut length will be on the back side of the paver and hidden when installed. Turn the paver on its side and adjust the saw blade to match the depth of the second cut. Leaving the block on its side, position the block and pass it through the blade. The two cuts should meet and complete the notch.

Chain Saw

Wood wall products are best cut with a chain saw. Bring extra blades, oil, and the maintenance kit if several cuts are required. Mark the tie/timber and place it so that the longer portion will have adequate support. Wear gloves, safety glasses, and ear protection. Use the foot to safely anchor the tie/timber or have a co-worker, standing at a safe distance, hold the piece. Set the saw on the ground to start it. Rev the saw to full speed and then slowly lower it into the material. Cut with the material close to the saw body. Avoid contacting the blade tip with any material. Move the saw with a slight rocking motion as it passes through the material. Make a square cut beginning on the top side, letting the shorter piece fall freely when cut.

Cutting Resilient Pavers

Rubber-like resilient pavers are difficult to cut using common cutting methods. Thin-bladed saws and knives work best. A pruning bow saw, hacksaw, or bread knife works best. Mount the paver in a vise to secure while making cuts.

Alternatives to Cutting Paving Materials

In any unit paving project, the cutting of pavers is inevitable. Few projects are laid out to match the dimensions of unit pavers perfectly. Cutting can be reduced, however, if factory-produced halves and edging pavers are available.

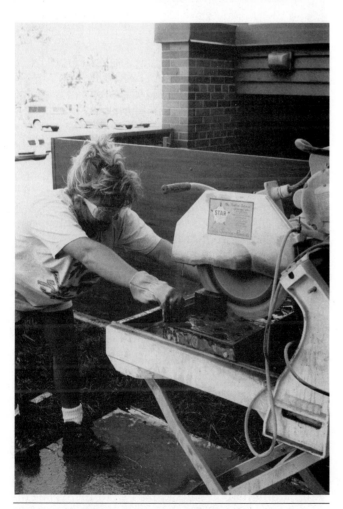

Figure 7–17 Cutting paving materials with a wet masonry saw. Cutting with a masonry saw is slow but accurate.

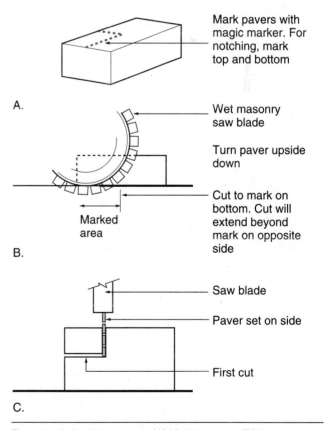

A.

Mark pavers with magic marker. For notching, mark top and bottom

Wet masonry saw blade

Turn paver upside down

Cut to mark on bottom. Cut will extend beyond mark on opposite side

Marked area

B.

Saw blade

Paver set on side

First cut

C.

Figure 7–18 Notching pavers. (A) Marking paver. (B) First cut for a notch. (C) Second cut for a notch.

CUTTING, DRILLING, FASTENING, AND CONNECTING WOOD MATERIALS

Cutting Wood Materials

Cutting is one of the basic carpentry techniques that is required to install wood improvements. Proper cutting begins with learning the basic cutting techniques. Four basic cuts are predominantly used in exterior carpentry—the cross cut (or square cut), miter cut, rip cut, and the bevel cut. (Figure 7–19). The square cut requires the carpenter to mark and remove a portion of the board at a right angle to the edge and face of the board. This cut is the most common one for exterior work and can be used for executing most types of joints. The miter cut runs at a right angle to the edge of the board, but at a determined angle across the face of the board. Miter cuts are used in joining railings, at corners, in rafters, and in other nonsquare portions of the structure, which require flat placement of a lumber

piece. The rip cut runs the length of a board parallel to its edge and at a right angle to the edge of the board. Any of the three previous cuts can be angled to the edge of the board to create a beveled cut. The bevel cut is used when overlapping materials to prevent warping and is used when lumber is placed on edge. Occasionally, usually in rafters, a complicated beveled miter cut may be required. This type of cut makes calculated angles to both the face and edge of the board.

Cutting with saws requires support for the piece of lumber. Rest the lumber on a flat, solid surface with the portion to be trimmed extending beyond the edge of the surface. If the trimmed portion is so long that it needs to be supported by hand while cutting, be sure the support does not bind the saw blade. The holder should not grasp the piece of lumber, but instead should place a hand below the piece and allow the wood to rest lightly. Most short pieces being trimmed can be cut without support and allowed to fall after being cut.

> **CAUTION**
>
> Read the manufacturer's instructions before operating power equipment.

Cutting with a Circular Saw. The direction of cut when using a **circular saw** should place the weight of the saw motor on the supported portion of the lumber. Position the body so that while sawing both the mark and the blade at the point of the cut can be observed. Verify that the power cord is not near the saw blade. The bottom plate of the saw should rest flat on the lumber being cut. Align the blade or directional mark on the guide with the mark on the lumber, start the saw, run the motor to full speed, then run the saw slowly forward along the mark and through the material (Figure 7–20). When cutting thick materials such as 4 × 4s, it may be necessary to mark on both the top and bottom of the lumber. Cut the top and turn the lumber over and make a second cut aligned with the first (Figure 7–21).

Notching wood materials requires the use of a hand saw or sabre saw to avoid overcutting the material. Notch by marking and cutting with a circular saw until the tip of the blade reaches the intersecting mark. When the blade stops, remove the circular saw and complete the cut using a sabre saw or a hand saw. Hand saws should be held so that the blade is cutting straight up and down when notching.

Combination cuts require additional planning and setup. Combination cuts utilize two of the basic cuts described earlier, such as a beveled miter cut. While not common in the landscaping field, joists, rafters,

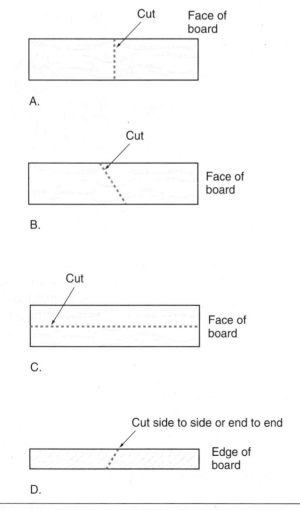

Fgure 7–19 Basic lumber cuts. (A) Square cross cut. (B) Miter cut. C) Rip cut. (D) Bevel cut.(

7–20 Proper cutting technique for a circular saw.

7–21 Cutting large-dimensioned lumber with two cuts.

and trim may be placed at angles that will require this specialty technique (Figure 7–22). Before executing the cut, double-check all measurements and angles to verify they are correct.

Cutting with a Sabre or Reciprocating Saw. If the cut involves the use of a sabre or reciprocating saw, use the same procedures and precautions as previously described for a circular saw. The reciprocating saw, which works like a large version of a sabre saw, is often used to trim pieces of lumber that have already been installed. Long blades for this saw allow access into tight corners, and even allow the blade to bend slightly to saw around corners. To reduce the heavy vibration caused by the reciprocating saw, hold the blade guide firmly against the piece being cut (Figure 7–23). Trimming of posts and unanchored wood members is best accomplished with a circular saw to reduce the chance of vibrating the member out of place.

Cutting with a Hand Saw. When using a hand saw, position the saw at the mark and hold it in position with the thumb against the saw blade. Begin with

7–22 Bevel/rip cut along length of dimensioned lumber.

short strokes until a groove is cut along the mark that holds the saw blade in position, then pull back the thumb and extend the stroke to the full length of the blade (Figure 7–24). Downward pressure on the saw is not required to cut—let the saw blade do the work. Any piece of lumber less than 1 inch thick, including trim and lattice, should be cut using a hand miter saw or other blade type with very fine teeth.

Drilling Wood Materials

Drills are used for boring holes and driving fasteners. When drilling pilot holes, select a wood bit for small holes or a spade bit for larger diameter holes. Insert the bit into the drill and tighten securely using the chuck key. Place a block of wood under the piece being drilled if the bit penetrates through the wood. Position the bit over the hole location and hold the drill at the same angle as the desired hole. Start the drill run motor to full speed and apply downward pressure (Figure 7–25). Deep or wide holes may require that the bit be pulled out of the hole every 2–3 seconds to clear the wood shavings out of the bit.

When using the drill to drive fasteners, install the appropriate bit for the drill. Tighten and secure the bit using the chuck key. Place the fastener on the drill bit and, while holding the fastener, position the tip at the location where the fastener is to be installed. Hold the drill and fastener at the angle desired for driving the fastener. Lightly hold the top portion of the fastener shank while slowly starting the drill. When the fastener "bites," or begins to enter into the wood, release the shank. Increase the drill speed, apply increasing downward pressure, and continue driving until the fastener is completely installed. It is important to maintain downward pressure on the drill to assure that the bit remains in contact with the fastener head, otherwise the bit will slip out and strip the head. Maintain a straight alignment between the drill and fastener to assure solid contact between bit and fastener head. When removing fasteners, reverse the drill direction and place the bit on the fastener head. Maintain downward pressure while removing the fastener. The friction may make fasteners hot for a few seconds after removal.

7–23 Cutting using a reciprocating saw.

7–24 Proper hand-sawing technique.

7–25 Drilling holes using a cordless drill.

JOINING, SPLICING, AND FASTENING MATERIALS

Joining Materials

Several methods for joining materials are available for the carpenter, with two primary ways of joining used for exterior applications (Figure 7–26). Selection of the proper method is based on the appearance and durability required of the joint. Connecting two lumber pieces often involves placing the square ends of pieces of lumber end-to-end in what is called a butt joint. The butt joint requires square cuts on the ends of each piece of lumber being joined. Butt joints are acceptable for structural applications and work in situations where the lumber is setting on edge or laying flat, and can be executed using direct nailing, **toenailing,** or many of the other connection techniques described in the following sections. The joint formed when cutting pieces of lumber with a miter cut and placing the angles edges together is termed a mitered joint. Mitered joints are typically used for appearance and practicality when pieces of lumber meet at a corner, or

to prevent warping along straight runs. Mitered joints can be executed with the lumber setting on edge or laying flat.

Other types of lumber joints that are available but seldom used in exterior carpentry include the mortise-and-tenon joint and the dado joint. While each of these joining methods is valuable to the carpenter in the proper application, they tend to hold moisture and work counter to good drainage recommendations when used in exterior stuctures.

Splicing

Strengthening joints in structural situations may require a special technique called splicing. When one piece of lumber will not cover the length required and two pieces must be connected to make a structural support, either the overlap or gusset method of splicing must be employed. Good planning of a project will minimize the number of splices that have to be made. If splicing is required, place the splice over a structural support (Figure 7–27). Unsupported splices run the risk of sagging over time.

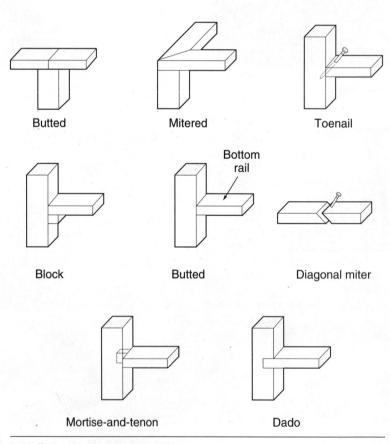

7–26 Typical lumber-joining techniques.

To overlap splice two pieces of lumber, extend each piece past the end of the other by 1 foot and bolt them together. If the structural member is composed of more than two pieces, the ends of the pieces can be staggered and overlapped. A minimum of four bolts are required to make a connection, and six to eight may be required for longer spans or larger members (Figure 7–28).

In situations where the pieces that are being spliced must maintain a straight alignment, an alternative to overlapping is installing a gusset. A gusset is a piece of dimensioned lumber that is bolted alongside the pieces being spliced. To install a gusset, cut a 2-foot length of 2 x treated lumber the same width as the pieces being spliced. Place the gusset flush with the spliced pieces, drill pilot holes, and install a minimum of six bolts—three in each spliced piece (Figure 7–28). Placing eight bolts in a rectangular pattern will provide additional strength.

Fastening Materials

Methods for fastening wood materials continue to evolve. The basic premise for fastening remains one of providing maximum strength with appropriate efficiency. Quality of workmanship has also entered into the picture as the use of power equipment increases in the carpentry field. Pneumatic nail guns and screw guns have appeared, adding speed at the expense of (according to many) quality. A contractor's selection of the appropriate fastening method remains a choice based not only on strength, but also on value and workmanship. (See chapter on materials for exterior carpentry.)

Nailing. The most common fastening method for lumber is nailing. Except for special situations, nails should be driven at a right angle to the face of the lumber. Position the nail at the location where it is to be installed, holding the nail lightly between the thumb and index finger. Grasp the end of the hammer handle and tap the nail until it is driven into the wood far enough that it stands on its own. Continue driving the nail, lightly at first, until the nail begins to enter the wood and then with full force, until the head is flush with the piece of lumber. Effective use of the hammer's leverage requires that it be grasped at the end of the handle and not near the head (Figure 7–29). To avoid missing or misdriving the nail, focus the eyes on the head of the nail during the entire driving process.

When fastening wood members that butt against each other, either end nailing or toenailing is used (Figure 7–30). A warning about connections of this type—end nailing is not a strong connection due to the nail entering the lumber parallel to the grain.

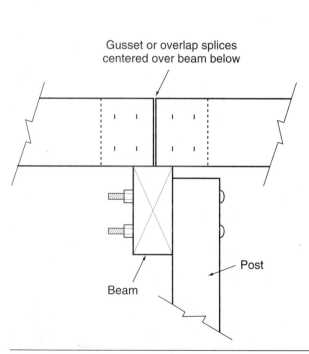

7–27 Splicing lumber over beams.

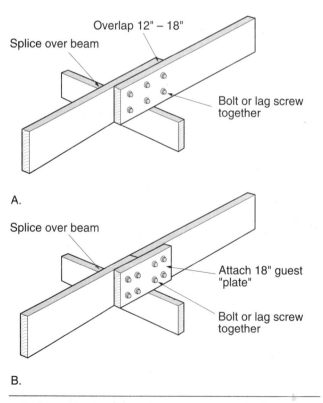

7–28 Splicing lumber. (A) Overlap splice. (B) Gusset splice.

7-29 Proper hand-nailing technique.

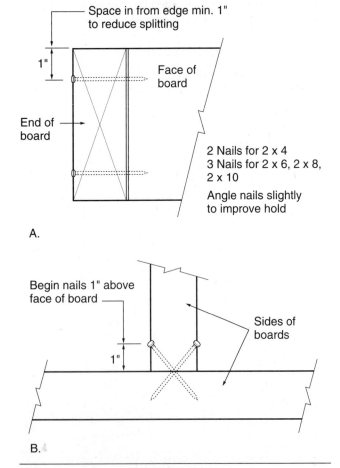

Space in from edge min. 1" to reduce splitting

1"

Face of board

End of board

2 Nails for 2 x 4
3 Nails for 2 x 6, 2 x 8, 2 x 10

Angle nails slightly to improve hold

A.

Begin nails 1" above face of board

Sides of boards

1"

B.

7-30 Nailing lumber. (A) End nailing. (B) Toenailing.

Review all alternatives for fastening wood members before selecting these choices. End nailing is accomplished by driving nails through the face of one piece of lumber into the end of the other. Slightly angling the nail will strengthen the connection. When a piece of lumber butts against a piece of lumber thicker than the length of the nails, toenail the pieces together. Toenailing requires driving the nail at a 45 degree angle through the face of one piece of lumber into the face of the second piece. To properly toenail, position the nail back from the second piece about one-third the length of the nail. Drive the nail lightly into the first piece at a right angle. When the nail bites into the surface of the first piece, tilt the nail to the 45 degree angle and complete driving the nail. Toenailing will require holding the first piece of lumber in proper position so that it does not shift when driving the nail. When improperly done, toenailing can also create a weak connection.

With thin lumber such as lath and lattice, nailing can easily split the wood. Examine the end of the nail and observe that the point is diamond shaped. Orient the long dimension of the diamond perpendicular to the grain of the wood to reduce splitting.

Screwing. Fastening materials using screws provides an extra measure of strength not offered by nails. As weather moves the wood around, nails tend to pull out from the wood pieces they are connecting. Screws remain in position despite minor structure movements. The variety of shapes, materials, and sizes makes screws a practical choice for aesthetic as well as structural connections.

Installation of screws is best executed by positioning the piece of lumber, temporarily nailing it if necessary, and drilling a pilot hole. The pilot hole should be slightly smaller than the diameter of the screw shank. The pilot hole and screw installation should be at right angles to the face and edge of the board. Some woods are soft enough to be fastened without the pilot hole. Regardless of wood type, drill pilot holes if working within 2 inches of the end of the board or within 1 inch of the edge of the board, if using screws with a shank diameter of 1/4 inch or more, or when using any carriage bolt. Holding the screw by the end, hand twist the screw into the pilot hole to start the connection. Continue twisting the screw into position using hand or power tools, lightly holding the smooth portion of the shank until the screw can stand in the pilot hole without assistance (Figure 7–31). Screw holders that eliminate the need to handhold the fastener are available. Attach the screw holder in the same manner you would attach a drill bit, then insert screws into the holder.

Bolting. Bolting makes very strong structural connections, but bolts are seldom used for surfacing or aesthetic situations. To install bolts, position the piece of lumber, temporarily nailing it in place if necessary. At the location where the bolt is to be located, drill a pilot hole that is slightly larger than the diameter of the bolt though all lumber pieces being connected. Galvanized carriage bolts that are 1 inch longer than the combined thickness of all lumber pieces being connected

should be selected. Drive the bolt through the hole (Figure 7–31). Whenever possible, the head of the bolt should be located on the side where the public will view the connection. Install a washer and nut on the opposite side and tighten. Check alignment and elevations before completely tightening the bolt.

Gluing. Gluing exterior connections is limited to locations that will not be exposed to moisture. Exterior glues, which are resistant to exposure, can be used for fastening caps and surface trim when other fasteners will not work.

Avoiding Problems with Connections

When making connections, consider alternatives that will reduce damage to the wood or structural failure. When installing connectors near the ends or edges of boards, there is a possibility of splitting the lumber. The best solution is to predrill pilot holes slightly smaller than the diameter of the connector. Dulling the end of the nail, or turning the nail so that the wide point of the diamond tip is perpendicular to the edge of the board, may also help.

Structural failure can result from using too few connectors or improper spacing of connectors in a piece of lumber. Too few connectors may not provide the resistance required to overcome forces attempting to pull the pieces apart. Improper spacing may weaken the lumber. Figure 7–32 identifies typical nailing and bolting patterns that should provide improved strength.

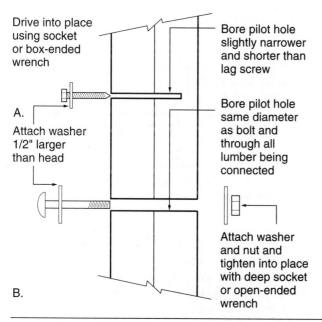

7–31 Installing lag screws and carriage bolts. (A) Lag screw. (B) Carriage bolt.

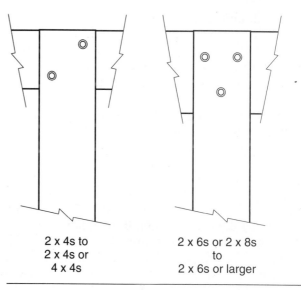

7–32 Fastening patterns that provide maximum stability.

Chapter 8

Construction Staking

Regardless of the size of a project, there will be a need to identify the location and elevation of the landscape elements to be constructed. Determining how to perform construction staking work will require a judgment regarding the complexity of the project and the level of accuracy required. It would be inefficient to hire a land surveyor if the project involves only a short walkway. It would be challenging to measure a large project with only a tape measure. Construction staking addresses techniques used in locating improvements on the site, establishing the elevation of improvements, and locating plant material. Each of these three construction staking requirements utilizes different approaches and should be performed in a logical order.

Locating improvements in the horizontal plane is typically performed first, followed by establishing the vertical elevation. These first two steps need to be completed before construction begins, and may require that improvements be measured and restaked several times during the project. Location of landscape plantings is typically completed later in the construction process, often after finish grading has been completed.

Projects that may require restaking several times benefit from the establishment of permanent reference points. Horizontal plane reference points depend on the type of location method used. Typical references are corners of structures, edges of roadways, fence lines, baselines, or other permanent improvements that can be accessed and will not be disturbed by the construction. If existing permanent improvements do not exist on a site, establish reference points by driving fence posts into the ground and painting them for visibility. At least three such reference points are required for most construction staking systems, and more locations will be beneficial. Vertical measurements should be controlled by a benchmark, or permanent marker, where the elevation is known. Benchmarks for projects can be set on existing foundations, fire hydrants, manhole covers, or other objects that will remain throughout construction, will not move, and can be seen from locations where a level will be set up.

LAYOUT EQUIPMENT

Layout of work requires the use of specialized equipment designed to make the work easier and more accurate. The choice of equipment is determined by the amount of layout you perform, the accuracy required by the projects, and the cost of the equipment. Typical layout equipment includes the following (Figure 8–1):

- Tapes. A tape measure is an essential piece of equipment for all contractors. Useful in measuring horizontal and vertical distances, handy to carry, and marked for levels of accuracy acceptable to almost every building trade, the tape measure remains the one indispensable piece of equipment for layout. It may be necessary for the landscape contractor to have tapes available that measure in both feet/inches and feet/tenths. It is also valuable to have at least one tape that measures distances of 100 feet or more. If measuring or marking by yourself, utilize a large screwdriver to anchor the zero end of the tape.
- Builder's levels. Builder's levels allow the contractor to determine slope, plumb, and in

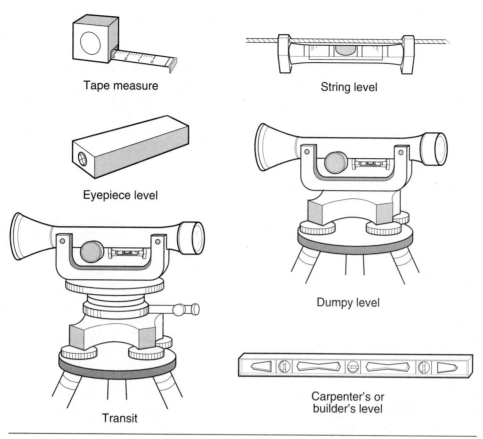

Figure 8–1 Layout Equipment and tools.

advanced models, angles, with a high degree of accuracy. The effectiveness of a level can be increased by selecting a piece of dimensioned lumber that is straight and resting the level on top of the board. This technique is common for determining grade for walks and paved areas.

- String levels. A string level is a simple bubble level that hangs from a line stretched between two points. String levels are handy for determining slope or level for paved areas, walks, decks, and other landscape elements that require a consistent grade between two points.
- Dumpy level. **Dumpy levels** are tripod-mounted focal instruments that provide a magnified eyepiece for the contractor to look through. The dumpy level has crosshairs that allow accurate measurement at long distances without serious distortion. The level and tripod are also adjustable, allowing the contractor to set up on a variety of terrains. Surveyor's rods, a vertical stick with measurements, or a tape measure are required to complete measurements with this level.

- Eyepiece level. The eyepiece level is a small, handheld version of the dumpy level. The limited accuracy of this tool limits its use to establishing rough grades for planting or turf areas and rough grading. A rod or tape measure is required to complete measurements with this instrument.
- Transit. **Transits** are dumpy levels that provide a means for measuring horizontal and vertical angles. Used by surveyors, such instruments are useful if working on a steep site or when making measurements from a single point. A process called stadia uses the eyepiece cross hairs to measure both level and distance in one step. Newer transits today utilize lasers that can measure height and angle over great distances with much higher levels of accuracy than optical equipment.

SPECIALTY LAYOUT AND MEASUREMENT TECHNIQUES

Following are suggestions that will make layout work easier and more accurate.

Measuring Short Slopes

Select a long, straight 2×4 and lay one end at the top of the slope. Place a builder's level on the 2×4 and adjust the board up or down until it is level. Using a tape measure, measure the distance from the bottom of the 2×4 to the toe of the slope. This distance is the fall of the slope (also referred to as the V in some formulas) (Figure 8–2). The run, or horizontal distance, can also be calculated by hooking your tape over the end of the 2×4. Holding the 2×4 level, measure along the board until directly above the toe of the slope.

Measuring Long Slopes

When slopes exceed lengths of 10 feet, elevation measurements using a straight 2×4 and level are difficult. To measure long slopes, utilize mason's twine anchored at the top of the slope using a screwdriver. Stretch the mason's twine to the bottom of the slope and hang a string level on the twine. Holding the twine level, extend a measuring tape from the twine to the

ground to obtain the vertical measurement of the slope.

If the slope is steep, this process may have to be repeated several times. For subsequent measurements, always move the screwdriver holding the twine to the location where the measuring tape contacted the ground in the previous measurement. Add all measurements to obtain the vertical dimension of the slope.

Using Offsets

Another component of project layout is establishing layout markings that are accurate but do not interfere with the construction. To accomplish this the construction markings made on a project may utilize the concept of offset. Offset places the stake or identification for the improvement at a consistent, predetermined distance and direction from where the improvement is to be actually installed. Typical offsets may place the construction stakes 5 feet outside of where the improvement should be placed (Figure 8–3). The

Figure 8–2 Determining slope by holding a board level and resting on the high point and measuring from the bottom of the board to the low point.

distance and direction are dictated by the project conditions, but should be consistent throughout the project. A common form of offsetting horizontal and vertical markings is the use of batterboards. Batterboards are ideal for many aspects of landscape construction, especially the building of decks, structures, and paved areas. Construction of batterboards is covered under the Location of Improvements section of this chapter.

Floating Dimensions

Dimensions that are labeled with a plus/minus sign in front of them are considered floating dimensions. Floating dimensions provide flexibility in locating the actual object being measured or accommodate the lack of accurate existing measurements. An object labeled with floating dimensions may also be located in another manner, such as being aligned with other objects or adjacent to a fixed object.

Layout of Right Angles with the 3,4,5 Triangle

Determining if an angle is exactly 90 degrees can be difficult in field construction. Use of a carpenter's square will help for small elements, but when laying out a large, paved area, the square may not provide the level of accuracy required. To perform this layout task, use a method known as the 3,4,5 triangle (math majors will recognize this as an application of the Pythagorean theorem used in triangle formulas) (Figure 8–4). Along the project edge you are turning the angle from, mark the point from the point where you need the perpendicular, or right-angled, line to begin (this may also be the corner of the project). From that point, measure 3 feet along the edge you are turning the angle from and make a second mark. From the beginning mark (or corner), stretch a tape in the direction you want the perpendicular line to run. Hold the tape at the 4 foot mark. Place another tape at the

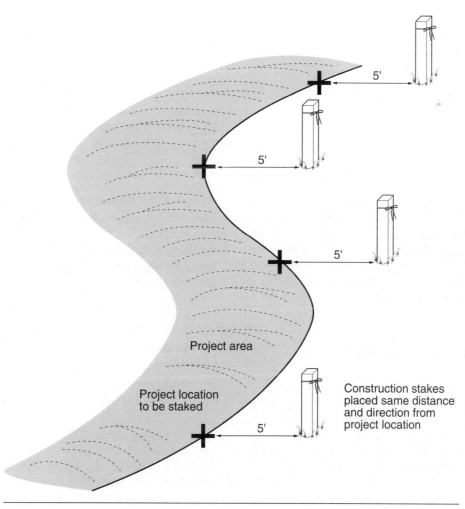

Project area

Project location
to be staked

Construction stakes
placed same distance
and direction from
project location

Figure 8–3 Staking with offsets to reduce restaking time.

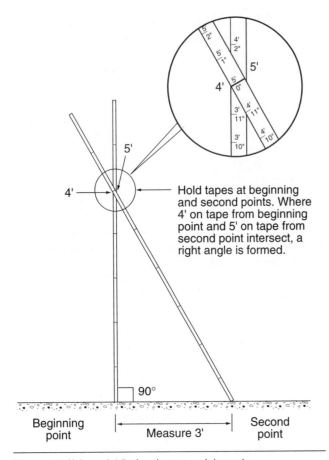

Hold tapes at beginning and second points. Where 4' on tape from beginning point and 5' on tape from second point intersect, a right angle is formed.

Figure 8–4 Using a 3,4,5 triangle to turn right angles.

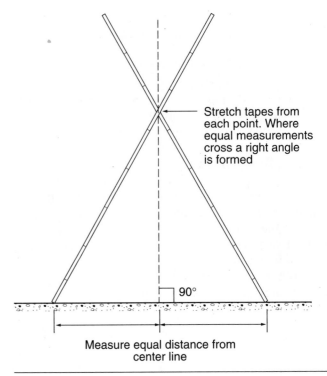

Stretch tapes from each point. Where equal measurements cross a right angle is formed

Measure equal distance from center line

Figure 8–5 Layout of a perpendicular line using the triangular method.

second mark and stretch the tape to 5 feet. Adjust both tapes until the 4 foot mark on the first tape and the 5 foot mark on the second tape intersect. Mark directly below that point. A line traveling from the first point marked (or corner) and passing through that third point where the tapes intersect will be at a right angle to the first edge. When **squaring** decks or paving, you may mark on the forms/joists. This technique will only require use of one tape to make the 5 foot measurement.

Layout of a Right Angle Using the Triangle Method

Obtaining a perpendicular, or right-angled, line to a baseline or any other straight line can be done using the triangle method (Figure 8–5). Measure two points an equal distance on either side of the point on the baseline from where the perpendicular line is to be located. The distance is not critical as long as it is the same on both sides, but minimum measurements of 3–5 feet are recommended. From each of these two points, stretch the tape measures in the direction of

the perpendicular. Select a measurement on one of the tapes and match it with the same measurement on the second tape. Mark the point where these similar measurements intersect. A line laid out from the original starting point through this intersection will be at a right angle to the baseline.

Diagonal Measurements to Check for Square

When laying out a landscape element that must maintain a square (or rectangular) form, check your work using diagonal measurements (Figure 8–6). Using a tape, measure from diagonal corners and compare measurements. If the measurements are equal, the shape is square. If one measurement is longer than the other, the long corner must be pushed in slightly and the short corner pulled out slightly until the measurements match.

Straight Lines

When street trees or other improvements are to be laid out in a straight line, using a **stringline** will save time over locating each individual plant. To obtain a straight line, place a stake at each end of the line and stretch a mason's line between the two stakes. Measurements can be made down the line to required improvements. It may be beneficial to offset the line 1–2 feet if it interferes with construction operations.

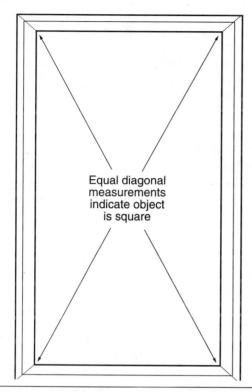

Figure 8–6 Checking for square using diagonal measurements.

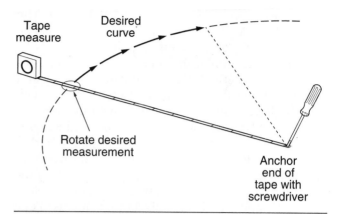

Figure 8–7 Layout of curves using a large screwdriver anchored at the center point and using a tape measure to locate the correct radius. Mark radius as tape is moved.

Locating Curves

Curves can be measured by locating three points—the beginning of the curve, the end of the curve, and the radius point. Use baselines or coordinates to locate the beginning and ending points. If the radius of the curve is identified but the radius point is not, locate the radius point by turning a right angle (see the 3,4,5 triangle) from the curve beginning point. Swing a tape anchored at the radius point along the alignment where the curve is to be located, marking as you go (Figure 8–7).

Ending points for curves can be marked using a deflection angle and chord length, if that information is provided . To utilize this method for locating curves, the beginning point of the curve must be located and a line extended beyond the beginning point for 2 feet. The deflection angle indicates the degrees to the left or right of this extended line where the end of the curve is located. Extend a second line from the beginning point of the curve that is at the correct angle to the first line. Measure the chord distance along this second line to locate the end point of the curve. This method does not locate the radius point for the curve.

Marking

Cutting and joining elements such as lumber, wall stone, and pavers can be simplified by marking the cutting or joining point. This frees up one hand to per-

form the work rather than to hold the tape. Using a "V" to mark points eliminates the confusion of determining which end of a mark is the correct measurement. When temporarily locating elements in the field, use white spray paint or base-path chalk to locate edges. A hose can also be used to temporarily locate edges of a curve. Experiment until you are satisfied with the alignment, then replace the hose with permanent markings.

Marking Center Line Locations for Digging

To aid in excavating holes, use paint to mark a large X, with the center of the X aligned with the center of the hole and the legs of the X extended beyond the excavation. If the legs of the mark are extended far enough, reference marks will remain after digging begins. These reference marks can be used to relocate the center of the hole at any time during the excavation.

LOCATION OF IMPROVEMENTS

Establishing the work area for a project should begin with the location of all elements on a horizontal plane. Among the typical items located during this process are corners of structures, edges of paved and planted areas, center lines for walkways and drives, and locations of points, edges, or center lines for all permanent improvements for a site. Site layout plans typically locate major objects on a plan, leaving specific dimensions to details. On sites that require substantial grading, it is not uncommon to locate only key improvements, such as structures and roadways, until major earth moving has been completed. Waiting until the rough grade has been generally established saves time restaking items that were in areas disturbed by the grading process. Locating elevations of improvements (described in the following topic), should be completed after major improvements have been located. The same stakes used to locate improvements can also be marked with any grade change required.

Location of landscape improvements requires selection of a method that will provide accuracy and efficiency. Projects that are performed by contract will typically have a site layout plan, which dictates the location method that must be used. One of the methods described in the following sections, or a hybrid of these methods, is common on projects designed by landscape architects or site design professionals. Projects that have no site layout plan will require selection of a method for locating improvements. Four methods—object dimensioning, baseline layout, grid layout, and bearings or survey lines—are typical for locating improvements for a project. The first three of these methods are the most common.

Complex layout problems should be contracted to a registered land surveyor. Simple layout can be accomplished by a contractor by combining the following methods with the techniques described earlier. Remember, however, that errors in staking performed by the contractor will require correction at the contractor's expense.

Grid Layout

If a project is covered with buildings and roadways, the grid layout method may be required to maintain accuracy. Grid layout utilizes a series of **Cartesian coordinates** (X and Y measurements) to locate improvements on a site. These coordinates can be located off a pair of baselines set up at right angles to each other. The beginning points and orientation of these baselines should be located on the layout plan.

To utilize grid layout, each point to be located will have an X and Y measurement that corresponds to the distance the point is from the X and Y baseline. To locate an object, select the point to be located and locate the measurement along the X baseline. From that measurement, extend a line perpendicular to the X baseline slightly beyond where you estimate the point is located. Locate the measurement along the Y baseline and, from that measurement, extend a line perpendicular to the Y baseline, which intersects the line extended from the X baseline. The intersection of the two lines is the correct location for the point in question. This process is repeated for all points that require location.

Baseline Layout

Baseline layout utilizes a straight line that runs through the entire site and is measured from a beginning point. Landscape elements are located by making a measurement down the baseline from the begin-

ning point and using a second measurement perpendicular to the left or right off the baseline. Baseline layout is common for small-to-medium sized projects that have a great deal of open space on the site. Baselines also work well if there is a straight curb line, roadway, fence line, or other existing baseline that can be utilized.

Object Dimensioning

Object dimensioning is utilized when the location of improvements can be measured off existing landmarks such as structures. For most small landscape projects, object dimensioning provides adequate accuracy.

Bearings and Survey Lines

Bearings and survey lines method uses lines measured a determined direction (bearing) and distance from a reference point to locate objects on the site. Using an established beginning point such as a property corner, objects are located by providing the bearing and distance from that point. Additional points along a boundary can be located by measuring a second bearing and distance from the previous point. Curves are typically located by interpreting curve data, such as chord and curve length or radius point. Layout of objects described in this manner requires a registered land surveyor to perform the staking.

Batterboards

Batterboards are an excellent tool for maintaining locations and grades for project corners. When properly constructed, batterboards are the only markings that are needed throughout the construction of a paved area, deck, or structure. Following are instructions for installing a batterboard for one corner of a project with an offset beyond the corner of the project being staked (Figure 8–8). The actual offset is determined by how much of the area outside the project will be disturbed by construction. Construct a batterboard using the following steps:

- Select three, pointed 2 × 2 stakes that are 4 feet long and two 1 × 6s that are 4 feet long.
- Stretch a stringline along the alignment of each side of the project being staked. Extend the stringlines 4 feet beyond the corners of the project in each direction.
- Drive the three stakes into the ground. The stakes should be in an L pattern, with each leg of the L paralleling one side of the project. The L should be located 2 feet beyond the stringlines and should straddle the stringlines.

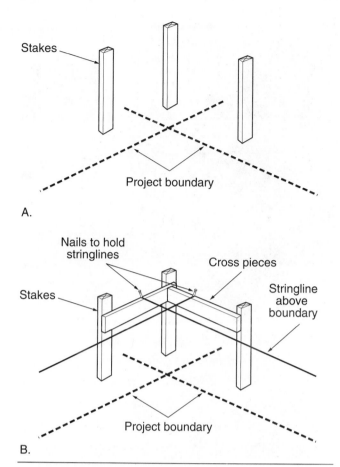

A.

B.

Figure 8–8 Constructing batterboards. (A) Installation of posts outside the project edges and straddling the boundaries. (B) Attachment of cross pieces and lines marking project boundaries. Height of the cross pieces can be set to project elevations if you want to locate boundaries with the same batterboards.

- Halfway up each stake along one side of the L, connect a 1 × 6 between the two stakes. Using deck screws allows connections to be made with minimal disturbance to the stakes.
- Connect the other 1 × 6 between the two remaining stakes.
- Place a second set of stringlines directly above the first set. Connect the stringlines to the 1 × 6 using a small nail.
- Measure to verify the stringlines are set at the project dimensions. Adjust lines left or right and reset if necessary.

ESTABLISHING ELEVATIONS

Once locations of improvements have been identified, the proposed elevations can be marked. Selection of a method for identifying elevations depends on the accuracy demanded by the project. Projects with no grading plan or with a single layout drawing may state elevations by indicating the direction and percentage of slopes. Another form of identifying elevations is to mark proposed heights of objects by labeling the object with a spot elevation. It is important that drainage away from structures and important improvements be maintained when establishing elevations. If grades are critical, consider hiring a registered land surveyor to perform the staking. As with layout staking, errors staked by the contractor are corrected at the contractor's expense.

Like the horizontal layout of a site, elevation measurements may require restaking as the project progresses. Often the site is staked once to allow the grading contractor to perform rough grading activities, and restaked or measured again when areas within the site are built. Individual portions of the project may be remeasured as they are built, such as grades for tile inlets, paved areas, and slabs for structures. Accuracy for rough grading should be within 2–3 inches. Finish grades should be accurate within 1 inch for lawn and planting areas and under .1 inch (one-tenth inch) for paved surfaces.

Offsets for construction staking of grades are sometimes unnecessary. Rough grading layout typically places a grade stake at the correct location and the grading contractor adds or removes soil around the stake until the new surface is at the correct bevel. Cut areas leave the stake on a mound of soil that is removed when the surrounding area is at the correct level. In fill areas, soil is piled around the stake until a mark on the stake is reached.

Batterboards and Stringlines

Vertical layout functions can be obtained from the batterboards described in the previous section if the proposed finish elevation of the corner being staked is known. Adjust the height of the 1 × 6 so that the stringline passes over the corner at the proposed elevation. The batterboard can now hold the stringline for the edges of the project in the correct location as well as at the correct elevation.

Slope Direction and Percentage

When a plan indicates no specific spot elevations, the contractor must rely on indications of slope direction. This method is most likely to be used when defining drainage patterns for lawns, but may also be used for paved surfaces. Using the slope arrows as a guide, grade the project with a consistent slope in the direction indicated. Set the elevations for structures, paved areas, and utilities first, followed by grades for lawn and planting.

Cross Grades

When plans indicate a slope direction but not a specific percentage, a straight 2×4 and carpenter's level can be used to stake the grade. Set the desired grade on one side of the construction area and then lay the 2×4 from across the project. Set the level on top and adjust the second side up or down until the level reads the general slope direction required. (see Figure 8–2).

Surveying to Determine Spot Elevations

When specific elevations are indicated for points of a project, it will be necessary to accurately measure and mark each of the points. Staking in this manner requires the use of a dumpy level to calculate the precise elevation. The surveyor's rod that is used must match any grading plan measurement increment. If feet and inches are used on the plan, use a rod that is calibrated in feet and inches. If decimals (feet and tenths) are used, use a rod that is calibrated in decimals. Math is easier to calculate using decimal measurements and most grading plans use decimal markings. Calculations that require angles as well as elevations should be staked by a registered land surveyor.

Level Setup. Select a location for placement of the level that has good visibility of the benchmark and all locations that will require construction staking. This site should be in an area as level as possible where construction traffic and activity will not disturb the instrument. Set the tripod by first spreading the legs and positioning them with the mounting plate at a comfortable height and as close to level as possible. When mounted, the instrument should be at eye level for ease of operation. Push the legs of the tripod securely into the ground to avoid accidental movement.

Remove the instrument from its case and secure it to the mounting plate. Some levels have a threaded connector that is attached through the mounting plate; others screw directly onto the plate. The telescope is leveled using the four thumb screws that support the instrument on the mounting plate. A small bubble level below the telescope indicates if the instrument is properly leveled. Turn the telescope so it is aligned with one pair of thumb screws (Figure 8–9). Turn the thumb screws in opposite directions to center the bubble. If the level bubble needs to move to the left, turn both screws toward the center of the mounting plate. If the level bubble needs to move to the right, turn both screws toward the edge of the mounting plate. With these motions, the screws are actually turning in opposite directions and cause the telescope

Figure 8–9 Leveling a survey instrument. Always turn the leveling screws in opposite directions.

to tip in one direction or the other. Continue adjusting until the level bubble is centered.

Turn the telescope 90 degrees to the right and repeat the leveling operation with the second pair of screws. When level in that direction, turn the telescope back to the original direction and repeat the leveling process. The telescope should now be level. Test by turning the telescope in all directions and verifying that the bubble stays centered. If the bubble is not centered, repeat the leveling process. The base of the thumb screws should always remain in contact with the mounting plate. If any screw lifts from the plate, the entire process needs to be redone.

Using the Level. When operating the level, use extreme caution not to bump the instrument out of place. Any movement will require resetting the level. Gently aim the telescope in the direction where you will be taking measurements. Look through the telescope using one eye and focus using the knob on the side (Figure 8–10). When properly aimed and focused, you should see the leveling rod (or tape) clearly.

Figure 8–10 Focusing a survey instrument. Use caution not to bump the instrument out of level.

Taking a Backsite. The first step in calculating elevations is to take a backsite. Have a second worker set the surveyor's rod on the benchmark. Look through the telescope, focus, and read the elevation on the rod. If you are unable to read a foot number, the rod operator may need to lift the rod until a foot reading is visible. Add the reading to the elevation of the benchmark to obtain the height of instrument (Figure 8–11).

Taking Foresites. The rod operator may now move to a location where an elevation measurement is needed. Drive a construction stake into the ground at that point (one may already exist from locating the improvements). Look through the telescope, focus, and read the number on the rod. Subtract the reading from the height of the instrument to obtain the elevation of the spot where the surveyor's rod is setting.

Marking and Reading Grade Stakes. Read the grading plan from the construction documents to determine the proposed elevation for the spot where the rod is setting. If the proposed and existing elevations are the same, the stake will be marked "0" (no cut or fill).

If the proposed finish elevation is higher than the existing elevation, subtract the existing elevation from the proposed elevation and mark the stake as a fill using the difference as the fill amount. Fill stakes are marked + X.X or F X.X, where X.X is the amount of fill necessary (1 foot of fill would be marked as +1.0 or F 1.0). If the proposed finish elevation is lower than the existing elevation, subtract the proposed elevation from the existing elevation and mark the stake as a cut, using the difference as the cut amount. Cut stakes are marked as –X.X or C X.X, where X.X is the amount of cut necessary (2 feet of cut would be marked as –2.0 or C 2.0). This process is repeated for each spot elevation that must be marked (Figure 8–12).

Moving the Instrument. In the event the instrument must be moved, take a foresite at a location that is visible from both the existing and new instrument locations. Have the rod operator stay in the same location while the instrument is moved to the new location and leveled. Take a backsite to the same rod location. Calculate the difference between the foresite and backsite. If the backsite is less than the foresite, subtract that difference from the old height of the instrument to obtain the new height of the instrument. If the backsite is greater than the foresite, add that difference to the old height of the instrument to obtain the new height of the instrument. Continue taking foresites and calculating elevations using the new height of the instrument.

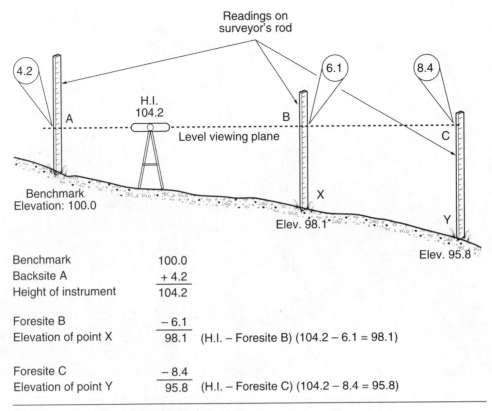

Benchmark	100.0	
Backsite A	+ 4.2	
Height of instrument	104.2	
Foresite B	– 6.1	
Elevation of point X	98.1	(H.I. – Foresite B) (104.2 – 6.1 = 98.1)
Foresite C	– 8.4	
Elevation of point Y	95.8	(H.I. – Foresite C) (104.2 – 8.4 = 95.8)

Figure 8–11 Performing elevation calculations.

Figure 8–12 Cut markings on a grade stake. This stake shows a cut of 1.6 feet below existing grade (existing grade is indicated by the markings at the bottom of the stake).

LOCATING PLANT MATERIAL

Locating plant material for a site is usually a process of marking the boundaries of planting beds and identifying tree and shrub locations. Locating trees and shrubs can be done with construction flags placed at the center point of the proposed planting location (Figure 8–13). White or green flags are recommended since they are less likely to be confused with standard utility markings. In projects with many varieties of plants, it is also beneficial to print the name of the plant on the flag. Planting bed edges can be marked with white paint indicating the edge between the bed and lawn or other types of plantings. Plantings within a bed may be marked using stringlines or paint lines placed on the ground, or reference markers placed along the edges of the planting bed. As with all markings, these indicators may need to be replaced several times during the course of a construction project. When available, utilize existing improvements as references for plant material location.

Figure 8–13 Flagging of proposed planting to review placement.

Success of a landscape project can often be determined by how well the contractor prepares for the work. Before turning a single shovel of soil, all preconstruction activities should be completed. Section 1 describes the construction process, legal requirements, interpretation of construction documents, safety issues, math calculations, estimating and bidding, basic construction techniques, and construction staking activities typically required before undertaking a landscape project.

Landscape construction should be approached as a process with several interlocking steps. To maintain work efficiency, these steps should be completed in an orderly and logical fashion that allows current work to build on, rather than disturb, work previously completed. While contractors may make adjustments to address particular project requirements, the typical steps to completing a landscape project include preconstruction activities, site preparation, rough grading including drainage and temporary erosion control measures, utility installation, retaining wall installation, paving, site structures and wood construction, finish grading, fencing and freestanding wall construction, and amenity installation. While not covered in this text, irrigation piping installation would occur during utility stages, and irrigation head installation before or after turf area establishment. Planting of trees, shrubs, and ground covers, and establishment of turf areas would occur following the completion of construction activities described in this text.

Landscape contractors must address legal and contractual concerns similar to those addressed by building contractors. Building codes, ordinances, and legal controls guide the installation of the portions of the project that may put the public at risk. In certain jurisdictions, permits or review of plans may be required before construction can be approved. Work that is bid typically has a set of plans and specifications that bind the contractor to the details of the project. Protection for the contractor, his or her clients, and the public is obtained in the form of insurance, bonding, and licensing. In addition to providing quality services, the contractor is also required to work within legal boundaries. Before beginning a construction business, contractors should be aware of common torts and activities that may lead to legal action against them.

Occasionally, landscape projects are designed as a general concept or an open-ended project, providing the contractor a great deal of flexibility in layout, detailing, and material selection. In many cases, particularly projects prepared by design professionals, the contractor must work from a specific set of instructions. Construction plans and specifications (specs) are prepared to guide the location, materials, and installation procedures used in completing landscape improvements. Plans and specs are part of the contract for construction and must be followed in order to satisfy legal terms of the contract. Various plans may be included depending on the design professional, but they typically address the preparation (or demolition) of the site, site layout of the improvements, grading and elevations of the improvements, utility installations, plant material installation, and details of critical entities of construction. Specifications provide detailed information regarding the materials and installation methods required for the project along with general information such as warranties, storage, and other topics related to the work. Interpretation of plans and specs requires that the contractor obtain and review a complete set of construction documents. The contractor must also understand the concept of drawing scale and be able to comprehend the variety of graphic standards, symbols, abbreviations, and textures used by design professionals to convey design ideas.

Safety is an overriding concern in any industry, and landscaping is no exception. While the risks of the landscaping industry may not be as obvious as those found in other professions, the potential for serious injury or death is just as real. Safety in the workplace should begin with prevention and educating the worker in safety precautions. Wearing proper safety clothing/equipment and awareness of risks on the job will reduce injury potential. Proper lifting of the materials used in landscaping is important in reducing muscle injury. Due to the amount of excavation in landscape work, utilities pose a significant risk. Calling for locates before digging is essential. The landscape contractor should also be aware of the potential hazards from working below grade, operating power equipment, and working with the variety of chemicals used in construction.

Mathematics is applied to landscape construction in virtually every phase of work. Whether calculating a bid or ordering materials, the contractor must master basic measurement and calculation techniques to succeed. Typical math applications that are found in landscaping work include calculating averages, item counts, linear measurements, perimeter measurements, area calculations, volume calculations, and weight conversions. A contractor must understand the formulas involved in these calculations and know when they are appropriate to use.

Landscape construction also requires an understanding of the estimating and bidding process, and calculating business operating costs such as labor, overhead, and profit. Understanding the techniques used by various types of firms to calculate and present project costs to potential clients can be equally as important to the success of a business as the work performed.

Basic construction techniques should be practiced and mastered before construction begins. Basic techniques range from digging to essential carpentry skills. Operation of specialized equipment is also required to successfully complete landscape projects.

Proper construction staking of the project begins the project on the right track. Staking the location and elevation of improvements is necessary to install landscape elements in the correct position. Contractors must be prepared to locate improvements for a project, or interpret a site layout plan, which utilizes object dimensions, baseline layout, grid layout, or batterboards. Establishing proper project grades requires understanding of batterboards, stringlines, contours and spot elevations, and slope calculations. The landscape contractor must also be prepared to locate, or interpret the locations, of plant material and planting beds for a project.

Section 2

Site Preparation

INTRODUCTION

The path to improving a site begins with properly preparing for construction operations. Generally, this step can be considered in two steps—preserving existing site elements and removing unwanted elements. Site preparation, or site demolition as this phase is termed in some contract documents, requires the same level of precision in workmanship as does any other phase of work. Failure to properly perform the steps of preparation may lead to reconstruction, hazards and penalties, or expensive replacement of landscape elements.

The chapters of this section address site preparation in the order in which it should be undertaken—preservation of existing site elements before removal of unwanted elements. Preservation begins with the identification of site access and storage areas, definition of construction limits, and securing of benchmarks and layout baselines. Preservation of existing site elements continues with the fencing or sheltering of plant material, structures, pavement, and utilities. The chapter on removal of unwanted site elements identifies techniques for removing plant material, pavement, minor utilities, and disposition of landscape waste.

Landscape contractors should carefully read specifications, and discuss the project with their general contractor, to determine if any or all of the site preparation work is prescribed in their contract. On large-scale projects, this aspect of the work is typically undertaken by the general contractor or the grading contractor. Smaller projects may require that the landscape contractor complete all site preparation prior to the start of grading and utility work.

PRODUCTIVITY SUGGESTIONS

Beginning a site preparation project requires that activities listed as preconstruction work be completed. Some specific suggestions that may be of benefit during this stage are:

- Have any questionable boundaries surveyed before beginning work.
- Locate a secure material storage area early in the project.
- Before beginning, review the site for elements scheduled for demolition that may be recycled.

Chapter 9

Preservation of Existing Site Elements

Few landscape bids include the price for replacing a 50-year-old bur oak damaged during construction or replacing a bay window struck by a skid loader. The cost of one construction mishap often exceeds the profit built into a project, requiring that proper techniques for the preservation of existing site elements be practiced throughout construction.

RELATED INFORMATION IN OTHER CHAPTERS

Information provided in this chapter is supplemented by instructions provided elsewhere in this text. Before undertaking activities described in this chapter, read the related information in the following chapters:

- Safety in the workplace, Chapter 4
- Basic construction techniques, Chapter 7
- Erosion control, Chapter 13

SITE ACCESS AND STORAGE AREAS

Defining the location of a site access may fall under the responsibility of the design professional, but is sometimes left to the discretion of the contractor. Selection of a site access could be as simple as finding a level spot to drive off an adjacent roadway, or could be made difficult by a remote, heavily wooded site with steep topography. Considerations for selecting a route to your work should include the following factors:

- Legal access. Access should not cross another landowner's property without permission. In certain communities, permanent and temporary access off-public roads must be approved.
- Visibility. Visibility for vehicles entering and exiting the site should not be restricted.
- Traffic. Traffic should be light during construction hours, since unloading may have to take place from streets.
- Level grades. Level grades make the transitions from street to access easier.
- Size and type of vehicles. Verify that openings are large enough for vehicles that will require access to construction areas.
- Utility locations. Overhead wires and fragile underground utilities should be avoided.
- Avoid construction. Select a route that will not be in the way of present or future construction.
- Minimal disturbance. Site disturbance should be kept to a minimum.

For sites that require a new permanent access (either walkway or drive), roughing out that new access may provide the best route to the construction site. If a construction site is to be completely disturbed, access may take place at any point until the late stages of construction. Projects that involve only landscaping may require that a route be defined across a lawn, with the lawn restored after construction is complete. Access for pedestrians and small equipment may be accommodated through the placement of sheets of plywood on the ground to reduce soil compaction and to protect turf. Regardless of the situation, expect your access to create a disturbance in the site that will require some level of restoration, ranging from aerating compacted soil to removal of temporary driveways.

Selecting the simplest route that will serve all construction is usually the best choice, even if it involves some work to prepare. Once in place, barricades or a temporary gate should be installed to reduce liability and security problems. The greater the risk of such problems, the more secure the perimeter should be.

Storage areas for work materials require many of the same considerations as site access, along with the following additional concerns:

- Delivery may take place from semi-trailers, requiring the storage area to be accessible from both a public street and the construction access.
- Security of the stored materials may be a problem in some environments. Fencing may be necessary to protect valuable supplies from theft.
- Temporary utilities may be required if plant material is watered or if materials are prepared (cutting, planing, trimming, etc.) in a storage area.
- Storage for plant material should provide the proper environment for plant health, including access to water and shade, if required.

Like construction access, selecting the simplest storage site usually works best for most construction projects.

CONSTRUCTION LIMITS

Where your work is to be performed is usually not in question. Areas where you are *not* to work can be confusing. Projects may have limitations on the areas that can be disturbed by your construction activities, and these boundaries are defined by the construction limit line. Identified in construction documents as a line around the project, limits should be located early in the project and fenced or marked to prevent violations of boundary restrictions. If the project does not have construction limits, verify that the limits are the property of the owner. Hire a land surveyor to locate property lines that are in question. If no limits are defined for your project, care should still be taken to avoid and protect sensitive areas, and concern given to how your work may affect areas outside your construction area. Chapter 2, Legal Requirements, identifies some of the off-site problems that could arise from your construction activities.

The effort made to lay out the horizontal and vertical controls for your site should be protected by clearly marking benchmarks, grade and layout stakes, and other reference marks that will be referred to

Figure 9–1 Baseline reference markings for project layout.

often (Figure 9–1). Protection could be as simple as spray painting with a bright fluorescent color to aid in identification. In more extreme cases, protection may require setting a fence post or metal pin. Multiple reference points ease the pain of an accidental disruption by providing several options for reestablishing destroyed references.

PLANT MATERIAL PROTECTION

For the landscape contractor, the need for providing plant material protection should be obvious. Unfortunately, many trades people are not familiar with plant culture and fail to practice proper protection techniques when working around plant material. Construction poses several perils to the mature plant species at a site, and in most cases the damage does not become apparent until several years after the construction project has been completed. For this reason, it is important that members of the landscape contracting industry become stewards of the site by practicing sound plant protection techniques. Familiarizing yourself with the identification and cultural requirements of plants helps you recognize which plants are most sensitive to construction, and which construction practices pose the greatest risk to plants. Using the concept of trying to keep protected plants in their same preconstruction environment throughout construction will help in reducing overt and latent construction damage.

Protection techniques vary depending on the species and maturity of plants, and the construction activity taking place. Overstory (shade) and understory (ornamental) trees are at the greatest risk of damage from site construction activities. This risk stems from the errant belief that if the contractor doesn't damage the trunk or canopy, the tree will not suffer damage. This belief fails in not recognizing that the root zone of a plant is as equally important as the trunk and canopy. The root zone of a plant extends from the trunk to at least the **drip line,** the imaginary line on the ground directly below the outermost foliage. Depending on plant species and environmental situations, many plants have root zones that extend beyond the drip line. Compaction, trenching, mixing of chemicals, excavation, storage of materials, and related construction activities within the drip line of a plant will damage its root system and eventually damage the remainder of the plant. Protection from damage must extend at least to the drip line of the plant, and in some cases consideration must be given to restricting activities outside the drip line (Figure 9–2). Because of their size and vigor, many shrubs are more durable when exposed to construction activities. However, shrubs should be afforded the same protection given to trees to enhance the chances of plant survival.

Compaction

Compaction from traffic and material storage within the drip line must be avoided. Fencing a plant at the drip line and leaving the fence in place throughout construction will divert traffic and activities away from the plant. If traffic within a drip line is a necessity, accommodate by building an elevated wood boardwalk, or by placing 8–10 inches of loose, organic mulch over the ground within the drip line. Water the

Figure 9–2 Fencing placed at drip line of a tree to protect the root zone.

mulch daily. These techniques will only reduce the compaction of root zone soil.

Root Zone and Canopy Damage

Avoid trenching within the drip line, since trenching will sever the root structure entirely from the plant. Reroute utility lines away from the area under the plant canopy when possible. If utility lines must pass within the drip line, bore the lines 3 feet deep to pass under the plant root zone. Excavation within the root zone creates the same type of disturbance as trenching, and filling within the drip line will suffocate many plants by blocking the path of gases and nutrients to the root zone. Tree wells (Figure 9–3) are recommended where possible to reduce the damage from excavation and filling. Tree wells should be constructed outside the drip line of the plant and should be designed so that the original grade and drainage patterns are maintained.

On some construction sites, the source of damage to plant material is well beyond the drip line. Contractors who take advantage of a shade tree to mix caustic chemicals run the risk of plant damage from spills and leakage. Even changes in drainage patterns from grading activities can lead to plant material damage.

Protection of plant canopies from construction damage can include covering the plant or temporarily tying the branches up/back to keep them out of harm's way. Secure the plants loosely and do not leave them tied for more than 2–3 days to avoid permanent damage. Placing sheets of plywood over turf will provide temporary protection in an area that may be subject to heavy traffic for a few days. Again, only leave such protection in place for 2–3 days to avoid permanent damage to turf areas.

STRUCTURE PROTECTION

Structures are provided a level of protection by their size and dominance of the landscape. Even with this inherent protection, structures are vulnerable to damage from landscaping operations when care is not taken. Windows can be covered with plywood to protect from shovels, gravel, and other flying objects. Corners of structures can be protected by placing wood fence posts on either side of the corner, particularly in locations where traffic is routed near a corner. Routing traffic away from a structure and placing storage areas apart from buildings also reduces the potential damage from equipment and activities.

Figure 9–3 Tree wells installed to protect existing tree root zone from fill.

Pavement Protection

When a construction project requires that foot or light traffic be routed across existing paved areas, care should be taken not to damage such surfaces. Rubber track equipment will do little damage to concrete and asphalt surfaces unless the pavement base is already in poor condition. Rubber deposits may be left on concrete after construction, and can be removed with water and scrubbing. Unit paved surfaces such as brick and limestone may be damaged by any equipment, and should be protected using the methods outlined for heavy traffic. Heavy traffic requires protection that will prevent cracking and settling. Simplest among the protection techniques is to route traffic around the paved area. If rerouting is impossible, mound 12 inches of soil over the pavement in the area where the traffic will be crossing. This will redistribute the weight and reduce the stress on the pavement. An alternative is to build a temporary bridge out of 4×4s and wood planks (Figure 9–4) This method will transfer the weight to the ground surrounding the pavement.

UTILITY PROTECTION

Damage can occur from heavy traffic passing over utility lines buried at a shallow depth. Particular concern should be given to shallow irrigation lines, shallow gas lines in copper or PVC pipe, and PVC or clay drainage tile. Protection can be provided in these instances by increasing the soil depth over the line or by placing 2×12 lumber in the tracks of the path over the utilities. Low overhead utilities should be avoided or temporarily relocated to reduce the risk of contact with vehicles and workers.

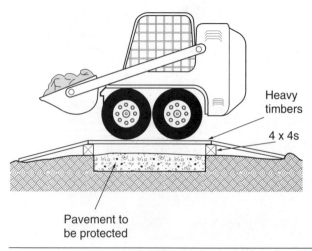

Figure 9–4 Temporary bridging to protect paving from construction traffic.

PERIMETER PROTECTION OF CONSTRUCTION SITES

Temporary fencing should be erected at all construction sites to protect the public from the hazards of a disrupted work area. Broken or removed pavement, open excavations, and unfinished surfaces are just a few of the potential danger areas on which an unsuspecting public could sustain injuries. Securely install fence posts and brightly colored utility fencing to warn the public of potential hazard (Figure 9–5).

Figure 9–5 Installing temporary fencing at a construction site to protect public from hazards.

PROTECTION OF ENVIRONMENTALLY SENSITIVE AREAS

Construction activities that deal with exterior environments are likely to face issues regarding protection of sensitive areas. Recognition of the importance of preserving delicate natural environments has led to approaches that mitigate the negative impact of construction on areas such as woodlands, wetlands, prairies, dunes, and wildlife habitats of all types. Probably the most important ingredient in the approach to sensitive areas is the awareness of the contractor. As workers who are engaged in the development of the environment, it is important that we be the first to recognize the need to value these areas and actively engage in preservation activities.

The step that is the most effective with sensitive area protection is to avoid the area with direct and indirect construction activities. Compaction of soil in a wooded area or even temporary draining of a wetland can do damage that cannot be repaired. Even minor changes in the environment around the area can change the characteristics of wetland or forest. If construction can avoid the sensitive area, the chances of preservation are good. Contractors from nonlandscape trades will likely lack the knowledge necessary to identify and preserve sensitive landscapes, so identification is an initial step to preservation. Fencing the area beyond the perimeter will assist in identification. Notifying all workers and inspectors will also aid in protection. Indirect construction can also play a significant role in damaging sensitive areas. Runoff from projects should be redirected if possible. Silt fence (see Figure 13–4) should be installed to prevent eroded soil from passing into a protected area.

In the event that construction must take place within the boundaries of a sensitive area, certain precautions can be taken to minimize the potential damage. Limit disturbing activities such as grading, utility work, and deliveries to a single corridor through sensitive areas. Provide temporary cover such as mulch or planking for the circulation routes through such areas. Suggest boring utilities under protected areas rather than trenching through them.

Chapter 10

Removing Unwanted Site Elements

Removing existing landscape elements seems foreign to the concept of landscape construction, but most projects require some level of demolition to make way for new elements.

RELATED INFORMATION IN OTHER CHAPTERS

Information provided in this chapter is supplemented by instructions provided elsewhere in this text. Before undertaking activities described in this chapter, read the related information in the following chapters:

- Safety in the workplace, Chapter 4
- Preservation of existing site elements, Chapter 9
- Erosion control, Chapter 13

PLANT MATERIAL REMOVAL

Techniques for removal of plant material differ depending on the type and size of plant being removed. When removing plants, it is important that the majority of the vegetative parts be removed. Uneven settlement will occur in paving or finish grades if large portions of stumps are left in place to decay. Excavating all roots is impractical, but root crowns and large feeder roots should be removed, and plant material waste should not be buried in areas where finish grade is critical.

Stripping sod and ground cover requires the scraping of a thin layer of plant growth off of a large area. Cut and remove the material by hand or utilize a sod cutter, and pick up debris with a skid-steer. A technique for larger areas is the use of a skid-steer with a toothed bucket to peel the sod away from topsoil. This technique requires an experienced operator and more fill due to additional topsoil being removed with the sod. Very large projects are stripped using large earth-moving equipment such as a paddle scraper, which can successfully peel the ground cover away from the topsoil over large areas.

Removing Shrubs

Shrub removal requires two steps to successfully remove plant debris. Canes or branches of shrubs are cut and removed, then the crowns or roots are dug out and removed. To facilitate the removal of shrub canes, tie twine around the canes and pull them into a bundle, then cut at the base using a chain saw. This technique works for all but a few spreading and prostrate shrubs, and speeds the removal and disposal process. Remove the crowns by hand digging around and under the crown, or by excavating a trench around the roots, hooking a chain around the crown, and pulling the crown out with a tractor. A backhoe will speed the process by excavating the entire crown in one operation. Prostrate shrubs can be removed in a single step by undercutting and lifting out with a skid-steer.

Tree Removal

Tree removal requires a systematic removal of plant parts, from crown to roots. Begin with the removal of as much of the crown portion of the plant as can be removed safely from the ground. Fell the remaining portion of the tree using the steps outlined in the following list. Remove and dispose of the remaining crown and trunk portions, then remove the stump. To safely cut down trees that do not meet the restrictions outlined in the Caution, follow the steps listed.

Safe felling of a tree requires the operator to be aware that even a small tree carries a great deal of weight and force when falling. Trees over 30 feet in height or 10 inches in diameter should be professionally removed. Trees located next to structures, utility lines, or other permanent improvements should be professionally removed.

Removal of trees requires use of power equipment. Follow manufacturer's safety instructions for equipment operation.

- Determine the direction for the tree to fall. The direction of fall can be controlled to a certain degree, but gravity will tend to pull the plant in any direction it is leaning.
- Verify that an area twice as long as the tree's height and twice as wide as the tree's spread is clear of any improvements and other large plants.
- Clear an escape path away from the tree and away from the direction planned for the tree to fall (Figure 10–1).
- At approximately 4 feet above the ground and on the side of the tree facing the direction the tree is planned to fall, cut a V-shaped notch one-third of the way through the trunk.
- On the side opposite and slightly above the V cut, begin cutting down at an angle toward the point of the V. Do not cut all the way to the V, and do not stop cutting until the tree begins to fall (Figure 10–2).
- When the tree begins to fall, remove and stop the saw and walk away along the planned escape route. For even a small tree, move 10–15 feet away from the tree to avoid the possibility of the tree "kicking back" toward the operator.

When the tree has been felled, cut the side branches off the main trunk. Use caution when completing this operation since the trunk may shift when a supporting branch is removed. Branches may be bent under the tree and snap back when cut. After removal of side branches, cut the trunk into manageable pieces that can be hauled away. Small stumps can be grubbed out using a skid-steer with forks or a backhoe. In some circumstances, a stump grinder can be rented to remove the aboveground portion. Large stumps require a great deal of excavating to remove the majority of the roots.

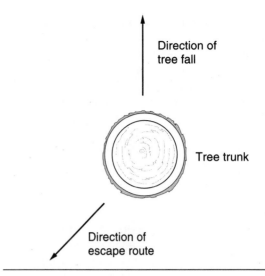

Figure 10–1 Plan view of tree removal showing direction of fall and escape route.

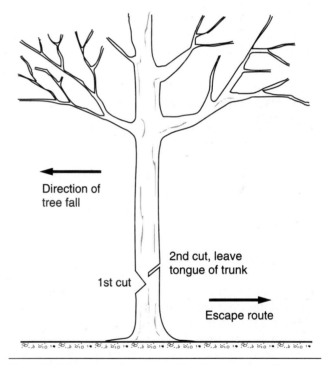

Figure 10–2 Elevation of proper notching locations for felling a tree.

PAVEMENT REMOVAL

Removal of pavement and other permanent improvements is a labor-intensive operation. Rental of specialized demolition equipment for this phase of work is highly recommended if large areas of pavement need removal.

Preparation for removal of pavement varies depending on the type of pavement being removed. If portions of the pavement are to be saved, cutting with a concrete saw is required before removal can occur. Unit paver surfaces require the removal of edge restraint material to allow access to the surface. Slab pavement usually requires that the slab be broken into pieces for removal and hauling. For small areas, this operation can be done by hand with a sledgehammer or pneumatic powered jackhammer. Larger areas will require that the pavement be lifted and dropped by a skid-steer bucket to break it into pieces, or presawing the slab into pieces and lifting it out with the skid-steer bucket. Larger equipment that will remove slab pavement without further reducing it into smaller pieces can be rented.

After the slab has been reduced to manageable pieces, locate a spot along the edge where equipment can work under the pavement to begin lifting it out. Excavate soil away from that portion of the slab to allow access under the slab. Work carefully in removing the old surfacing. Caution should be used so that pieces of the pavement are not forced against structures, footings, or other improvements that need to remain intact. Working at right angles to, rather than toward, structures can reduce damage to existing footings or foundations.

Footings and foundations that are buried to frost depth will require the use of a backhoe to excavate a working area around the footing. They are then broken using the backhoe bucket and removed.

UTILITY ABANDONMENT AND REMOVAL

Unearthing and removing old cables and pipes will reduce confusion for future excavation projects. Most utilities for a site that are abandoned are disconnected, capped, and left in place. This operation should be accomplished by a technician certified to perform such operations, particularly for gas, electric, fiber optic, potable water, and sanitary sewer lines. Physical removal of abandoned utilities is recommended if possible.

Storm water drain lines and tiles are sometimes encountered by landscape construction operations. If the lines are active, they should be reconnected or rerouted to avoid new construction. Lines that are no longer active should be capped or plugged to prevent subsurface drainage problems. Such lines can be capped with a PVC cap placed over the end of the pipe, or plugged by placing mortar or **bentonite** into and over the end of the abandoned line.

BURIAL AND ABANDONMENT OF UNWANTED SITE ELEMENTS

There is a temptation to cover over existing site elements rather than to remove them. This practice is not recommended for the following reasons:

- There is always uncertainty regarding utility lines uncovered later.
- Buried site elements can restrict plant growth. Roots for trees and shrubs may not be able to reach groundwater levels, and sod may dry out faster over an old improvement left in place.
- Future excavation is made more difficult and expensive when old footings or slabs are uncovered.
- Settlement of topsoil will take place at different rates over old site elements left in place.

In the event that on-site burial of waste is planned, be certain that the waste pit is in an area free of any existing or planned structures, pavement, or utilities. Burial pits should be covered with a minimum of 5 feet of soil to provide adequate base for plantings.

RECYCLING AND WASTE DISPOSAL

Methods for waste disposal in landscape construction run the range from landfills to creative recycling and reuse of by-products. Whether the method is environmentally sensitive or insensitive, landscape waste is an issue for which landscape contractors must prepare. Temptations to landfill all waste are now counterbalanced by restrictions on placement of organic waste in landfills and markets for recycled material. Opportunities to reduce waste from landscaping operations requires additional effort and planning because the quantities of recyclable materials can be small for a single project. Recycling may require stockpiling of materials or taking several small loads to the recycling plant. Many of the by-products of site preparation can be utilized in other aspects of landscaping if the contractor is willing to make the effort to reuse and recycle.

Use caution when removing, transporting, recycling, and disposing of hazardous materials. Identified in the following paragraphs are suggestions regarding how to dispose of materials that are removed from the landscape.

Reuse of Landscape Materials

Piecemeal adaptive reuse of old landscape materials has been taking place for decades. Following are a few creative suggestions for reuse of hardscape material:

- Stone wall material can be reused. This material is timeless and has several applications as walls, paving, steppers, edging, or lining swales.
- Bricks or pavers can be reused as edgings or steppers.
- Concrete pieces can be used for wall material or ground and spread as a loose aggregate surfacing. Verify that all reinforcing materials have been removed before reuse as a surfacing material.

Recycling of Landscape Materials

Healthy, woody plant material can be recycled by the landscape contractor using a chipper to create shredded and chipped mulch. Quality and reuse value of this mulch will vary depending on the parent material. Material composed primarily of woody plant parts can be used for landscape beds, while material with small branches and foliage may be acceptable only for field mulch or **composting** purposes. Diseased or insect-infested plant material should be burned or buried, the choice depending on the disposal technique recommended for the affliction. Sod that is stripped from the site can be composted along with miscellaneous plant waste from the site.

In appropriate quantities, most landscape materials have the potential to be recycled. Metals from fences, wood from decks, plastic and metal from edging, concrete and asphalt paving, crushed bricks, wiring and piping, and many other materials unearthed by the landscape contractor can be taken to various centers that will recycle the products. Limitations on engaging in a wholesale recycling process will be the quantities of waste generated and the time required to separate and deliver the materials.

Landscape Waste As Fill Material

Certain waste soils can be used as fill when properly placed and compacted. This filling procedure can be done on site or transported to an off-site facility. Clean soil without debris is satisfactory for this procedure. Soil contaminated with wood, paving, or other nonsoil debris will deter trenching and future underground work in the fill area and should not be used. Decomposition of large quantities of organic material will create instability in the soil surface, eliminating sod strippings as fill material.

Landfill Disposal

An option that requires a minimum of effort, but can be expensive to the contractor and the environment, is placing waste in the local landfill. Many localities have strict regulations on what materials can be placed in landfills, including limitations on plant material. Typical use of the landfill requires that the contractor haul and place the waste material, paying fees based on the tonnage of material disposed. Organic materials, including green plant matter, are sometimes composted if the solid waste facility has the processing equipment.

Transport of Waste

When waste is hauled over public roadways, ensure proper enclosure is provided so that spillage will not occur. This may require use of vehicles with tailgates or enclosed cargo areas, covers, or tie down of loose plant waste. Review with municipalities and state Departments of Transportation for regulations regarding proper hauling of waste

Initial work for a landscape site includes preparing for the construction that is planned. Any site elements that are to remain should be identified and protected, and those elements that are not incorporated into the design may need to be removed. Section 2 identified construction techniques basic to these steps and standard site protection and demolition activities.

Preservation of existing site elements can cover a wide range of activities for a large project, or can be a simple marking of plant material for a small project. Typical considerations when preparing a site for construction include identification of safe and legal access to the site and secure storage of materials. Most projects will also require identification, and in some cases, fencing the limits of the construction. If layout was performed using benchmarks or baselines, those reference points should be clearly marked to avoid future disruption. Landscape contractors should maintain an awareness of the impact of construction of existing plant material. Plants to be preserved should be protected from compaction, trenching, and other construction activities within, at least, the drip line. Permanent improvements such as structures, pavement, and utilities also require protection from construction damage. These preservation techniques may range from marking to fencing the improvement. Environmentally sensitive areas also require special attention when construction activities are in proximity of wetlands, prairies, woodlands, and similar protected plant and wildlife areas. Avoiding the sensitive area provides the best measure of preservation, but in cases where avoidance is impossible, protected corridors and perimeter protection may be required.

Removal of unwanted site elements may take the form of cutting out plant material to demolition of old site elements. Attention to safety should be a priority when removing old site features, particularly large plants, structures, and utilities. Burial or abandonment of unused site elements is not recommended in most cases due to the impact such actions may have on future work. Proper disposal, including consideration of reuse and recycling of landscape waste, is required for any demolition project.

Section 3

Grading, Site Drainage, and Erosion Protection

INTRODUCTION

Water and soil play two of the leading roles when considering the many elements of the typical landscape. A contractor's ability to properly manipulate these two elements often determines the long-term success or failure of a project. Soils vary significantly in terms of composition, texture, stability, and fertility. Moving soil without consideration of placing the proper amount and type for the function it must serve in the landscape may lead to premature settling of improvements or poorly performing plants.

Water's role in the landscape is probably more extensive than that of any other element. Controlling water in its various roles can reduce the impact of such landscape problems as drought, frost heaving, **erosion,** settling, and ponding. The interaction of water and soil creates the potential for erosion, one of the most devastating landscape problems. Improper manipulation and inadequate protection of the surface of a site is a common problem in landscaping projects of all sizes.

The chapters of this section explore the steps of the grading process, methods for effectively draining water from a site, and protecting a site from common forms of erosion. Included is a review of the grading process, which is often split into rough grading steps that occur during the initial steps of the construction

process, and finish grading, which typically occurs near the end of the process. A task performed in concert with grading is development of drainage systems to effectively remove unwanted water from a site without flooding, eroding, or otherwise disturbing the surface improvements of the site. This challenge is addressed in the development of surface drainage patterns and, if necessary, the addition of subsurface drainage structures.

Techniques are presented for grading surface slopes and **swales** to provide positive drainage throughout a construction site. Subsurface systems such as tiles, french drains, and storm sewers are also introduced as more comprehensive methods for dealing with large quantities of storm runoff and water problem areas. Erosion protection concentrates on protecting the perimeter of a site from runoff, providing surface protection from sheet erosion, and reducing channel problems from gully erosion.

WORKING WITH SOIL

Essential to successful grading and erosion protection is the development of an understanding of the types of soils on the site and the suitability of those soils to perform the functions required by the landscape plan. Soil types have an impact on whether a soil can be reused on the site, whether it must be disposed of off site, whether it can be built upon, or how erodable it might be. It would be impossible to define the characteristics of all soil types in this text, but one basic trait of soil that provides clues to its workability is its composition. Soil is composed of three basic particle types—silt, sand, and clay. Each soil type is a combination of these particle types, and the percentage of each determines the soil's stability and suitability for landscape tasks.

A general rule is that soils high in silt are suitable for planting areas but are undesirable for built areas, while clays are suitable for built areas but are undesirable for most planting areas. Based on the percentage of sand, soils may be unsuitable for both. Sandy clays are possible substructure soils, while sandy loams, or silts, are suitable for plantings. Soils with high sand or silt content are the most likely soils to be highly erodable. Topsoil (dark soil found near the surface of a site) is typically rich in organic matter and, in most situations, is capable of supporting plant growth. Specific information regarding the suitability of site soils can be obtained from a soil survey book, soil conservation service, or for critical areas, from a soils engineer. Reviewing and understanding this information will improve the success of any landscape project.

PRODUCTIVITY SUGGESTIONS

Grading a site is one of the many landscape activities that is weather sensitive. Effective completion of construction requires practice of techniques that counter the problems associated with weather. Following are some suggestions for reducing the impact of unworkable weather:

- Avoid working soils high in clay or organic matter content when they are wet.
- When working in dry conditions, lightly wet the soil before compacting. Dry soil shifts rather than compacts.
- If rain is anticipated, do not strip sod and surface vegetation. The drying process takes longer if a site has been exposed to the elements.
- Consider how water will drain from the site at all stages of the grading operation. Avoid creating low areas where water will pond and drying will be slow.
- Provide temporary erosion protection throughout the grading process.
- Maintain a covered stockpile of dry topsoil that can be accessed when an exposed site is still wet.

Chapter 11

Site Grading

Most landscape projects require moving soil to accomplish design goals. A simple concept of grading could be described as removing soil from where it is not needed and placing it where it is desired. Grading a site requires background knowledge about soils and the grading process, the ability to calculate quantities of cut and fill, and judgments about the size of the project and what approach should be taken.

Grading is accomplished through a process called cutting and filling. Cutting is the removal of unneeded or undesirable soil from a location. Filling is the placement of appropriate soil where it is needed. While this concept seems simple, conditions exist that complicate the process. If soil with plant matter, such as turf, is reused when filling, it would later decompose and settle. The desire to save topsoil and place it in the area where it will benefit plants adds complexity to the **cut** and **fill** process. Respreading topsoil and smoothing the finish grade complicate the idea of cut and fill even further.

The amount of soil moved and the types of landforms required by the landscape plan will determine whether the project can be accomplished by the landscape contractor using typical landscape equipment, or whether the project should be subcontracted to a grading contractor with special equipment and expertise in large earth-moving projects. Determining the size of project that can be effectively graded with a skid-steer and dump truck is a matter of experience. It is not uncommon for a landscape contractor to undertake projects that require up to 50 cubic yards of earth moving (roughly enough soil to cover the floor of a two-car garage 3 feet deep). Because of the greater efficiency of heavy equipment, many landscape contractors subcontract larger projects to grading special-

ists. When considering the grading portion of a project, the decision should properly match a contractor's equipment and capabilities.

RELATED INFORMATION IN OTHER CHAPTERS

Information provided in this chapter is supplemented by instructions provided elsewhere in this text. Before undertaking activities described in this chapter, read the related information in the following chapters:

- Safety in the workplace, Chapter 4
- Construction staking, Chapter 8
- Erosion control, Chapter 13

STEPS TO GRADING A SITE

Small projects offer the possibility of using a simplified grading approach of pushing the soil and sod away from a project and pushing them back when construction is completed. When major grade changes are required, a process that accommodates the orderly completion of all steps is necessary (Figure 11–1). Unless other specialized equipment is identified, each step can be accomplished by a skid-steer on a small site or by a bulldozer or scraper on a larger site.

Strip Sod

Because vegetative matter will decompose over time, ground covers such as sod should be removed separately and either disposed of, composted, or stockpiled separately from topsoil for use in noncritical areas. The site should be stripped to a depth of

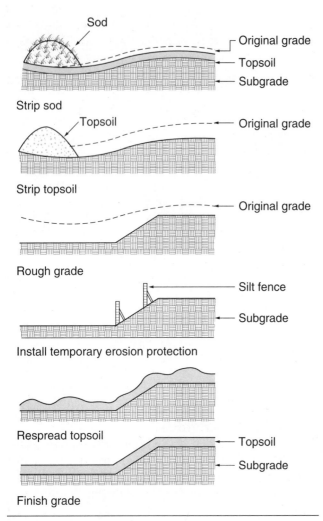

Figure 11–1 Steps of the grading process.

approximately 4 inches to remove the most dense layer of vegetative matter. In some areas, this depth may be decreased or increased based on the type and maturity of the covering plant material. Any areas where tree stumps were located should be excavated with a backhoe to a depth of 24 inches in a 10 foot diameter circle centered on the stump.

Strip Topsoil and Stockpile
Following the removal of vegetative matter from the site, strip enough topsoil to cover all areas where topsoil is to be respread in the final grading step. Include an additional 15% in this stockpile to cover settling, erosion, and compaction losses. When calculating areas to receive topsoil, include all turf and ground cover areas as well as planting beds. A shortage of topsoil indicates that topsoil will need to be imported to cover planting areas. Place this topsoil in a stockpile that is located away from construction areas, yet is easily accessible both for immediate needs and at the end of construction.

Excess topsoil can be stockpiled or, if suitable, used to establish subgrade on the site. Most topsoils are stable enough to serve as the **subgrade** for **berms,** lawn areas, and other areas without permanent improvements. It is not recommended that soils high in silt content be used as subgrade for paved areas or structures because of the possibility of soil settlement.

Rough Grading
Rough grading involves the cutting and filling of the site to establish proper subgrade elevations. These elevations are obtained by subtracting the required topsoil layer, pavement and pavement base, or other nonsoil subgrade improvements from finish grade. Subgrade establishment does not require the precision of finish grading, but should be as accurate as possible to reduce grading work later in the project. Productivity can be increased in some situations by loading dump trucks for hauling rather than hauling soil in the bucket of slow earth-moving equipment. When using the skid-steer to excavate, use a bucket with teeth allowing penetration into the surface and making excavation easier.

Cutting will take place in all areas where the current grade is higher than the desired subgrade. This soil is often directly moved to areas that require fill. When filling, it is recommended that soil be placed in 6 inch lifts (layers) and compacted before more fill is placed. This provides more stability than a site that is compacted only after all fill is placed. Water may have to be sprayed lightly on lifts to assist compaction.

During this process, the contractor should also be aware of area soils that are unsuitable for construction. These areas appear wet, spongy, foul smelling, or differently colored. If such areas are encountered, the entire depth of unsuitable soils should be excavated and replaced with suitable soils. If the depth of unsuitable soils exceeds 24 inches, a soils engineer should be consulted to recommend a solution. The area should also be reviewed as to the cause of the soil problem. **Tile** installation may be required to remove excess water. Providing stable subsoil is important to the long-term stability of all paved areas and structures.

Installation of Temporary Erosion Protection
In most projects, the grading work halts after rough grading so that walls, paving, and other site improvements can be constructed. Some projects may be small enough to allow the grading process to proceed from beginning to end, but as the size and complexity of a project increases, grading must be completed in stages. If grading halts after rough grading, temporary erosion control measures (see Chapter 13) should be installed.

Respread Topsoil

When the majority of heavy traffic over the site has ceased and all walls, pavement, and structures have been installed, grading operations can continue. Topsoil from the stockpile is removed and spread over the site to the required depths. Depths of topsoil vary depending on project specifications, but lawn areas typically require a minimum of 6 inches up to 12 inches. Planting beds may require a minimum of 12 inches up to 24 inches. The topsoil is laid in and roughly spread to the finish grade required by the site grading plan.

Finish Grading

The final step prior to seedbed preparation or planting is to smooth the surface to the exact grades stated on the grading plan. This work is typically done with a grader or tractor-mounted blade. Hand work may be required in tight areas to accurately obtain desired grades. In cases in which no grading plan exists, finish grading will need to maintain proper elevations next to paved areas and permanent improvements, and proper drainage direction and slope. At this stage, soil is often lightly compacted with a drum roller. Further surface manipulation, such as **gilling** or cultivation, is performed as part of the preparation for turf and areas. Permanent erosion protection should be installed immediately after finish grading to protect the site.

Chapter 12

Site Drainage

Much of landscape construction today is directed toward addressing the issue of water. Irrigation systems combat drought; footings and pavement design strengthen improvements against frost; and grading, tiling, and other drainage work are implemented to reduce flood and water damage. Of all of the elements that can create difficulties on a site, water is the element that can single-handedly wreak havoc. Whether the problem is an excess of water from flooding or poor drainage, a shortage of water from drought, or frozen water that heaves pavement or ices walks, proper management of the many potential impacts of water on a site will minimize construction and long-term problems. In this chapter, we examine ways to control water by draining it away from improvements on a site.

Two general concepts are incorporated into making site drainage improvements. The first is the concept of draining water from the site on the surface. Surface drainage consists of shaping the ground to direct runoff where desired. When preparing a site for proper surface drainage, the intention is to slope the ground away from important features that can be damaged by water, such as structures, walls, and paved areas. In addition, the contractor attempts to avoid low areas that will collect water and make spaces unuseable after rains or irrigation. This method of removing water is typically the most cost-effective and easiest to construct. It is applicable on most sites to some degrees and, in many sites, is the only technique needed to drain the site.

Basic to the idea of surface draining a site is the principle of maintaining a low point, or series of low points, that are lower than structures and important site fea-

tures. These low points are sometimes referred to as "**free outs**" or "swale high points." These low points guarantee that if rains continue to flood stage, water will flow away before entering a structure. This principle is easy to maintain on sites with ample topography, but is a challenge on sites that are flat. Sites that limit the surface drainage possibilities or make locating low points difficult require drainage systems that are more complex than surface treatments.

When the volumes of water are great, or the site itself is not conducive to surface drainage, subsurface drainage systems must be employed to maintain positive drainage. Subsurface systems involve collecting the water on the surface and then letting gravity draw it into a piping system for removal. While more expensive and time consuming to construct than surface drainage, subsurface systems are effective in draining difficult sites.

Also discussed at the end of this chapter is the concept of storm water detention for a site. Many projects incorporate the idea of retaining rainwater for a period after a storm to minimize the impacts on downstream areas. Construction and finishing of the **detention** area may become the responsibility of the grading or landscaping contractor.

RELATED INFORMATION IN OTHER CHAPTERS

Information provided in this chapter is supplemented by instructions provided elsewhere in this text. Before undertaking activities described in this chapter, read the related information in the following chapters:

- Safety in the workplace, Chapter 4
- Basic construction techniques, Chapter 7
- Site grading, Chapter 11

SURFACE DRAINAGE

In any proper grading and drainage scheme, positive surface drainage should be the first consideration. Surface drainage can best be described as draining water away from locations where it will be a problem using surfaces that are shaped and sloped. More specific techniques involve the grading of swales, or shallow ditches, and the use of diversion berms to direct runoff where the designer desires. Surface drainage is the predominant method for sites that are small and/or have limited amounts of runoff. As a site increases in size, surface drainage may be limited to directing the runoff to more permanent drainage structures such as tile or storm sewers. Another factor that may limit the effectiveness of surface drainage is the amount of runoff. Sites that are largely covered by structures and pavement limit the amount of water that will be absorbed into the ground, leaving large amounts of water that must be removed to protect structures and occupants. Surface drainage methods can be quickly overwhelmed during storms with large amounts of rain on such sites.

Grading Slopes

A simple technique that is essential to good site drainage is to slope surfaces away from permanent improvements. Structures, walkways, patios, walls, and other intensive use areas of the landscape benefit from having the ground slope away for a short distance, allowing water that runs off of these structures to stay off the improvement. This shaping process should be addressed during the grading of the site, especially during the rough and finish grading steps.

The percentage of slope graded for a site is dictated by construction drawings or, in the absence of a grading plan, is determined by standards and site conditions. Standards for most sites recommend that a minimum slope of 2%–3% be maintained across turf areas to provide positive drainage. Pavement areas drain well if a 2% slope can be maintained. Athletic fields and sporting surfaces have required slopes recommended by the authority governing that type of activity. Grade around buildings should be maintained 6 inches below the finish elevation of the structure, and the ground outside the structure should fall an additional 6 inches in the first 10 feet away from the building.

Grading Drainage Swales

Drainage swales are shallow ditches designed to carry water through and away from a site. If a site has adequate slope and area to accomplish drainage without swales, none should be utilized. While swales are drainage channels, they are typically constructed with side slopes that are gradual enough to mow. As with all surface drainage, the general slope of swales can be completed during rough grading, and the detailed slope completed during finish grading steps. The highest point of a swale should be at least 6 inches lower than the finish elevation of any paved area or structure. From that high point, a minimum of 2% downstream slope should be maintained, with grades of up to 3%–5% allowed for short distances. While the side slopes are maintained at grades that allow mowing, water volume or occasional steep slopes may require permanent channel erosion protection in some cases.

Swales should be utilized where they will be most effective. Placement around structures and through low areas of lawns allows the swale to perform the function of collecting water from downspouts and lawns and transporting it away (Figure 12–1). When a hillside with high runoff abuts an improvement, a swale should be placed between the hillside and the improvement to intercept runoff. Water carried in swales can be directed to storm sewer **inlets** or emptied into drainage channels, streams, or unused lawn areas. This latter choice will create wet areas that are unuseable for periods of time.

Grading Diversion Berms

Occasionally, a site will have a slope that directs runoff toward structures and other improvements. If a swale cannot be constructed to intercept this runoff, a diversion berm may be an alternative. While a swale lowers the surface elevation, the diversion berm mounds soil upward to direct water (Figure 12–2). By mounding soil a short distance from an improvement, and continuing that mounding around the improvement, water can be directed to the point below the improvement.

Diversion berms should be constructed during the grading of the site. To be effective, the berm must create a small swale on the upslope side to carry water, and must continue around the improvement being protected until the highest point of the berm is level with the improvement's finish elevation. Caution should be taken when using diversion berms near structures. The space between the berm and the structure will continue to drain toward the structure, and if

Figure 12–1 Drainage swale constructed to direct runoff away from improvements.

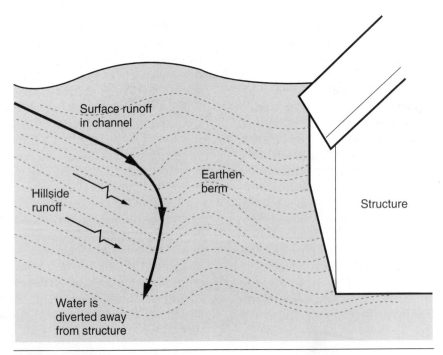

Figure 12–2 Diversion berm to direct surface runoff.

water from downspouts, roof drainage, or other large quantities of runoff land in this zone, the structure could flood. If this is the situation, consider using a tile or storm sewer system to handle drainage.

SUBSURFACE DRAINAGE

While the first consideration for site drainage should be surface drainage, certain sites require construction of subsurface drainage systems. Situations that require subsurface systems rather than surface drainage systems include sites that have variable grades that do not allow the construction of swales, sites that temporarily flood, sites with high volumes of runoff, sites that have a high percentage of the coverage by structures and/or paved areas, and sites where it is critical that runoff be kept away from structures.

Three approaches to subsurface drainage are available, with the first two systems considered within the scope of the landscape contractor's work. The first approach is the use of a **french drain** to draw excess water off the surface and store the water until it percolates into the surrounding soil. The second approach is the use of an underground tile to transport runoff to an outlet point. This system requires that the surface be graded properly to direct water to the system collection points, typically a surface inlet. Subsurface water can be removed from a site with the use of a perforated tile system. Perforated tile systems are available that collect water without the use of surface inlets. The third approach is the construction of storm sewers, a system of drainage structures and underground piping engineered to collect runoff at surface low points and carry the water to an outlet point.

Construction of French Drains

French drains are gravel trenches covered with a thin layer of soil and ground cover. The gravel in these trenches stores water in the open spaces between each stone (Figure 12–3). French drains are considered a closed drainage system, requiring no outlet point. Proper application of a french drain would be placement in a low area that collects water or in locations where poor soils slow the percolation of surface water.

Begin by using the formula identified in the Chapter 5 to calculate the length of french drain required. When the length of french drain has been calculated, mark the location of the trench on the site. If the site will not accommodate a single trench, the required length may be split into separate trenches placed at least 36 inches apart. To be effective, the majority of the trench should be located directly under the area that floods. Excavate a trench 12 inches wide and 42 inches deep along the lines marked. In

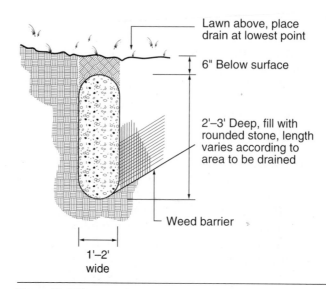

Figure 12–3 Cross section of french drain.

turf areas, the sod may be removed and set aside for reuse. Save approximately 10% of the soil excavated from the trench for cover.

For each trench, cut two lengths of 48 inch wide landscape fabric 2 feet longer than the trench length. Place the first piece of landscape fabric up one side of the trench. Fold the top over the outside edge of the trench. It is not necessary to cover the entire bottom of the trench. Repeat the process with the other piece of landscape fabric on the opposite side of the trench. Fill the trench to within 6 inches of the top of the trench with 1–2 inch diameter washed river rock. Use care when filling the trench that the landscape fabric is not disturbed. Fold the remaining landscape fabric over the top of the river rock. Backfill the trench with the soil set aside earlier and compact. If sod was saved, replace over the fill.

The average life of a french drain is approximately 5–10 years. After that time, the storage capacity of the pores between the river rock will be reduced from silt that has worked into the drain. A faster alternative is to construct the french drain without placing landscape fabric around the river rock. Such an alternative will have a typical life of less than 5 years due to faster siltation.

Installation of Tile Systems

Tile systems can be used to drain both surface water and subsurface water. Surface water is drained away through the use of inlets that allow water to drain into a pipe and be carried to an outlet point. Tile also can be used to intercept the water table (subsurface water that is moving up toward the surface) and reduce damage from frost. Installation of each of these systems is described in the following sections.

Tile Systems to Drain Surface Areas.   Installation of a tile system to drain surface areas begins with the flagging of inlet points and an outlet point. Inlets should be located in paved areas and lawns, near roof downspouts, and in any location where significant runoff is expected. Between inlet locations, a simple network of connecting tile lines should be marked leading to the outlet point. This network should avoid uphill runs.

Beginning at the outlet point, excavate an 8–12 inch wide trench along the entire length of the collection system (Figure 12–4). The trench should be deep enough that the grade can fall at a 1% rate from the highest inlet to the outlet point. Beginning at a point just before the highest inlet, lay 4 inch diameter *nonperforated* tile that exits at the surface and drops to the bottom of the trench. At this beginning of the system, cut the tile flush with the surface and install a cap (Figure 12–5). This cap will serve as a cleanout in the event the system becomes plugged with debris. Continue running tile along the bottom of the length of the trench. Large numbers of inlets or heavy runoff areas will require increasing the tile size to 6 inches or larger.

At each location where an inlet is planned, cut the tile with a carpet knife and insert a T connection. Insert a vertical tile **(riser)** (Figure 12–6), running from the T to approximately 12 inches above the surface for connection to an inlet. Continue laying tile and inserting Ts until the entire system is completed. Premanufactured Ts, angles, and Ys are available to make branching and connecting systems easier. Apply duct tape around any connections and joints to reduce problems from connections pulling apart. Backfill the system to the surface, lightly compacting the backfill after every 12 inch layer.

Fiberglass inlets with flanges that fit into corrugated plastic pipe are premanufactured for tile systems (Figure 12–7). To install inlets, excavate the finish grade around each location where a riser was placed. Cut the riser at this elevation and place the inlet into the end of the riser. Position the riser flush with the surrounding ground. Backfill the hole and adjust the finish grade to create a low point at the inlet.

Tile Systems to Drain Subsurface Water.   Tile used to intercept subsurface water is laid out in a pattern that covers the entire area that requires drainage (Figure 12–8). Tile line locations are marked out in a parallel line pattern with a connecting tile that runs perpendicular to these lines at the low end. A single tile leads from this connecting tile to the outlet. Spacing of the parallel lines is determined based on soil type, rainfall, and severity of existing water problems. Spacing of the parallel rows may range from 5 feet apart to 40 feet apart, depending on the conditions.

Figure 12–4 Trenching and placing plastic drain tile.

The more severe the conditions, the closer the spacing. An engineer should be consulted if there is uncertainty about spacing of tiles for a particular application. Inlets are rare on this type of tile system.

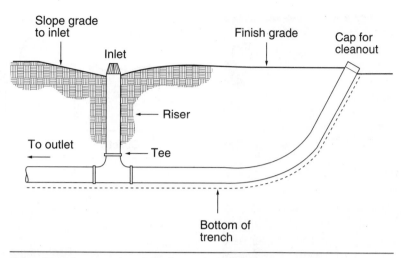

Figure 12–5 Cross section of tile riser and cleanout.

Figure 12–6 Tile riser before being trimmed for inlet installation.

Excavate and place tile in the same manner as that used for surface drainage tile systems, with the exception that *perforated* tile is used. To reduce the potential of silt filling a perforated tile, backfill with coarse backfill material such as pea gravel or utilize a perforated tile with a **geotextile sock** wrapped around the tile. If the tile system is intended to protect structure footings from frost, the tile must be slightly deeper than the bottom of the footing.

Storm Sewer Systems

Sites that require engineered storm sewer systems also require a specialized contractor for the construction of those systems. Heavy equipment requirements and hazardous trench work place this type of drainage system beyond the expected scope of the typical landscape contractor. Landscape contractors will, however, be required to grade surfaces around storm sewer inlets and possibly connect tile systems they

Figure 12–7 Tile inlet.

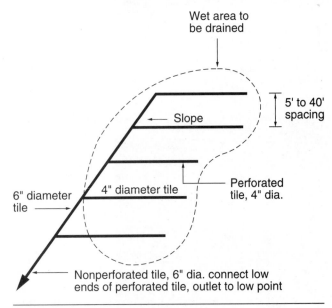

Figure 12–8 Plan view of tile layout for intercepting subsurfafce water in a wet area.

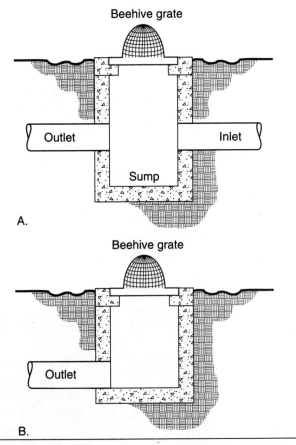

Figure 12–9 A. Cross section of catch basin and B. drain inlet.

install to the storm sewers. Because of this relationship, the components of the storm sewer system will be reviewed.

Storm sewers are permanent subsurface drainage systems composed of two major parts: inlets and piping. The water collection for a storm sewer is performed by inlets, which are underground concrete (either precast or cast-in-place) boxes with openings in the top. Water is directed to the inlets on the surface and then falls through these openings. Different types of inlets are used, based on where the water is being

collected. **Catch basins** have a **sump** in the bottom of the inlet designed to catch debris and sediment that fall into the inlet. **Curb inlets** are concrete boxes with metal frames shaped to match roadway curbs lines. Openings along the curb allow water to enter the inlet from the side. Once inside the inlet, water flows into clay or concrete piping and is conducted to outlets at streams or rivers. This pipe is typically laid at a minimum slope of 2% to maintain flow (Figure 12–9). Surface grades that lead to storm sewer inlets should be maintained relatively flat. Steep slopes adjacent to the inlet may lead to erosion. It is also important that the adjacent finish grade be flush with the inlet to avoid water ponding by the inlet.

With approval from the storm sewer owner, tile systems may be connected to inlets as an outlet point for small drainage systems. This requires that a hole be knocked into the inlet with a sledgehammer and the tile pushed through the hole into the inlet. The space around the hole should be filled with mortar so that soil does not wash into the inlet (Figure 12-10). When this outlet method is used for tile systems, verify that the tile line can be emptied no lower than halfway down in the inlet. Tile entering near the bottom of the

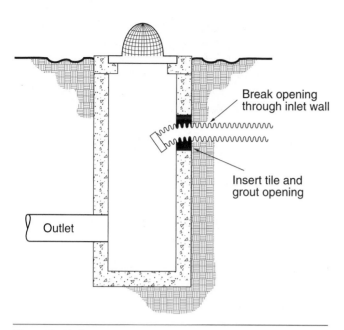

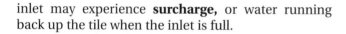

Break opening through inlet wall

Insert tile and grout opening

Outlet

Figure 12–10 Connecting drain tile to existing storm sewer inlets.

Figure 12–11 Treatment of drain outlet with 4–8 inch rip-rap to prevent erosion.

inlet may experience **surcharge,** or water running back up the tile when the inlet is full.

Treatment of Outlet Areas

Outlet areas for surface and subsurface drainage systems require special treatment to prevent erosion. End points of swales, channels, tile systems, and storm sewers typically carry large volumes collected from the site (or several sites), and the water at these locations also is moving quickly. The combination of these two factors makes the locations prone to severe soil erosion. To reduce the impact of erosion, it is suggested that each of these locations be treated with rip-rap.

Rip-rap is large stone used to line drainage swales for erosion protection. Composed primarily of 4–8 inch rock, rip-rap reduces erosion by slowing the water as it strikes the surface of the rock. The weight of rip-rap holds the surface contour in position so that it is not washed away like soil or smaller stone. The rip-rap at outlets should be placed over geotextile fabric centered on the outlet. The geotextile should be placed in a square with dimensions twice the width of the outlet (Figure 12–11). Clients should be advised to monitor this area periodically to watch for erosion problems.

With tile systems, consideration should be given to covering the outlet opening with landscape fabric to prevent small animals from moving into the pipe. Cut a large piece of fabric and wrap around the pipe end.

Duct-tape the fabric onto the pipe to prevent water from washing the fabric off.

TEMPORARY STORAGE OF STORM WATER

Storm water detention is based on a very simple concept. Before a site is developed with paving and structures, most of the rainwater that falls on the site is absorbed, or percolates, into the surface rather than running off. After the site is developed, a majority of the water runs off the site rather than percolating into the ground. Storm water detention plans design a pond on the site that will hold storm runoff from a developed site and release it at the same rate as an undeveloped site. If the detention of storm water is properly practiced in developed areas, flooding that occurs from runoff during heavy storms should be reduced.

Constructing the detention areas sometimes falls under the responsibilities of the landscape contractor. Detention ponds on the site can be placed in turf areas and are occasionally placed on paved parking areas when open space is restricted on a site. Turf-lined ponds must be built to exact dimensions in order to detain the proper amount of water. When grading for a detention pond, the rim elevation (elevation where the water will reach its highest level) is critical. Low points in the rim elevation will allow water to exit the detention structure rather than being retained. Most detention ponds also have an overflow to accommodate rains that exceed design capacities. This overflow may be a grassed or paved channel similar to those discussed in the erosion control chapter.

Ponds also require an outflow pipe in which the size of the opening (called an **orifice**) is determined by extensive calculations. Whatever the type of orifice (plastic or metal), it may require that openings in the surface be covered to reduce the outflow. When sizing the orifice for detention, the most common method is to have a metal plate welded over the required percentage of a metal grate or to glue a covering over plastic grates. Calculations of the size of the detention area and the orifice opening should be left to the design professional.

Chapter 13

Erosion Control

One goal of most landscape designs is to provide positive drainage throughout a site with stable surfacing that will remain in place during all conditions. Unfortunately, to change a site from existing conditions to the shapes and forms desired requires various levels of disruption to the site. The time period between existing conditions and finished product provides a window of opportunity for one of the most challanging problems facing the landscape contractor—site erosion. Improvements to sites such as structures and paved areas may increase the amount of water exiting a site, complicating the goal of maintaining stable surfacing. When construction is complete, the potential for erosion still exists on slopes that remain unprotected. Stabilizing these slopes and channels is an important part of the erosion protection concept.

Erosion is the removal of soil from a site due to the action of water and wind. Conditions that exist during the landscape construction process, particularly site grading, make most sites vulnerable to the processes of erosion. Several factors influence the process of erosion, including soil type, vegetative cover, soil particle size, slope of the site, exposure of the site, and the amount and speed of water and wind over the site. Changing or disrupting these factors is the focus of most erosion protection methods. Control methods typically involve covering the soil, reducing or diverting the amount of water and wind passing over disturbed soils, anchoring the soil with plant roots, or slowing the water and wind speed to reduce their capacity to move the soil particles. At each construction site, an important technique for reducing erosion is to cause minimal disturbance to the existing vegetative cover.

The techniques presented in this chapter will use the methods just identified to address erosion on three fronts. Erosion controls are identified for the perimeter of the site, the exposed surfaces of the site, and for drainage channels that carry runoff water away from the site. Techniques are identified as temporary or permanent control methods, with use and installation methods provided for each. Unfortunately, the majority of erosion control measures available are aimed at the problem of water erosion and have minimal effect on wind erosion patterns.

The decision to protect the site from erosion may be a required contractual issue or a judgment by the landscape contractor to provide control measures. Contract documents may spell out specific measures that must be taken to protect the site from erosion, but occasionally provide only a clause stating "contractor shall protect site from erosion." Keeping soil on the site protects a valuable resource and reduces cleanup efforts during the project. Erosion protection can also provide protection from liability by preventing the disruption of adjacent properties from outwash soil. In either situation, whether contracted or through the independent judgment of the contractor, erosion protection provides a valuable preventative service to the landscape project.

As with all steps of landscape construction, calculations need to be performed to order materials and complete the steps identified in the following sections. Refer to Chapter 5 to review methods of linear, area, and volume measurement required by this chapter.

RELATED INFORMATION IN OTHER CHAPTERS

Information provided in this chapter is supplemented by instructions provided elsewhere in this text. Before undertaking activities described in this chapter, read the related information in the following chapters:

- Safety in the workplace, Chapter 4
- Basic construction techniques, Chapter 7
- Site grading, Chapter 11

PROTECTING THE SITE FROM PERIMETER EROSION

Protection of the site perimeter is utilized primarily to prevent soil eroded off a site from washing onto neighboring properties. Two techniques have been utilized with varied degrees of success—straw bales and silt fence. Silt fence is an erosion control product composed of landscape fabric attached to posts or stakes. The stakes are driven into the ground with the fabric suspended vertically between stakes. Water strikes this fabric and slows, dropping its load of sediment. Both techniques rely on slowing the velocity of water leaving the site to the point where the sediment is dropped on the upstream side of the protection. Both provide an 18–24 inch basin on the upstream side of the protection for sediment storage. Both techniques are more effective if the existing ground cover vegetation is left in place for at least a 10–20 foot strip along the entire perimeter of the site.

Placement of this type of protection in the proper location greatly increases its effectiveness. Areas to be treated are those below disturbed areas of a site, particularly locations downstream from disturbed areas. Lower areas of sites and channels (see channel protection later in this chapter) are other potential areas for protection. Treatments of this type should be considered temporary and removed after construction and when permanent erosion control is functioning properly.

Maintenance of the perimeter protection is also critical to proper functioning of the systems. Whichever method is selected, the sediment in front of the protection should be periodically removed to allow water to run through, and not over, the protection. Maintaining the protection in an upright position is also important. Bales that have tipped over and silt fencing that sags reduce the effective area of protection.

Installation of Straw Bale Perimeter Protection

Straw bales have been used for many years to hold eroded soil on a site. While inexpensive, bales are prone to washout and have a limited life. Their use should be limited to short-term projects in which maintenance of bales can be performed. Materials required for straw bale perimeter protection include enough bales (measured along the longest dimension) to cover the perimeter and 2–3 foot wood or metal stakes for each bale placed.

Begin installation by marking out the locations where perimeter protection will be provided. Clear weeds and debris for a 2 foot wide path along the perimeter identified for protection. Level the surface within this cleared path, since the bales must make continuous contact with the surface to prevent leakage under the bales.

Beginning at the lowest point along the perimeter, place a bale on edge and push stakes into the top of the bale 12 inches from each end and centered in the bale. Drive these two stakes into the ground until the stakes are flush with the top of the bale. Place a second bale snug with the end of the first bale and repeat the staking procedure. Continue placing and staking bales until the entire perimeter to be protected is lined with bales.

Installation of Silt Fence Perimeter Protection

Use of silt fence to reduce outwash at the perimeter is the most effective protection method available. Materials required for silt fence perimeter protection include enough length of the premanufactured silt fence with stakes to cover the perimeter that is to be protected.

Begin installation by marking out the locations where perimeter protection will be provided. Clear weeds and debris for a 2 foot wide path along the perimeter identified for protection. In a straight line along the low side of this cleared strip, excavate a 6 inch wide by 12 inch deep trench along the entire perimeter (Figure 13–1). Use of a trenching machine for large projects will save time and labor.

Beginning at the low point of the site, place the bottom of the silt fence in the trench. Drive the wood stakes into the ground leaving 24 inches of stake above the top edge of the trench. With each successive stake, stretch the fabric taut between the stakes. Tuck fabric into the trench and backfill. If more than one length of fabric is required, overlap the first length of fence with the second by 5 feet. Fasten both lengths together using nylon ties. Complete the installation by backfilling and compacting the trench.

An alternative to silt fence with preinstalled stakes is to hang weed barrier from an existing fence at the perimeter of the site. The bottom edge of this installation should be buried in the same manner as the silt fence. If using an existing fence for support, excavate a

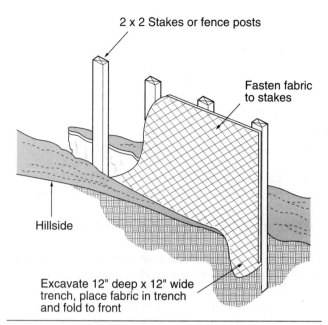

2 x 2 Stakes or fence posts

Fasten fabric
to stakes

Hillside

Excavate 12" deep x 12" wide
trench, place fabric in trench
and fold to front

Figure 13–1 Installation of silt fence with wood stakes.

trench along the upstream (high) side of the fence. To fasten to an existing fence, push nylon ties through the silt fence 2 inches from the top of the fabric and wrap around the fence wires. Stretch the fabric taut and place a tie every 2 feet.

PROTECTING THE SITE FROM SURFACE EROSION

Once the plant material cover has been removed from a surface, the process of erosion begins. Water droplets hit the surface and dislodge a particle of soil. Several particles are eroded away, and rills, or small valleys, are formed. If left unchecked, the rills eventually form a network of gullies and soil is systematically eroded away. Steep slopes and/or highly erodable soils speed the process. To reduce this potential problem, surface erosion protections are designed to either hold the soil in place with plant roots or cover the soil to limit the dislodging of soil particles. The best protection methods combine both of these techniques.

Protection methods used to reduce surface erosion include establishment of **cover crops, mulching,** covering sites with erosion preventing mats, and structural coverings of steep slopes. Each of these methods can be used as both temporary and permanent control measures. Most effective among these controls are ones that combine establishment of a cover crop with mulching, erosion mats, or permanent structural controls. Areas best suited to these protection methods

include any surface that has been disturbed, but particularly disturbed areas with slopes greater than 2% (2 feet of fall in 100 feet of horizontal distance) and highly erodable soils.

Using Cover Crops to Reduce Erosion

Roots of plant material provide an effective method for holding soil in its place. Seeding a disturbed site with a quick-growing crop will both buffer the impact of water drops and hold the soil. Selection of a cover crop requires knowledge of locally available crops that germinate quickly. Common selections include oats, wheat, annual ryegrasses, and buckwheat. The most serious limitation of this method is the time required for plant material to germinate. This limitation suggests a combination of this method with a method that provides immediate abatement of erosion.

Many designs rely on vines, herbaceous ground cover, and woody plant material as a permanent cover to protect slopes from erosion. Since these plants require establishment time, it is necessary to combine their planting with temporary erosion control measures. Suggestions for minimizing the effects of erosion during the establishment period include closer spacing of the plants, use of erosion mats, and placement of mulch.

Mulching for Erosion Protection

Providing an inert covering over a disturbed surface reduces the impact of water drops on soil particles. Using mulch in this way reduces the energy with which the water strikes the soil, limiting the disturbance of soil particles and the potential for erosion. Mulching is a labor-intensive process with limited effectiveness on steep slopes, and functions best when combined with seeding of a cover crop under the mulch.

Mulching begins with selecting a suitable mulch material for the project site. Common choices are straw, wood chips, and shredded wood, with the former a more cost-effective solution for construction protection. Spread the mulch evenly over the entire disturbed area by hand or mechanical means (spreader). This treatment requires periodic remulching during construction. Expect some minor erosion and gullying that will need to be repaired prior to respreading topsoil. If gullying becomes too serious a problem in some areas, consider channel protection for those areas.

A new variation of mulching that has proven to been more effective and less labor-intensive is **hydromulching.** This process, which requires special equipment, mixes mulch with water and a **tackifier** (sticky substance). The mixture is then sprayed on the surface of the site. Hydromulching is typically accomplished by feeding the materials into a trailer-mounted hopper and spraying evenly over a site using a large hose. If desired, seed can also be mixed with the mulch, although the germination and survival rate of seed is reduced when this mixture is used. Hydromulching allows use of more durable mulches and reduces the labor required to evenly spread the mulch over the site. Newer hydromulch formulas, called bonded fiber mulches (BFMs), chemically bond the materials together and provide a durable, papier-maché-like coating over an exposed area.

Installation of Erosion Control Blankets (Erosion Mats)

Erosion control blankets (ECBs) or erosion mats, are a layer of biodegradable materials sandwiched between two layers of lightweight netting. Mats of this type are intended to be anchored to a disturbed surface that has been seeded, either below or on top of the mat,

and left to degrade over time. The mat material, often wood fibers, excelsior, straw stems, or other wood by-products, degrades over one to two seasons and allows any seeded ground covers below the mats to grow. The lightweight netting is ground up by mowers or imbedded into the ground cover, which eventually covers the site. While mats present a higher cost for erosion control than mulching, they do provide better cover and a more stable form of protection for slopes up to 5%.

Installation begins with the identification of the area to be covered. Prepare the entire seedbed below the area where the mat is to be placed. Starter fertilizer and seed may be placed on top of the mat, but placement below the mat provides better success for seed germination rate. Beginning at the high side of the site, excavate a trench 6 inches deep and bury the top edge of the mat. Staple the fabric to the bottom of the trench. Backfill and compact the trench. Roll out the mats in the same direction that water will run across the surface (top to bottom)(Figure 13–2). Verify that there is good mat-to-soil contact. Where two mats are adjacent to one another, overlap the mats 12 inches.

Figure 13–2 Placement of erosion mat along length of a channel. Note that the mat is placed with the long dimension running the same direction as water flow.

Mats can also be overlapped if there is excess material in odd-shaped areas. If mats are placed on a sloped surface, overlap by placing the mat from the higher side over the mat on the lower side. Using metal staples or stakes (usually provided by the mat manufacturer), secure the mat every 12 inches around all edges and at overlap joints (Figure 13–3). Place staples in staggered rows every 18 inches for the length of the mat. It is advisable to remove all stakes and staples that can be located after the ground cover is established and prior to the first mowing. This will reduce the chance of mowers "throwing" loose stakes as projectiles and will protect mowing equipment from damage.

Structural Erosion Control On Steep Slopes

On steep slopes, it is sometimes necessary to install a more durable alternative for reducing surface erosion. Two choices of structural controls are available that have proven reliable in many situations—slope retaining block and rip-rap.

Slope Retaining Block. Interlocking precast concrete block can be used to structurally stabilize a steep slope. These blocks are similar to retaining wall block, but are installed with a steeper **batter,** or backward lean, and open voids between units. Slope retaining blocks anchor a slope by creating a loose wall that holds the soil at its natural **angle of repose** and allows plant material to establish in the voids. To install slope retaining block, excavate a 12 inch wide by 6 inch deep trench along the entire length of the slope. Place the base course in this trench. The base course is laid by leveling the blocks side to side in the trench. Slope the base toward the hillside to provide the batter recommended by the manufacturer (typically 2 inch batter for every 1 foot height of wall). Blocks in subsequent courses should be placed straddling the two blocks below. Backfill with soil and continue adding courses until the top of the slope has been reached. Ground cover plants may be added in the voids as the wall is being constructed or after the installation is complete.

Rip-Rap. In cases where plant material is not a reliable or practical choice for embankment erosion

Figure 13–3 Anchoring erosion mat using metal staples.

protection, permanent protection can be obtained by placing rip-rap over landscape fabric. Install the landscape fabric running in the same direction as the slope and staple in place with sod or fabric staples. In instances where there is significant water running onto the slope from above, excavate a 6 inch deep trench along the top of the slope and bury the top edge of the fabric in this trench. Staple the fabric to the bottom of this trench, backfill, and compact. Beginning at the bottom of the slope, place rip-rap of at least 4–8 inch diameter over the fabric. Material choices may include rounded or angled stone, or other locally available stone. Rip-rap must be large enough and heavy enough to counter the force of water washing down the slope. Carry the rip-rap 18 inches beyond the edges of the fabric at the top and sides to help hold the fabric in place.

PROTECTING THE SITE FROM CHANNEL EROSION

Unless a site is extremely flat, water will collect in lower areas to drain away. This runoff is often heavy enough to form channels in areas with high volumes of runoff. If the channels are disturbed or without protection, severe erosion can result. To provide temporary protection from erosion, techniques such as silt fences are used to slow down the runoff and trap soil washed away from higher elevations. Permanent protection is provided by the use of erosion mats to help establish well-rooted ground covers on the surface of the channel. In locations where water volumes and speeds are great, paved surfaces or rip-rap (loose rock surfacing) are used.

Determining the proper technique to use, as well as proper location of the treatment, varies greatly from site to site. If the site is still under construction and more grading or construction activities are planned in the channel area or erodable areas above the channel, temporary silt fences along the entire channel route are recommended. Spacing of the silt fences depends on the slope and volume of water, but a standard is to place the fences 50 feet apart for a channel that has a slope of 2%–3% or drains an area that is primarily covered by impermeable surfaces. Steeper slopes and greater permanent coverage will require closer spacing. In channels with a very flat slope, it is possible to place the erosion mats as described in the surface erosion section.

Channels in areas where all construction is completed may be addressed with one of the permanent solutions. Establishment of a ground cover is acceptable if the flow in the channel is intermittent enough to allow seeding to germinate. Guidelines stated in the previous paragraph also provide limits for when ground covers may be expected to be functional in stopping erosion. Maximum slopes of 2%–3% and drainage areas that are less than 50% permanent cover approach the upper limits of channel erosion control using living plants. Channels above those limits should lead to consideration of rip-rap or paved channels to limit erosion.

Installation of Silt Fence in Channels

Installation of silt fence for channel protection is done in a manner very similar to silt fence installation for perimeter protection. Materials required for channel silt fence will include lengths of the premanufactured silt fence with stakes that are 50% longer than the channel width (e.g., 10 foot wide channel requires 15 foot fence lengths). Begin installation by marking out the locations where fences are to be located. Silt fence should be placed across the channel perpendicular to the direction of the water flow. Clear weeds and debris for a 2 foot wide path along the path identified for the silt fence. Excavate a 6 inch wide by 12 inch deep trench along the path.

Beginning at the edge of the channel, drive in place at the bottom of the trench the first stake of one of the previously cut silt fence lengths. Leave half of the stake/fabric above the top edge of the trench, and place with the fabric on the upstream side. Stretch the silt fence out along the length of the trench, driving wood stakes into the ground. With each successive stake, stretch the fabric taut between the stakes. The silt fence should be completed using a single piece of material. Complete the installation by backfilling and compacting the trench (Figure 13–4).

As with silt fence used for perimeter protection, this application must be maintained by periodically digging the trapped soil out of the reservoir. Failure to clean out in front of the fence may allow water to flow over the top of the fence, making the fence nonfunctional. If the reservoir fills up rapidly, consideration should be given to placing more fencing between existing installations, or addressing surface erosion above the channel with a temporary control.

Installation of Erosion Mats in Channels

Erosion mats can also be used for protection of channel surfaces. Installation is the same as the installation on a slope described in the previous paragraph, with the following exceptions. Beginning at the low end of the channel, roll out the mats in the same alignment that water will follow in the channel. Where two mats are adjacent to one another, make certain the mats overlap 12 inches. Overlap by placing the mat from the higher side over the mat on the lower side and

Figure 13–4 Silt fence installation.

securing with staples. Using metal staples or stakes (usually provided by the mat manufacturer), secure the mat every 12 inches around all edges and across the center of the mat. Place staples in staggered rows every 12 inches for the length of the mat. In instances where there is significant water running into the channel from above, excavate a 6 inch deep trench along the top of the slope and fold the edge of the mat in this trench. Staple the mat to the bottom of the trench, backfill, and compact. It is advisable to remove all stakes and staples that can be located after the ground cover is established or prior to the first mowing. Riprap should be placed at either end of erosion mat in areas where high water volumes are present.

Installation of Rip-Rap

Rip-rap erosion protection is accomplished by lining a channel with large, loose stone that cannot be picked up and washed away by the water passing through the channel. This permanent installation performs the functions of slowing the water as it strikes the stone and holding the soil beneath the stone in place. A variation of this application is to place landscape fabric below the rip-rap before it is placed.

The success of rip-rapping is entirely based on selecting the proper size of stone. For most small channels, 4–8 inch stone will provide adequate weight

to resist erosion by passing water. Larger channels may require larger stone to perform this same function. Consult a design professional if there is uncertainty regarding what size stone to select. Place the stone beginning at the lower end of the channel and evenly cover all surfaces (Figure 13–5). To be effective, the stone must extend from the bottom of the channel to the top, with no gaps in coverage.

Use of rip-rap for channel protection requires periodic maintenance by the owner. Sediment dropped in the channel creates an ideal environment for grasses and weeds to germinate. Manual, mechanical, or chemical weed control will be needed to control growth in the channel.

Installation of Paved Channels

Paved channels provide the most reliable protection against channel erosion. Unfortunately, they are also the most expensive and difficult to construct. The reader is referred to Chapter 23, Concrete Paving, for details regarding preparation, forming, and pouring cast-in-place concrete. Following are some special considerations for review before selecting and pouring such a permanent erosion protection technique.

Pouring slabs for channels that will carry large volumes of water requires special preparation and finishing to assure an effective installation. An initial

Figure 13–5 Swale protected with 4–8 inch rip-rap.

consideration is making sure that the channel can be paved without being disturbed by drainage through the channel. If flows are not intermittent, methods must be explored to divert or stop the water to allow construction. Since the slabs that form the channel may not be completed in a single pour, providing joints that are solid and impermeable will be necessary. Any water that works under the slab will erode and cause failure of the channel. The surface of the channel should be "distressed" or textured to slow the water as it moves through the channel. This can be done by dragging a rake across the finished concrete perpendicular to the direction of water flow. Other methods of distressing the surface are available, but may require that the slab be thickened by 1–2 inches to avoid weak areas.

Treatment of outflow areas of all channels should be considered when providing channel protection. The area where water passes out of a channel is a potential area for significant erosion. The grade in this area should be flattened out and covered with rip-rap to prevent the channel protection from being undermined (see Chapter 12, treatment of outlet areas).

Section 3

Summary

Creating attractive landforms and effective site drainage requires grading the soil on the site and protecting it from erosion. Section 3 discussed the process for grading a site, installation of basic drainage structures, and temporary and permanent erosion protection measures. Each of these components works together to create interesting site features, effectively remove unwanted water from a site, and keep valuable soil in place.

The grading process is a logical removal, sculpting, and replacement of the layers of soil on a site. Initial steps in grading a site include stripping sod and vegetative matter from the surface, stripping and stockpiling topsoil, rough grading to approximate elevations desired, and placement of temporary erosion control. At the conclusion of rough grading, the process is suspended until drainage systems, utilities, irrigation piping, walls, paving, and structures are in place. When intermediate construction activities are completed, the grading process resumes with the respreading of topsoil and grading to desired finish elevations. Additional smoothing and shaping of the finish grade may be undertaken before establishing turf areas. Permanent erosion control measures follow either finish grading or completion of turf areas.

Managing water draining from the site can be addressed by surface drainage or through the use of subsurface drainage systems. Surface drainage involves the shaping of grade to direct water to desired locations. On sites with little existing topography, the construction of drainage swales or diversion berms may be required to direct surface runoff. When surface techniques do not provide effective drainage, subsurface systems such as french drains, tiles, and storm sewers may be required. French drains are temporary storage trenches that drain and hold small amounts of water in low, undrainable areas. Tile systems involve the trenching and laying of pipe that will intercept water and carry it to an outlet point. Storm sewers work in the same manner as tile systems, with the exception that water is collected at concrete structures set at low points throughout the site and carried in large, nonperforated concrete pipes. In areas where flooding is a problem, temporary storage of storm water runoff in detention ponds may be required. Detention ponds are open basins into which storm water is directed and that have an outlet sized to slowly release detained water.

Keeping valuable soil on the site is the function of erosion protection measures. Erosion can be addressed at the perimeter of the site, over surfaces of highly erodable areas, and along channels that carry runoff. Perimeter protection is accomplished by burying straw bales or silt fence at the construction boundaries and other required locations. Measures typically used over surface areas include establishing fast-germinating cover crops, mulching, placement of erosion mats, and rip-rapping of steep embankments. Channel erosion protection ranges from installation of erosion mats and silt fencing to the rip-rapping and paving of high-volume channels.

Section 4
Site Utilities

INTRODUCTION

While drainage systems are the primary major utility for which the landscape contractor has responsibility, references to other aspects of utility work common in the landscape are made in this section. In addition to drainage systems, preparation of the site for work by an electrical contractor and installation of direct current (DC, or low-voltage) lighting systems are most likely to be addressed by the landscape contractor.

Because separate utility contractors are typically retained to perform other utility work, only reference information is provided with regard to sanitary sewer, potable water, gas, and cable/fiber optics.

Whether engaging in a simple or complex utility installation for a site, caution is always urged when working with underground operations. Call for locates before digging and use extreme caution with the equipment and trenches involved in utility work.

Chapter 14

DC Site Lighting and Related Electrical Work

In this chapter, the installation of DC electrical systems is presented, along with a summary of site-related AC electrical installations that may become the responsibility of the landscape contractor.

RELATED INFORMATION IN OTHER CHAPTERS

Information provided in this chapter is supplemented by instructions provided elsewhere in this text. Before undertaking activities described in this chapter, read the related information in the following chapters:

- Safety in the workplace, Chapter 4
- Basic construction techniques, Chapter 7
- Related site utility work, Chapter 15
- Site amenities, Chapter 35

DIRECT CURRENT ELECTRICAL SYSTEMS

Exterior low-voltage electrical systems are used primarily for powering decorative lighting. Spotlights, stair lights, low-level walk lights, uplights, and downlights in plant material are examples of direct current (DC) lighting applications that can be installed by the landscape contractor. Installation of basic DC lighting **fixtures** varies slightly from manufacturer to manufacturer, but the steps undertaken for system assembly are similar.

Locating Electrical Sources and Controllers

DC electrical systems require an alternating current (AC) outlet to provide power. Typical AC sources are 120-volt outlets located in garages, basements, or exterior locations. Identify an outlet in a weatherproof location where a transformer can be placed. It is preferable to select an outlet that is a ground-fault circuit interrupt (**GFCI**) circuit. GFCI circuits are designed to shut off electrical power to the circuit when voltage fluctuations from short circuits are sensed, providing protection from accidental shocking.

Most **transformers** are designed so they can be mounted on a post or wall with a grounded plug for an outlet. The transformer reduces the AC current so that it is useable by the low-voltage system. A low-voltage electrical cable plugs into the transformer and must be fed to the locations where lights are located. If the transformer is inside, drill a hole through an exterior wall and feed the lighting cable through to the outside. Plug the hole with insulation or steel wool and fasten the cable to a nearby stud with a wire staple.

Some DC lighting systems have **controllers** that turn on lights at dusk or allow the user to control the on and off times. These controllers may be built in with the transformers or are separate units placed along the cable. Either method requires that the controllers be located where the owner can gain access to the equipment. In the case of photocells, controllers must be located in an area that receives ample sunlight. This may necessitate that transformers with built-in controllers be located outside a structure, and that the cord that connects to the AC outlet be fed through the wall to the outlet.

Fixture Installation and Cable Connection

Fixtures require some level of assembly according to the manufacturer's instructions. Locate the fixtures in the areas where they are to be placed before stringing the cable. Lay the cable from the transformer to the are where the first fixtures are located. Provide ample cable to allow for adjusting the fixture locations after they are connected. Typical fixture assembly and connections are performed as follows (Note: This assembly procedure is general and may not apply to every style and manufacturer of light fixtures. Review manufacturer's instructions before assembling any light system):

- Insert bulb into lamp base socket (Figure 14–1).
- Attach lens to lamp base. Lenses will snap or twist into the base (Figure 14–2).
- Run a loop of cable through the mounting stem bracket (Figure 14–3).
- Connect the lamp base to the cable by pressing the cable onto the metal prongs projecting from the lamp base (Figure 14–4). These prongs puncture the cable and make contact with each conductor in the wire. One metal prong must make contact with the wire inside each side of cable. Use care not to bend the prongs.
- Attach a threaded cap, which will hold the cable in place (Figure 14–5).
- Slide the mounting stem bracket into place over the threaded cap (Figure 14–6).

- Insert the mounting stem into the mounting stem bracket (Figure 14–7).
- Gently push the fixture into the ground (Figure 14–8).

Ground-level lights may have the stem and fixture in a single unit, allowing the installer to connect the light and insert the mounting stem into the ground in a single operation. Mounted lights may require that the fixture be snapped or bolted to a mounting base.

Continue installation of fixtures at each location along the cable where a light is planned. DC cables can be placed directly on the ground. Typical installations place the cables through planting beds covered by mulch. Shock danger is minimized if the cable is cut, but this shallow installation does create more opportunity for the cable to be severed by construction and gardening activities. If severed, or if additional cable length is required, splicing kits are available that clamp the ends of two cables together to complete circuits.

Testing System

After all the fixtures have been installed, plug in the transformer and test the system.

Figure 14–1 Inserting bulb into lamp base socket.

Figure 14–2 Attaching lens to lamp base.

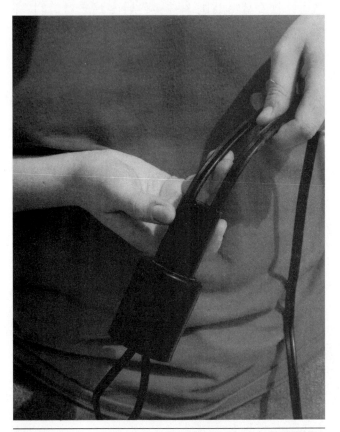

Figure 14–3 Running loop of cable through mounting stem.

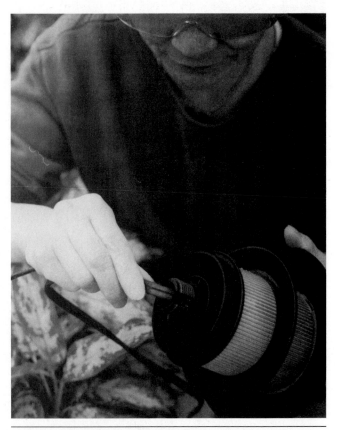

Figure 14–4 Connecting cable to lamp base. A metal prong must penetrate each side of the cable for the light to work. Do not bend the metal prongs.

Figure 14–5 Attaching threaded cap to lamp base.

Figure 14–6 Sliding mounting stem over threaded cap.

Figure 14–7 Inserting mounting stem.

Figure 14–8 Gently pushing fixture into the ground.

ALTERNATING CURRENT ELECTRICAL SYSTEMS

High-voltage electricity is used to power exterior lighting systems used for decoration, safety, and security. In addition, exterior electrical systems are used to power pumps, fountains, and other landscape systems that utilize motorized equipment. Installation of AC electrical systems is a task for a licensed electrician, with the scope of assistance by the landscape contractor limited to site preparation.

Exterior lighting systems can be as simple as a security light mounted on a pole or as complex as systems utilizing numerous lights and multiple lighting styles. While DC lighting is limited to small areas and simple uses, AC electrical systems are capable of illuminating large areas such as parking lots, walkways, and building facades. AC lighting also utilizes a variety of lighting types such as mercury vapor, metal halide, and high-pressure sodium to accomplish lighting tasks.

Common lighting uses associated with site AC systems include pedestrian walkway lighting, area lighting, decorative lighting, and utility lighting. Pedestrian walkway lighting includes low-and medium-height lights used for illuminating walkways and entries. Area lights are higher-mounted lights used for filling parking lots and open areas with light. Like DC light systems, AC lighting can also perform decorative functions. Floodlights for buildings, uplights and downlights for plants, and spotlighting for signs require high-voltage systems when high light levels are required.

Landscape contractors should not be required to complete AC electrical tasks such as pulling cables, supplying power, making electrical connections, or related circuitry work. Preparation of footings for light fixtures, installation of fixture standards, or installation of empty **conduit** for future landscape lighting are tasks that may fall under the scope of items assigned to the landscape contractor.

Fixture Footing Installation

Preparation of the site for AC landscape lighting applications may require installation of footings for **light standards** (posts on which lights fixtures are mounted). Properly installed exterior lighting requires frost footings to prevent damage to the fixtures from heaving (Figure 14–9). Installation of footings requires expertise in excavating and forming of footings and

Figure 14–9 Poured concrete footing to which site electrical fixture will be bolted.

placement of internal conduit. Manufacturers prepare extensive instructions for both the completion of footings and installation of standards, which should be reviewed prior to beginning such work. Chapter 35, on site amenities, covers footing preparation for common types of amenities that require electrical service.

Installing Standards

Standards are placed over the anchor bolts inserted into the footing. They are then plumbed and connected with lock washers and nuts. Caps may be provided to cover the bolted connections, or connections may be completed inside the handhole at the base of the fixture (Figure 14–10).

Conduit Installation

Placement of empty conduit that is to be used later by the electrician to pull cables is another task that may be assigned to the landscape contractor. Conduit installation requires the trenching, connection, and placement of weatherproof PVC or metal conduit to locations where future electrical service is planned. Check construction documents or local codes for allowable types of conduit and proper burial depths. If no plans or codes are applicable, bury **Schedule 40** conduit a minimum of 18 inches deep. Trenching for conduit and piping of all types should be executed after the rough grade has been established and deep utilities such as drainage, water, and sewer service have been completed. Trenching before paving and finish grade have been established reduces disturbance of completed parts of the project.

Leave a stub of conduit projecting 2 feet above the finish grade at each location where a fixture or other electrical service is located. Cap or duct tape over the open end and mark the location with a colored flag for visibility.

Figure 14–10 Handhole at the base of a lighting standard. Electrical connections are made inside this handhole after the fixture has been installed.

Chapter 15

Related Site Utility Work

Utility construction for site development can involve numerous installations. Utility systems for any site can include service lines to structures for water, sewer, gas, electric, phone, TV, and telecommunications. In addition to bringing service to the site, the site itself may have a network of utility lines for site lighting, irrigation lines, and utility services to outlying structures. While landscape contractors are limited by codes, labor unions, and safety issues in the scope of work that they are able to undertake as part of their landscape work, they must maintain an awareness of the timing and location of utility work on the site.

> **CAUTION**
>
> Installation of many utilities is restricted to licensed specialists in most areas. Because of its special nature, landscape contractors should engage only in the utility installations for which they are qualified and certified. Significant health and safety hazards exist with the improper installation of utilities listed in this section.

RELATED INFORMATION IN OTHER CHAPTERS

Information provided in this chapter is supplemented by instructions provided elsewhere in this text. Before undertaking activities described in this chapter, read the related information in the following chapters:

- Safety in the workplace, Chapter 4
- Basic construction techniques, Chapter 7
- DC site lighting and related electrical work, Chapter 14

WATER SYSTEMS

Exterior water systems can include irrigation, external **hydrants,** and supplying potable water to outdoor living areas. In many projects, the type of work required for this section is completed by a plumber or irrigation specialist, but the landscape contractor may be called upon to complete a hydrant installation or a supply line.

> **CAUTION**
>
> Improper installation of water fixtures can cause contamination of public and private water systems. Have the installation inspected by the health department before operating the system.

Hydrant Installation

Exterior hydrants provide convenient sources of water for landscape maintenance. Freezeproof hydrants that can be connected to threaded PVC, copper, or galvanized pipe fittings are available. It is assumed that the water line is buried below frost depth, otherwise the installation will not function properly in cold weather. Install by first turning off the water that supplies the line on which the hydrant will be connected (Figure 15–1). Excavate a 3 foot diameter pit over the water line in the location where the hydrant is to be located. Excavate around and 1 foot below the water pipe and place pea gravel backfill at the bottom of the pit up to the water pipe level. Cut out a section of pipe and insert a T connection. Run a short supply line from the T and connect the hydrant to this supply. Turn the water back on and check the system for leaks and proper function. Fill around the connection with another 6 inches of pea gravel and backfill the pit.

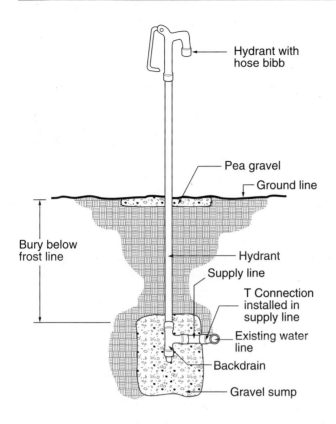

Figure 15–1 Elevation of freezeport hydrant installation.

On the surface directly below the hydrant spigot, or opening, place an 18 × 18 inch precast concrete block or 2 CF of pea gravel. This provides a "splash block" for the hydrant and allows access across a nonmud surface.

PHONE, TV, AND TELECOMMUNICATIONS SYSTEMS

Most people underestimate the impact of phone, TV, and telecommunications systems on the landscape. Any deck or gazebo can be "wired" to the information highway with simple preparation during the construction phase of landscaping. Installing a phone line or **coaxial** cable to an outdoor structure expands the idea of the virtual office into the virtual garden. Installation of private phone or TV lines requires specialized contractors. Installation of empty conduit is recommended even if service lines are not initially placed.

Please note that utilities may have regulations regarding who makes the connections to their service lines, and may have a charge for additional outlets connected to a single service. Inform clients that there is also a security issue with phone and TV connections in exterior applications. Without locking covers on access points, there is nothing to prevent someone from plugging a phone into the client's line and dialing unauthorized numbers.

GAS SYSTEMS

Gas systems can be either natural gas or liquid propane (LP). Gas used in exterior applications is typically reserved to supply lines to outlying structures, for outdoor cooking, and occasionally for lighting or firepits. As with other utility systems that pose health and safety risks, specialized contractors should be contacted for the installation and connection of gas systems.

PLACEMENT OF CONDUITS FOR FUTURE USE

When installing permanent landscape features, particularly those adjacent to buildings, it is advisable to consider installation of empty conduit under the feature that can be used for future use. Conduit can be placed under walks, patios, decks, and structures to avoid the necessity of digging up these elements at a later date to install utilities. Consider placing two oversized conduits (4–6 inch) through which electrical, gas, water, and other lines can be pulled in the future. Conduit can be placed through a basement wall or left just outside the wall if there are concerns about moisture or insect entry to the structure. The conduit should be capped to prevent moisture from filling the pipe and possibly flowing back into the structure. Sloping the conduit away from a structure is also advisable.

While a majority of the utilities designed to serve a site are installed by specialized contractors, the landscape contractor may be requested to perform minor utility tasks and coordinate activities with other utility installers. Section 4 described those utility tasks related to landscape work.

DC lighting systems are one utility element that is often assigned to the landscape contractor. Installation of the DC system requires the identifying of a suitable electrical source, connecting any controllers, laying cable, and connecting fixtures. Due to the versatility and reduced risk of DC electrical currents, such systems can be used to serve several decorative lighting purposes. AC electrical work is limited to laying empty conduit, pouring footings for exterior lights, and installing of lighting standards.

Water utility installations prepared by the landscape contractor are limited to the installation of yard hydrants and irrigation systems. Yard hydrants are often buried to frost depth to allow use during cold weather, while irrigation systems are placed at shallower depths, with the system being winterized prior to the onset of cold weather. Consult a reliable irrigation manual for a detailed description of the installation of the various irrigation components.

Wiring for the various communication mediums available may be accomplished by specialized contractors or, in smaller projects, specialized contractors assisted by the landscape contractor. Placement of the conduit, cables, and service entries to landscape structures can be accomplished by the landscape worker. Due to the safety risk, gas service connections should be delegated to a plumbing professional.

Section 5

Landscape Retaining Walls and Stairs

One of the most dominating functional and visual elements of the landscape is the retaining wall. Whether used as a single element or several walls grouped together in **terraces,** the immediate impact of this vertical element creates a significant change in a landscape. Walls also serve a variety of functional purposes for the homeowner, from mitigating erosion to creating space. Primary among the functions of the retaining wall is the leveling of grades to create more useable space in the landscape. On sites that have a significant amount of slope, normal use of open areas for lawns, gardens, play areas, or circulation routes may be restricted because of the steep grade. Retaining walls can create level areas that will accommodate these activities.

Contemporary construction materials and techniques make walls a versatile and effective landscape element. Walls of almost any form, color, and material can be blended with other landscape elements to accomplish design goals. Despite these advances, walls are still an expensive and time-consuming landscape element to construct, so give consideration to all possible solutions when addressing grade problems.

The chapters of this section cover construction techniques used for building walls with several common materials available to the landscape contractor. Chapters cover the installation techniques used for tie/timber wall construction, segmental wall unit (precast concrete) construction, dry-laid stone wall construction, and gabions. Not included in this are retaining walls constructed of cast-in-place concrete or mortared veneeer materials. Both cast-in-place concrete and mortared masonry veneer retaining walls are impermeable materials that require specialized drainage systems placed behind the wall to reduce he risk of failure due to water pressure. In addition, the skills required and risk involved in forming and pouring such installations typically place these walls beyond the scope of the landscape contractor. Nonretaining, freestanding wall construction, including mortared walls, is discussed in Chapter 33.

PRODUCTIVITY SUGGESTIONS

The process for constructing walls and stairs can be approached a number of ways. The following suggestions may provide general ideas from which a project can be started:

- Always begin wall construction at the lowest point (Figure SI–1).
- Work from lower walls to upper walls when terracing.
- Build cast-in-place concrete stairs first and then build the wall up to the stairs.

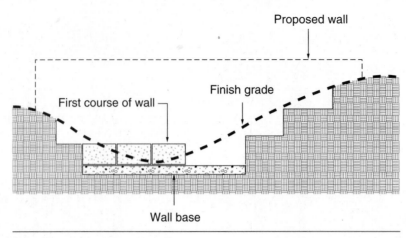

Figure SI-1 Beginning wall at low point.

- Locate stair and/or corner locations and plan material layout to end evenly at these points.
- To avoid delays caused by weather, install base courses for several projects rather than beginning and completing one project at a time. Bad weather will interfere with base construction, but has minimal impact with stacking of subsequent courses, or horizontal layers of wall material.
- The bottom course of all walls should be buried below grade. This will enhance the stability of the wall. There are accepted manufacturer's and engineering standards for bottom course burial depths.
- Excavation of a wide trench along the wall alignment will ease access problems, allow use of a skid-steer for excavation, and provide more room for working.
- Never use pea gravel as a base or backfill materials for walls. It does not create enough friction between stones to maintain stability.

Chapter 16

Materials and Installation Techniques for Retaining Walls

Wall material selection is dictated by the construction document prepared for a project. If the choice is to be made by the contractor or the client, a review of available materials will help with proper selection. Regardless of the type of wall material selected, site preparation is similar. Proper base preparation, providing for drainage behind the wall, and planning wall stabilization are required for walls that are taller than 1 foot, and should be considered on all walls.

RELATED INFORMATION IN OTHER CHAPTERS

Information provided in this chapter is supplemented by instructions provided elsewhere in this text. Before undertaking activities described in this chapter, read the related information in the following chapters:

- Safety in the workplace, Chapter 4
- Basic construction techniques, Chapter 7
- Concrete paving, Chapter 23

SELECTION OF WALL MATERIALS

In many construction projects, the choice of wall material is made by the designer based on engineering and aesthetic considerations. In the event the opportunity arises to recommend wall materials to clients, review with them the positive and negative aspects of the many materials available. The following sections identify several common retaining wall materials that are available to the landscape contractor.

Railroad Ties and Treated Wood Landscape Timbers

Wood railroad ties are often the least expensive wall material, but availability of quality ties has been limited since the late 1980s. Ties provide a gray to dark brown color for the wall and blend well with naturalistic design themes, but poor quality ties will provide an inconsistent look to the wall surface. Workability of the ties can be very difficult, considering their weight, irregular lengths, and irregular dimensions. Consistent wall building requires modular materials, and railroad ties require a great deal of sorting to match dimensions. Using either a chainsaw or cutoff saw is the most efficient way to trim ties, but the creosote, wood hardness, and buried nails and metal end ties are extremely hard on this equipment.

A more costly substitute for railroad ties is treated wood **landscape timbers** (Figure 16–1). Manufactured to consistent dimensions and lengths, timbers overcome the inconsistency problems presented by ties. Timbers also present a consistent wall surface color and texture. Timber weight is slightly less than that of ties, and the cost for timbers is significantly more than the cost of ties. Do not confuse landscape timbers with the small edging and planter timbers sold in lumber centers. Landscape timbers are typically sold in 6 inch × 6 inch or 8 inch × 8 inch dimensions, in 6 foot to 8 foot lengths, and have squared edges. The much smaller edging timber has rounded edges and is not suitable for walls.

Figure 16–1 A stacked wood tie retaining wall with vertical anchors.

Segmental Wall Units

Many manufacturers now produce **segmental precast concrete units** used for constructing walls (Figure 16–2). Similar in surface area to a concrete block, these wall units come in a variety of forms, shapes, and installation methods. Segmental units create an appearance of a rough-surfaced stone wall and, combined with the variety of colors, provide an attractive and consistent wall surface. Wall units can be heavy, ranging from 30 to 80 pounds per unit, but they weigh less than timbers and are more conducive to individual workers. Installation requires careful layout and base course preparation, with most subsequent layers stacked. Corners and stair incorporation can be difficult. Cost is in the mid- to slightly high range when compared to other wall materials. Two common types of segmental units are **pinned wall units,** which utilize metal or fiberglass rods to stabilize between layers, or **lipped wall units,** which utilize a protrusion on the bottom of the unit to integrate layers. Anchored units, or thin concrete fascia blocks with precast anchors placed into the hillside and pinned together, are also available.

Dry-Laid Stone

A wall material that has been used for years and is considered a classic is **dry-laid stone** (Figure 16–3). As the name suggests, this natural wall material is installed without mortar between the stones. Considered a highly aesthetic wall surface, the stone wall can be constructed of regular or random thickness material and with a variety of stone types. Costs vary, but are typically high due to availability and quarrying costs, high demand in construction, and intensive labor for preparation and installation. Stone is a workable material, since most cut stone weights are within the lifting abilities of a worker and are available in reasonably consistent dimensions. Stability of the material may be a concern on walls that require a great deal of height.

Figure 16–2 Segmental precast concrete unit retaining wall. (Courtesy of Gary Pribyl, Kings Materials)

Figure 16–3 Dry-laid stone retaining wall.

Gabions

A more contemporary solution to constructing retaining walls is the use of **gabions** (Figure 16–4). Gabions are heavy-duty rectangular wire baskets that are placed in position and filled with large stones. The size and weight of the gabions make it a very stable wall installation. Varying the size and type of stone inside the gabion produces different aesthetic effects, but the presence of the wire basket will always be a detraction. Because of their size, gabions require special equipment to place and fill, but the cost is in the lower range of wall materials.

WALL LAYOUT

Prior to construction, the location of the front of the wall should be identified. Placing stakes and a string-line, or painting the ground along the alignment, will help in identifying potential starting points and any problems that may be encountered. Locations of stairs and corners should also be identified. This layout will guide the location of the rough trenching necessary to establish an approximate grade for the wall. Offset temporary markers away from the excavation so that the wall location can be reset after rough trenching.

Figure 16–4 Gabion retaining wall one course high.

EXCAVATION

Excavation may be required in locations where walls are to be constructed. Sod and topsoil should be excavated from an area 2 feet in front of and 2 feet behind the entire wall alignment. Excavate to a depth that is approximately 8 inches lower than the desired finish grade along the base of the wall (Figure 16–5). This depth is necessary because the bottom, or base, course in any wall is placed below the finished grade. When excavating, any unstable materials encountered should be removed and replaced with **granular base.**

BASE PREPARATION

Wall construction begins at the lowest point of the wall. After rough trenching is complete, replace the stringline that identifies the front of the wall. In the rough trench and along the wall alignment, excavate a trench to be filled with **base material** that is 6 inches deep and 12 inches wider than the wall material being used (Figure 16–6). The bottom of this trench should be level from end to end. When the

depth of this level trench exceeds the thickness of the wall material being used, the wall should step up one level. When walls encounter embankments, this stepping may occur rapidly. Walls should never step up more than one course at a time, and the course on the bottom should always be covered by grade in front of the wall.

Fill this trench with free-draining angular 3/4 inch to 1 inch crushed stone. Level the base material and compact using a **vibratory plate compactor.** Placement of a 1/2 inch to 1 inch layer of finer granular material on top of the crushed stone will ease the leveling of the base course of wall material. Crushed stone that passes a 3/8 inch sieve is suitable for this second layer.

Flowable Fill

A stable alternative to compacted granular base material in wet locations is a medium-grade flowable fill. Flowable fill is a very thin concrete mixed with fly ash that will flow throughout a prepared trench. Consult premixed concrete suppliers for a suitable flowable fill

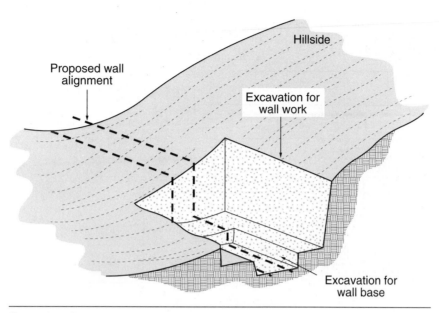

Figure 16–5 Preparing an excavation for retaining walls.

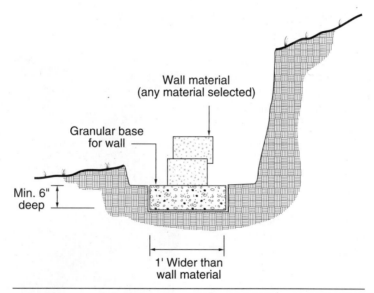

Figure 16–6 Granular base material for wall. Trench should be 1 foot wider than wall.

formulation. To install flowable fill, excavate a base trench a minimum of 24 inches wide and 12 inches deep along the alignment of the proposed wall. Place a grade stake in the trench and mark the elevation desired for the top of the base material. Begin pouring of flowable fill at one location along the trench and allow the material to fill the trench to the desired grade. Flowable fill is liquid enough that it will seek its own level without screeding or finishing. Allow the flowable fill to harden before beginning wall material installation.

Grade Beams

In areas with highly unstable soils, a design professional may require that walls be placed on a grade beam. Grade beams are reinforced concrete footings supported by concrete piers sunk to stable subgrade (Figure 16–7). While this application is uncommon and expensive, it is an effective way to support a wall. Construction of grade beams begins with excavation and pouring of piers in locations and dimensions determined by the design professional. Reinforcing rod should be left projecting out of the top of the piers

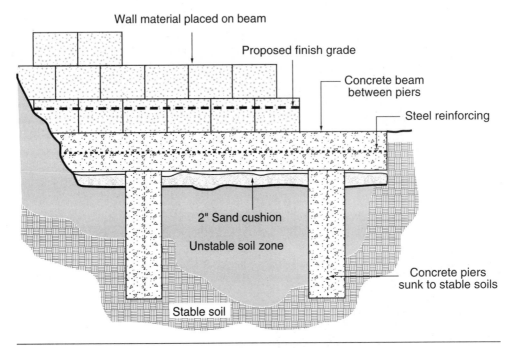

Figure 16–7 Cross section of grade beam used to support walls in poor soils.

to anchor the footings. Under the alignment of the footings, excavate and place a minimum of 2 inches of sand (or depth stated by design professional). This will accomodate frost pushing up against the bottom of the footing. Form the footing and place steel reinforcement that is tied to the stubs projecting from the piers. A typical grade beam is 1 foot square, but dimensions vary depending on the situation. Verify that the tops of the forms are level and at the correct subgrade elevation. Pour the footing and remove the forms when hardened. Place drainage tile behind the wall as described in the following section, backfill with **free-draining angular backfill,** then proceed with wall construction.

DRAINAGE BEHIND WALLS
Wall failure can often be traced to a buildup of **hydrostatic** (water) **pressure** behind the wall. The forces exerted by water behind a wall are strong enough to collapse even reinforced installations. Release of water from behind a wall is partially accomplished by drilling holes through the base course of the wall every 6 to 8 feet **(weepholes)**, or by water naturally passing through the open wall joints (narrow openings between wall materials).

While weepholes and open wall joints are beneficial, supplemental drainage is recommended in any type of wall that exceeds 2 feet in height. This is typically accomplished through the use of drainage tile placed behind the base course of the wall (Figure 16–8). A

Figure 16–8 Placement of socked drain tile behind retaining walls.

4 inch perforated **geotextile socked tile** placed in this location along the entire length of the wall should aid in the removal of excess water. In order to work properly, free-draining angular backfill should be placed above the tile to allow water to seep down to this tile. Proper material selection for backfill behind the walls is critical in reducing hydrostatic pressure. A 12 inch zone directly behind the wall should be backfilled with a 3/4 inch to 1/2 inch clean, free-draining, angular granular material. This material will provide adequate compaction without disrupting water movement downward to the drainage tile. The tile must be sloped to the low point of the wall. If the low point does not allow the tile to run around an end of the wall, the wall material will have to be cut or notched to allow the tile to pass through the wall and empty the collected water. Engineered walls that exceed 4 feet in height may have additional layers of tile along the length of the wall to expedite collection of water.

Landscape Fabric

Certain wall materials may have open joints that are large enough to allow backfill material to pass through the wall. Designers may opt for a solution that places landscape fabric behind the compacted fill material to reduce the movement of soil through the wall openings. In most wall construction projects, this approach is not recommended because it can increase the hydrostatic pressure behind the wall and counter the benefits gained from drain tile and free-draining backfill. For precast unit walls, this material is not recommended because it also disrupts any wall stabilization material placement (see following paragraphs).

If required by landscape plans and specifications, landscape fabric placed behind the wall should run from the base to the second-from-the-top course of the wall (Figure 16–9). The fabric is run the long direction along the wall and placed in the trench behind the wall material before the first backfill. The fabric is draped over the backside of trench until backfilling behind the wall reaches the top. The fabric is folded back over the backfill and tucked under the top course of wall material. Wall construction and backfilling is then completed.

COMPACTION BEHIND WALLS

Remaining backfill beyond the first 12 inches directly behind the wall may be original or imported soil. Backfill material should be placed in **lifts,** or layers, of no more than 6 inches before compaction. Compaction methods should be selected carefully to reduce the chance of wall collapse. For a distance of 5 feet behind the wall, compaction should be performed with a vibratory plate compactor (Figure 16–10). Beyond 5 feet, the compaction may be performed by heavier equipment. Avoid overcompaction immediately behind the wall to prevent displacing wall material.

WALL HEIGHTS AND STABILIZATION

As a wall increases in height, the potential of failure increases. Drawn downward by gravity, and compounded by water pressures, the forces that can topple a wall must be countered by measures that will

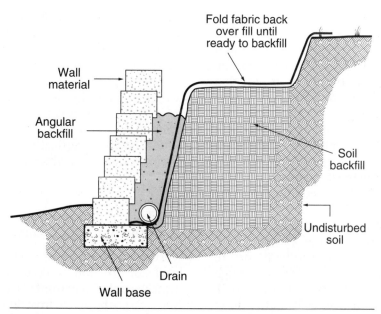

Figure 16–9 Placement of landscape fabric behind retaining walls.

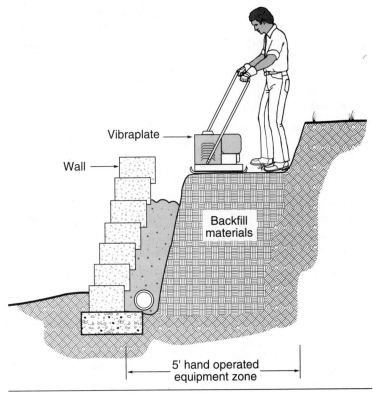

Figure 16–10 Compaction of fill behind retaining walls.

support the wall. Typical measures used to combat wall failure include burying and/or pinning the base course, **batter,** placement of **buttresses** in front of tie/timber walls, wall anchors, and **terracing.** Detailed techniques of anchoring are explained in the chapters on specific wall materials.

Burying of First Course

To improve wall stability, burying the base of the wall provides the first measure of stability. Actual burial depth depends on engineered specifications provided by the design professional or manufacturer. At a minimum, the first course of the wall is buried, and in lower height wall applications the depth of one course may be adequate. With all wall materials, excavate the base trench low enough to place at least one full course of material below the finish grade on the front side of the wall. Another formula used for calculating burial depth is to bury the base course 1 inch below grade for every foot of wall height, with a minimum of one course below grade. Many of the wall materials, particularly the unit wall materials, can be pinned to the subgrade using 24 inch rebars driven through holes in the material and into the base. Both of these measures anchor the base of the wall and counter pressures pushing out at the base.

Batter

Batter is the backward leaning or stepping of a wall (Figure 16–11). Batter can be built into a wall by tilting the base course slightly backward (approximately 1/4 inch fall from front to back). This will cause all subsequent courses to lean in the same direction when placed. Batter can also be built into the wall by setting the front of each subsequent course back from the front of the previous course a small amount. This process, called **step-back batter,** helps anchor the wall but keeps the materials level. Standards for step-back batter range from 1–2 inches for each foot of wall height. Most walls incorporate one of these two types of batter, particularly along straight sections and outside curves of the wall. A few wall materials may have a non-battered vertical face on inside curves due to the materials from which they are constructed.

Anchoring

In addition to batter, many wall systems incorporate some sort of anchor to combat overturning.

Verticals. One type of anchor used in timber or tie wall construction is **vertical** timbers, or buttresses, which are placed in front of a wall (Figure 16–12). These verticals are buried to frost depth and placed at joints in the wall in order to hold back the wall materials.

Figure 16–11 Retaining wall showing step-back batter. (Courtesy of Gary Pribyl, Kings Materials)

Figure 16–12 Vertical tie anchor for a stacked tie retaining wall. Note that the vertical is placed at the joint between wall sections.

Deadmen. Another type of anchor that is used in timber and tie wall construction and occasionally in stone walls is a **deadman.** Deadmen are horizontal anchors that are connected to the face of the wall and run back into the hillside. The friction from contact between deadmen and the soil behind the wall counters the forward pressures.

Deadmen can be constructed by placing a tie/timber with a cross piece into the hillside or by using rods or cables that are run through the face of the wall to an anchor in the hillside. Tie/timber deadmen are typically nailed to the wall face, while rods or cables are connected to metal plates attached to the face of the wall.

Geogrid. A newer version of anchoring that is similar in principle to the deadman is geogrid. **Geogrid** is commonly used in segmental unit wall construction. Geogrid is a gridded polyethylene or polyester fabric that is laid in contact with backfill behind a wall. The front edge of the geogrid is laid between courses of a wall or hooked over the wall pins and spread out in a flat sheet behind the wall (Figure 16–13). The friction between the backfill and fabric provides similar results to those of a deadman. Geogrid is rated according to the strength and direction of resistance. 2T geogrid, which can be laid in either direction, is adequate for most short wall applications. Stronger

Figure 16–13 Installation of geogrid anchoring behind precast concrete retaining wall.

geogrids typically have a higher number (e.g., 5T) and must be placed with proper orientation to have maximum design strength. Proper orientation places the loose strands of the geogrid parallel to the face of the wall and the bonded solid bands perpendicular to the wall face.

Terracing

One effective technique for reducing wall failure is to limit the height of a wall. While design professionals can design walls of extreme heights, without the benefit of their expertise, the maximum wall height should be 4 feet. In many landscape situations, this wall height is ample for accomplishing design goals. If there is a need for higher walls, either terrace the site or consult an engineer. Terracing is the construction of a series of short walls, each stepped back a calculated distance from the previous wall (Figure 16–14). By using terracing, a 12 foot wall can be constructed as three 4 foot walls without extensive engineering. Terracing is of limited usefulness in areas where there is extensive fill or in locations that have limited space.

Figure 16–14 Terracing of segmental precast concrete retaining walls for stability. (Courtesy of Gary Pribyl, Kings Materials)

Terracing should be constructed beginning with the lower wall first, then grading an area behind the wall, and constructing the next higher wall. The grade between the walls should slope forward with a fall of 6–12 inches. Spacing between the walls depends on whether the area leveled to create the second wall is above a slope. Generally, the spacing between the walls should be no less than twice the height of the taller wall. If the wall is built on fill or has heavy traffic above, consult a design professional to determine the proper method for stabilization.

ENDING WALLS

Ending retaining walls without an abrupt stop requires some method of transitioning the grade. Methods used range from tapering the grade in front of a wall to stair-stepping the top of the wall to meet the grade (Figure 16–15). Choice of a method depends on whether the grade behind the wall is falling or moving upward behind the wall.

Tapering Grade

The easiest method for transitioning grade is to construct the retaining wall, backfill behind the wall, and slope the grade from the higher elevation to the lower elevation in front of the wall at each end. Select the steepest slope possible that will allow maintenance and minimize erosion. For a wall 4 feet in height, using

10 feet of horizontal distance in front of the wall to bring the grade down is common. Place fill carefully so that the wall is not disturbed. Slope will need to be tapered in front of wall to accommodate grade change.

Stepping Wall Down

When the grade behind the wall is dropping, the transition to ending the wall can be done by gradually stepping the top of the wall down. Execute this transition by ending the top course and extending lower courses a few units beyond that ending point. Continue stepping down until the wall matches grade or can be ended in an existing grade.

Return Walls

If no grade exists close enough to use for a transition, a small section of perpendicular wall should be constructed. This **return wall** should carry the full height of the wall until it can be stepped down or ended in an existing grade.

PREVENTING EXCESS RUNOFF OVER THE TOP OF A WALL

When a wall is constructed near the base of a slope or near a paved area, the potential for excessive runoff passing over the wall can be a problem. To reduce the potential for wall failure from this runoff, grade a shallow swale above the top of the wall that runs the entire length of the wall (Figure 16–16). This swale should empty water out at a storm sewer inlet or carry the water around to the front side of the wall. An alternative to a drainage swale for paved areas is to install a curb that directs water to a location away from the wall.

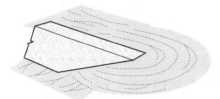

A. Tapering grade

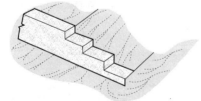

B. Stepping wall down

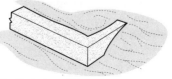

C. Return walls

Figure 16–15 Ending retaining walls.

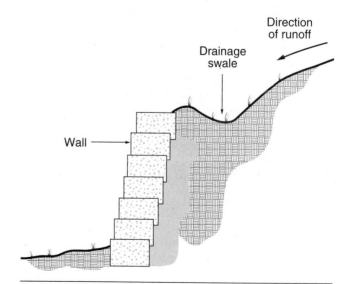

Figure 16–16 Drainage trench at top of retaining wall to intercept runoff before it can run over the top and face of the wall.

Chapter 17

Tie and Timber Retaining Walls

A standard in home landscaping has been the presence of retaining walls made of railroad ties. Such walls are versatile in performing functional duties and in adding a natural look to the landscape. With railroad ties suitable for wall construction becoming more expensive and difficult to locate, newer timber materials have been introduced that produce aesthetic effects similar to those produced by the railroad tie. Whichever material is chosen, proper installation should produce a wall that will serve the dual purpose of retaining hillsides and adding a natural feel to the landscape.

> **CAUTION**
>
> Many timber and tie walls are lasting fewer years than originally anticipated.

This chapter describes the construction techniques utilized to install walls constructed of either ties or landscape timbers. A variety of wall types are available with wood retaining walls, with the primary types being vertical placement, horizontally stacked ties/timbers, or horizontally staggered ties/timbers.

RELATED INFORMATION IN OTHER CHAPTERS

Information provided in this chapter is supplemented by instructions provided elsewhere in this text. Before undertaking activities described in this chapter, read the related information in the following chapters:

- Safety in the workplace, Chapter 4
- Basic construction techniques, Chapter 7
- Materials and installation techniques for retaining walls, Chapter 16
- Stairs, Chapter 21

PLANNING THE PROJECT

Additional preparations are required when using railroad ties as building material. Ties need to be separated according to thickness and trimmed to a consistent length. Matching the ties by thickness eases construction by allowing rows of a similar thickness. Layout of the wall location, including locating stairs, corners, and ending points, allows planning the project for minimal cutting.

VERTICAL POST OR TIE/TIMBER WALL CONSTRUCTION

For walls under 4 feet in height, ties/timbers and even round posts can be placed vertically in a trench and reinforced to create a retaining wall (Figure 17–1). This method is considered easier to construct than horizontally placed wall systems, but lacks stability over the long term.

Site Preparation for Vertical Walls

Initial layout and rough trench excavation is similar to the method described in Chapter 16. Rather than excavating a trench for base material, a trench 12 inches wide with a depth equal to the height of the wall should be dug along the entire length of the alignment. A maximum trench depth of 4 feet and maximum wall height of 4 feet should be used for this installation technique. Placement of 4 inches of granular fill in the bottom of the trench is optional.

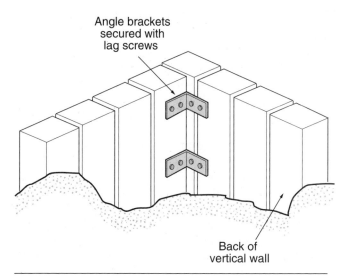

Figure 17–2 Corner connection on back of vertical timber/tie wall.

Figure 17–1 Tie retaining wall with all ties buried vertically. Elevations of tops are staggered for interest.

Installation of Wall Material

Place the first segment of selected wall material into the trench. The wider trench should allow the vertical to be placed with a backward lean that will serve as the batter. Continue placing vertical segments, making sure each fits snugly against the previous segment. After installing several segments, verify that the batter is the same for all pieces, then partially backfill the trench in the front of the wall.

On the backside of the wall, anchor a treated 2 × 6 with galvanized lag screws to each vertical near the top of the tie/timber. This 2 × 6 should be placed horizontally and connected to each of the vertical segments. When constructing curved walls, a strip of rust-resistant metal can be substituted for treated wood. Select a metal such as steel or aluminum edger and drill holes at each tie/timber location. Connect with 4 inch lag screws as just described. As a vertical wall

approaches an inside or outside corner, the batter will have to be reduced gradually until the corners are vertical. Corners may be secured using 8 inch galvanized angle brackets and lag screws on the backside of the wall (Figure 17–2).

Drainage tile should be placed near the bottom of the wall on the backside, and the back and front sides of the wall backfilled. If the top of the wall is irregular, trim the tops of the wall material to the proper height. For decorative purposes, the tops can also be left with varied heights. For a more stable installation, consider backfilling the trench at the base of the ties/timbers with concrete rather than granular material or compacted soil. Anchoring a vertical wall is accomplished by the batter and burial of the base of the wall. Additional anchoring typically proves ineffective.

Interlocking stairs cannot be constructed with vertical tie/timber walls. If stairs are necessary, consult the **butt stair** installation instructions or consider pouring concrete stairs before the wall is built.

STACKED TIE/TIMBER WALL CONSTRUCTION

Stacked tie/timber walls utilize the construction material placed in courses stacked directly on each other to form panels with a single vertical joint between each panel. Vertical ties/timbers or buttresses are placed in front of this common joint, overlapping ties in each panel. Stacked walls are typically built using a tilted base course batter. This type of wall also requires trimming of ties/timbers to a consistent length.

Placement and Leveling of First Course

Installation of ties/timbers begins with preparation of the site as described in Chapter 16. The first course of ties/timbers is placed on the compacted base material in the trench. Place a tie/timber in the trench and, using a carpenter's level, check it for level end to end, and for a half bubble batter toward the back of the wall. If the batter is not correct, lift the tie/timber and remove or add base material as necessary. Ties and timbers can be leveled by placing small piles of granular base near each end of the timber. Once grade is established, granular base material can be gently forced under the timber using a shovel. Recheck the tie/timber and repeat until correct in both directions. If the project requires exact elevations for the bottom and top of the wall, use a survey instrument to verify elevations at this point. If the elevation is incorrect, add or remove base to adjust to the correct elevation.

Continue placing ties/timbers the length of the first course, making sure that the each tie/timber has the same level and batter as the previous one. All ties/timbers in the first course should be similarly placed. If a corner is encountered, trim a tie/timber to the correct length (Figure 17–3). If an opening is placed in the wall for stairs or for other reasons, level across the opening to assure that the wall is level on both sides of the opening. Place drainage tile behind the wall at this point and, if the tile does not outlet at the ends of the wall, leave an opening at the low point for the tile to pass through the wall.

Placement, Connecting, and Backfilling of Subsequent Courses

Subsequent courses are stacked on top of the first course and backfill is placed behind the course (Figure 17–4). In instances where the lower course of the wall steps up, complete the lower course to where the step occurs. Prepare the granular base for the upper course so that it is level with the top of the lower tie/timber. Set a second course of ties/timbers straddling this base and the lower course (Figure 17–5). Ties/timbers are anchored to the course below by drilling three 3/8 inch pilot holes through the top tie/timber and driving a 12 inch spike or piece of #3 **rebar** through each hole. The three pilot holes should be located 6 inches in from each end and in the center of the tie/timber. All holes should be centered between the front and back of the tie/timber. Use a

Figure 17–3 First course installation for stacked timber wall.

Figure 17–4 Placement of subsequent courses for timber wall.

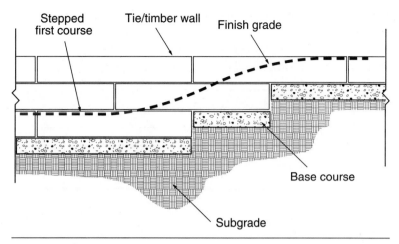

Figure 17–5 Stepping wall first course up a hill.

heavy-duty drill with a long bit to auger the holes, and a sledge to drive the nails. Compact the backfill every course after the courses are connected.

When turning corners, the end of each tie needs to be trimmed to match the batter of the adjacent wall section. Corners of stacked walls require butting one panel against the adjacent panel. Wall panels butted together require vertical ties/timbers on each side of the corner for stability (Figure 17–6).

Placement of Vertical Ties/Timbers

Following completion of the second or third course, auger 1 foot diameter holes in the ground in front of the wall centered on the joints between ties/timbers. The recommended depth for the holes is frost depth, but they should be at least two-thirds the height of the wall. (Note that this will limit the height of the wall to the length of the tie/timber minus the amount buried.) Use caution not to disturb the courses when auguring the holes. When all courses of ties/timbers have been placed, the vertical ties may be set in the holes and leaned back against the wall overlapping the joint. Backfill and tamp around the vertical. For decoration, the tops of the verticals may be trimmed at an angle.

STAGGERED TIE/TIMBER WALL CONSTRUCTION

Staggered tie/timber walls differ from stacked walls in that the courses are staggered so that the vertical joints do not line up (Figure 17–7). Whereas stacked walls rely on batter and vertical members for anchoring, staggered walls can be built with a step-back batter and deadmen for anchoring. Utilizing the step-

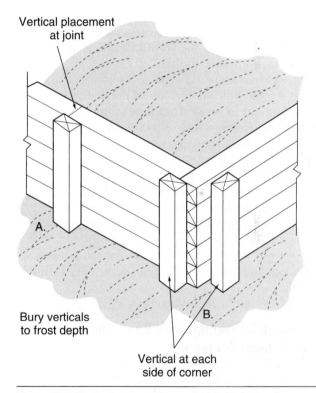

Figure 17–6 Vertical supports for stacked tie/timber wall. (A) Vertical support placement for stacked tie/timber wall. (B) Placement of vertical ties to reinforce corners.

back approach allows the corners of this wall system to be easily overlapped.

Placement and Leveling of First Course

Installation of ties/timbers begins with preparation of the base trench as described in Chapter 16. Place the first course of ties/timbers on the compacted base material placed in the trench. Place a tie/timber at one

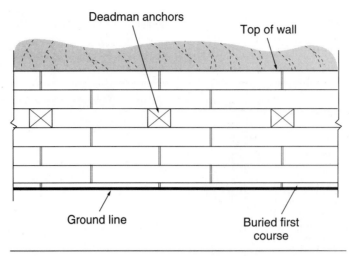

Figure 17–7 Staggered wall pattern for tie/timber walls.

end of the trench and, using a carpenter's level, check it for level end to end and front to back. If the level is not correct, lift the tie/timber and remove or add base material as necessary. If the project requires exact elevations for the bottom and top of the wall, use a survey instrument to verify elevations at this point. If the elevation is incorrect, add or remove base to adjust to the correct elevation. Recheck the tie/timber and repeat until correct in both directions. Continue placing ties/timbers the length of the base course making sure that the next tie/timber is flush with the previous one and has the same level. If a corner is encountered, trim the tie/timber to the correct length. Place drainage tile behind the wall at this point and, if the tile does not outlet at the ends of the wall, leave an opening at the low point for the tile to pass through the wall.

Placement, Connecting, and Backfilling of Subsequent Courses

Subsequent courses are placed on top of the first course by centering the top tie over the joint between the two ties/timbers below. Backfill is placed behind the course and compacted after each course. In instances where the wall steps up, complete the lower course of wall to the point where the step occurs. Prepare the granular base for the upper course so that it is level with the top of the lower tie/timber. Set a second course of ties/timbers straddling the base and the lower course. Do not interrupt the staggered pattern when the wall steps up. If necessary, trim a tie/timber to maintain a consistent pattern.

Ties/timbers are anchored to the course below by drilling four 3/8 inch pilot holes through the top tie/timber and driving a 12 inch spike or piece of #3 rebar through each hole (Figures 17–8 and 17–9). The

pilot holes should be located 12 inches from each end and 36 inches from each end (Figure 17–10). All holes should be centered between the front and back of the tie/timber. Use a heavy-duty drill with a long bit to auger the holes, and a sledge to drive the nails. Compact the backfill every course after the courses are connected.

When a corner is encountered in a staggered wall, overlap each new course over the previous course (Figure 17–11). Drill a pilot hole in the center of the overlap area and drive a 12 inch nail or rebar through the pilot hole.

Installation of Deadmen

Deadmen should be installed every fourth course of a tie/timber wall. Measuring from the end of the wall, install the deadmen every 8 feet (Figure 17–12). To install a deadman, cut an opening in the appropriate course of the wall equal to the width of the deadman. In this opening, set the end of a tie/timber running perpendicular to the face of the wall. The length of the deadman should extend back into the hill between 4 feet and 8 feet. It is sometimes necessary to trim the deadman if the full length interferes with an obstacle behind the wall. Excavating a trench for placement of the deadman is also sometimes necessary. At the hill end of the deadman, place a 3 foot tie/timber crosspiece parallel to the face of the wall and under the deadman. Connect the deadman to both the wall and crosspiece by drilling a 3/8 inch pilot hole and driving a 12 inch nail through the pilot hole into the tie/timber below. When the subsequent course is placed over the deadman, fasten that tie/timber to the deadman using a 12 inch spike through a pilot hole.

An alternative to using tie/timber deadmen is to use the drilled cable deadman system noted in Chapter 16.

Figure 17–8 Drilling holes in timber with a cordless drill for 3/8 inch × 10 inch spikes.

Figure 17–9 Driving spikes in timber with a 2 pound sledge.

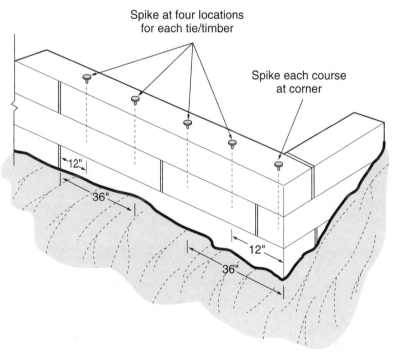

Figure 17–10 Spiking pattern for staggered tie/timber walls.

Figure 17–11 Interlocked square corners with a stacked timber wall.

This system requires less work than that required for the wood deadmen, but leaves an unsightly metal plate or connector on the face of the wall. Anchoring a vertical wall is best accomplished using a drilled deadman placed every 6 feet horizontally. Deadmen should be placed two-thirds up the height of the wall. Threaded bars or flanged wall anchors make acceptable deadmen. Drill a 1 inch diameter hole through the wall and slide a bar or cable through the hole. Anchor the bar or cable in the hillside and to the wall with 12 inch diameter plates threaded into both ends of the bar. Manufactured flanges can be used to anchor the bar into the hill.

CONSTRUCTING STAIRS WITH TIE/TIMBER WALLS

Selection of a stair construction method is based on the type of wall constructed. Stacked and vertical wall construction methods require solid cheek walls, which will not accommodate the overlap of wall material necessary for interlocking stairs. Thus, only the butt stair method may be used for those wall types. Staggered tie/timber walls can accommodate either interlocking stairs or butt stairs. See Chapter 21 for directions on installing butt and interlocking stairs.

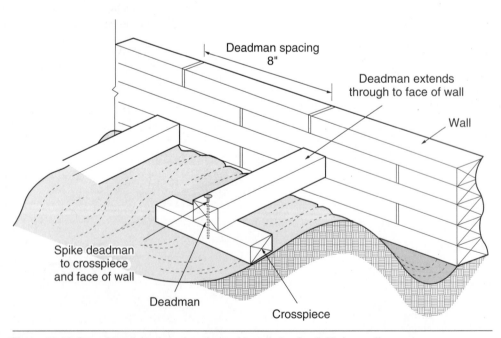

Figure 17–12 Rear view of deadman location and installation for tie/timber walls.

Chapter 18

Segmental Retaining Walls

Utilization of segmental walls in the landscape construction industry has increased dramatically. Experimentation with concrete wall blocks has grown into an industry standard for construction of attractive, stable wall systems. Precast concrete units are one of the few wall materials that can be engineered for placement in vertical installations. In addition to the stability of the product, a variety of surfaces and colors has enhanced the aesthetic qualities of the wall material. This chapter presents the installation of segmental wall units that utilize pins, lips, and anchors as stabilization and battering systems.

RELATED INFORMATION IN OTHER CHAPTERS

Information provided in this chapter is supplemented by instructions provided elsewhere in this text. Before undertaking activities described in this chapter, read the related information in the following chapters:

- Safety in the workplace, Chapter 4
- Basic construction techniques, Chapter 7
- Materials and installation techniques for retaining walls, Chapter 16
- Stairs, Chapter 21

PINNED AND LIPPED SEGMENTAL WALL CONSTRUCTION

Pinned wall units rely on metal or fiberglass pins to tie one course of segmental units to the next. Pins provide a very strong connection between courses that will not slip or be dislodged by pressures behind the wall. Pins also provide a very strong connection between the wall and geogrid anchoring. Lipped wall units rely on a thickened segment on the bottom of the block to position the subsequent courses behind units on lower courses. Lipped units are faster to place but do not provide as strong a connection between courses. Lipped units also provide a weaker connection to geogrid anchoring.

Most manufacturers recommend placement of segmental wall units level since a set-back batter is designed into the systems. In many pinned wall systems, the pins provide a means of automatically determining the setback from one course to the next. In most lipped units, the lip creates an automatic setback from the lower course. A few manufacturers may recommend that the blocks be tilted backward to create a batter for the wall. Review the literature provided by a wall material supplier to determine what method it recommends for installation.

Placement and Leveling of the First Course

Chapter 16 provides important information regarding the preparation of the base for unit placement. As with all walls, first course installation should begin at the lowest point of the wall. Measuring the distance between the beginning point and corners or ending points may allow for slight adjustment of the unit placement so that cutting is minimized. Once the base has been compacted and leveled, the first block can be set. Verify which side of the block is the top side. For pinned units, the top side should have openings for placing pins and the bottom should have openings to receive pins. For lipped units, the bottom has a thickened lip. This lip should be broken off with a hammer for all blocks used on the base course. For some block types an alternative to knocking the lip off is to turn the bottom course upside down and backward for the base only. This will place the lip up and at the front of the base course.

One of the more difficult tasks in constructing walls is the setting and leveling of base stones. To speed the leveling process, a 1/2 inch to 1 inch thick layer of fine granular material (select stone that passes a 3/8 inch sieve) over the compacted granular base is recommended. This layer provides a more workable material when making fine leveling adjustments. Place the block and check for level front to back and side to side with a carpenter's level (Figure 18–1).

If the block is not level, make minor changes by tapping it lightly with a rubber mallet. Adjustments to level may also be made by tipping the block and using the fingertips to scratch material away if the block is high, and sprinkling handfuls of fine granular material under the block if it is low. These methods allow minor height adjustments and create less disturbance than striking with a rubber mallet. Units can also be adjusted by twisting the block. Major adjustments require lifting the block and adding or removing base material as required.

After the first block is set, place the second block next to it and verify that the edges are in contact and the tops are flush. Check the second block for level in both directions and adjust as necessary. Use caution not to disturb adjacent blocks that have already been set, and verify that the edges of the blocks are still in contact with each other after adjustment. Repeat this process along the entire length of the wall first course, checking periodically using a level long enough to cover three blocks. It is imperative that the first course be flush and level, since any irregularity will be compounded as later courses are added. Place any drain tile behind the first course and cut or leave an opening for the tile through the wall if drain tile is not routed around the ends of the wall. Backfill and compact the void behind the wall with a free-draining, angular material. If a landscape fabric is being utilized, it should be installed at this time (see Chapter 16 for instructions). Place backfill and compact.

Placement, Connection, and Backfilling of Subsequent Courses

After backfilling, sweep any base material and debris from the tops of the blocks. Prior to placement of further courses for pinned units, pins need to be inserted in openings on the top of the base course blocks (Figure 18–2). Review manufacturer's instructions to determine into which holes the pins should be placed. Some block styles have pinhole locations that will create vertical walls and others that will create a set-back batter. Set pins along the entire length of the wall. Lipped units do not require any pinning preparation. Blocks for subsequent courses should be placed straddling the joint between the two lower blocks, with half on one block and half on the other. Continue placing blocks until the entire course has been set (Figures 18–3 and 18–4. Backfill behind the wall and compact, then repeat for each course until the top of the wall is reached. If geogrid is required, it needs to be placed between courses.

In locations where the base of the wall must step up, fill the void beyond the end of the last block with base material and compact. This base will continue under all blocks on the upper courses, similar to the base below the lower course. Level the base flush with the top of the lower course, place a pin in the last block (for pinned units) on the lower course. Place a block

Figure 18–1 Leveling and placement of first units for precast segmental units.

Figure 18–2 Fiberglass pin placement in precast segmental units.

Figure 18–4 Section of precast segmental unit wall showing backfill and drain tile.

Figure 18–3 Unit placement for precast segmental units.

straddling on the lower block (over the pin) and the base for the upper course. Check for level and adjust if necessary.

Different types of corners require different treatments. **Concave,** or inside, **corners** can typically be built by keeping the front edges of the blocks in contact and fanning out the back of the units. **Convex,** or outside, **corners** will also require that the block edges be in contact, but may require trimming of back portions of units with a cutoff or wet masonry saw to fit the required curve (Figure 18–5). Lipped units may require removal of a portion of the lip to accommodate a tight radius. Most manufacturers have technical publications that indicate the minimum curve at which radii can be built without serious modification of blocks. Expect major cutting with radii smaller than 5 feet.

Inside and outside right-angle corners can be difficult to construct with segmental wall units. Solid units can be split by hand, hydraulic block cutter, or by cutoff saw with partial units placed in the wall. Hollow units can be split, but placement at a corner may expose the inside of the unit. In most cases, the partial unit should be placed not at the corner, but one or two blocks in from the corner. Some manufacturers produce special corner units to address the exposure and trimming problems encountered at right-angle corners. If hollow cores must be exposed, mix mortar to match the wall color and fill the open cavity. Wall adhesive should be used between courses at the corner, especially when partial units disrupt the normal placement of pins or interlocking lips.

Placement of Geogrid

Manufacturers or design professionals must calculate if a wall requires anchoring with geogrid fabric. These specifications should indicate the course in which the geogrid should be placed, and the width of the fabric behind the wall. If the wall is tall enough to require geogrid, it will need to placed between the appropriate courses before the wall can be built any higher. Since geogrid comes in rolls, it can be cut easily to the width required using a knife or fine-toothed saw. The required orientation must be maintained for various

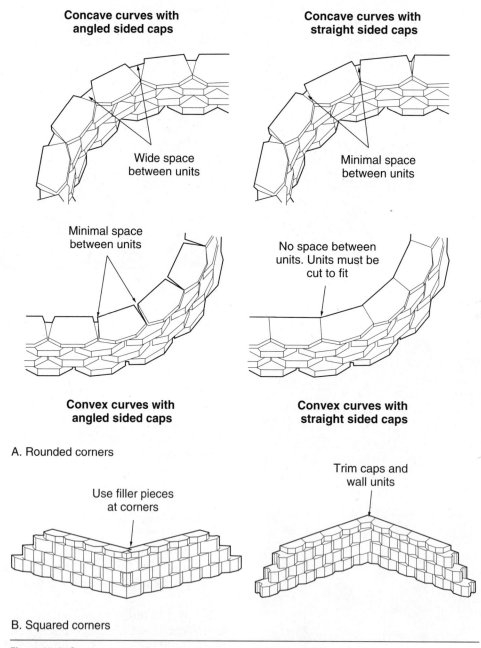

A. Rounded corners

B. Squared corners

Figure 18–5 Corner construction with precast segmental units. (A) Rounded corner construction with precast wall units. (B) Squared corner construction with precast wall units.

types of geogrid in order to obtain the designed strength. Proper orientation is determined by the direction of the bonded and loose strands. Bonded strands are load bearing and should be placed perpendicular to the face of the wall. Loose strands should run parallel to the face of the wall.

After compacting the backfill for the course below the geogrid, roll the geogrid out flat behind the wall along the entire length of the area to be reinforced. The backfill area may have to be widened to accommodate the width of the geogrid. Slide the geogrid to within 2 inches of the front of the wall units. Pull the material back slightly if any will be exposed when later courses are added. For pinned units place the pins for the next course so that they pass through one of the grids (Figure 18–6). Lipped units are set in place with the geogrid resting between courses. Place the next course of blocks and then stretch the geogrid away from the wall to remove any slack. Backfill and compact over the geogrid.

Capping the Wall

Most segmental units require a cap to cover open voids and spaces between the block. When the wall is nearing completion, cap installation can be started (Figure 18–7). All cap units should be installed with pins or adhesive. Clean the top of the wall and apply adhesive generously prior to placement of caps. Pinned units require short pins be used to secure the cap. Push down on the cap after placement to ensure good contact with adhesive. Some systems provide caps with angled ends that can be alternated for straight wall sections, or placed with all angles facing in or out to cap curved walls.

To avoid exposing open cavities at the top of a wall that steps down, either cut a cap to fit the step or replace the unit on which the voids are exposed with two cap units stacked vertically. Bond the cut cap or stacked caps with adhesive.

Figure 18–6 Geogrid placement between courses of precast segmental units. Geogrid should be hooked over the fiberglass pins before placement of following course.

Figure 18–7 Cap installation for precast segmental unit walls.

Adhering and Bonding of Precast Concrete Wall Materials

When wall capstones or corner units require bonding to maintain stability, thoroughly clean all surfaces to be bonded. Generously apply a flexible adhesive and place the two pieces together. Scrape any excess adhesive from the joint. Because it maintains its flexibility, use of a one-part urethane sealer is preferable to using subfloor or construction adhesives.

ANCHORED UNIT WALL CONSTRUCTION

In addition to pinned and lipped segmental wall systems, some manufacturers produce a wall unit that utilizes precast anchors that are connected to a face piece and protrude back into the hillside. These anchors are covered with backfill and provide stability for the wall. Anchored units are lighter in weight but also require more labor to assemble. The thin face piece provides a thinner profile for the wall, and the reduced anchors for the top course make planting behind the wall easier. Geogrid is installed in the same manner with anchored walls as it is for pinned wall systems.

Installation of anchor units is similar to that of pinned units in most aspects, except that each face piece has two rectangular precast anchors that are connected to the front with metal U pins. These anchors protrude from the back of the face piece and form a V shape. The anchors are then pinned together with a third U pin and the cavity is filled and compacted. Anchors for subsequent courses, rest on the anchors for previous courses (Figure 18–8). As the wall approaches the top courses, the number of anchors is reduced. The top and second-from-the-top course have only one anchor per face piece. A cap is pinned to the top course to provide a finished appearance.

STAIR CONSTRUCTION WITH SEGMENTAL WALLS

Stair construction with segmental units is accomplished in a variety of ways. Most manufacturers have engineered a solution to the stair issue, either by designing special stair units, producing a combina-

tion of base unit and cap units, or by utilizing standard wall units for stair construction. Most stair solutions utilize the butt method of placing stairs between cheek walls, but some interlock with cheek walls. Stair solutions with precast units typically work best when the stair width is a multiple of the units used to construct the stairs. This requires careful spacing between cheek walls to assure that an even number of units will fit. See Chapter 21 for directions on installation of butt and interlocking stairs.

Figure 18–8 Anchored precast segmental units showing anchors in place. Openings on tops of anchors are for U-shaped pins, which hold anchors and face blocks together. (Courtesy of Lucy Hershberger, Forevergreen Landscaping)

Chapter 19

Dry-laid Stone Retaining Walls

Successful installation of a dry-laid stone wall is as much a skilled craft as it is a construction technique. The varied types of stone available to the landscape contractor require patience and vision in arranging patterns that are structurally sound and attractive. **Ashlar** is a term that refers to stone that has been cut, or cleaved, in modular lengths and widths with a consistent thickness to aid in placement. An alternative to ashlar is **rubble,** which is stone possessing natural edges that have not been trimmed. Techniques for laying stone are discussed in this chapter. Chapter 32 covers the construction of freestanding walls.

RELATED INFORMATION IN OTHER CHAPTERS

Information provided in this chapter is supplemented by instructions provided elsewhere in this text. Before undertaking activities described in this chapter, read the related information in the following chapters:

- Safety in the workplace, Chapter 4
- Basic construction techniques, Chapter 7
- Materials and installation techniques for retaining walls, Chapter 16
- Stairs, Chapter 21

PLANNING THE PROJECT

Stability of dry-laid stone walls is a primary concern. Most stone used today is quarried and cut into widths of 4–8 inches. This narrow width limits the weight and surface area between courses, which are factors that help in countering forces that can push the wall forward. As a result, dry-laid stone walls sometimes need both leaning batter and set-back batter to provide stability. Placement of geogrid is difficult and use of verticals in front is unattractive and impractical. Stone deadmen can be placed, but they are of limited value since they cannot be attached to the wall or to a crosspiece in the hillside. These factors combine to limit the effective structural height of dry-laid stone walls without being engineered by a design professional.

Sorting of stone prior to placement speeds the process and enhances the stability of the construction. Laying several pieces out so that length can be quickly observed allows for fast sorting and selection of stone. Preparation of the site and laying of the first course is similar to that of other wall materials, but the courses above the first require a different approach.

Following are suggestions that add to the quality of a dry-laid stone wall:

- Utilize longer stones for the base course and stairs for increased stability.
- Avoid vertical joints that run more than two courses. One of the strengths of dry-laid stone comes from the overlapping of joints (Figure 19–1).
- Maintain spacing between vertical joints that is at least as wide as the thickness of the stone.
- Interlock all corners when possible.
- Avoid the use of short or small pieces of stone near corners. Larger pieces space the joints farther apart and improve the stability.

Figure 19–1 Staggering of vertical joints in stone walls improves stability.

- Cut the stone pieces a consistent thickness to eliminate one of the variables that makes laying stone wall a challenge. The worker is able to concentrate on length and joint placement and does not have to be concerned with matching both the thickness and length dimensions.

DRY-LAID STONE WALL CONSTRUCTION

Placement and Leveling of the First Course

Prepare for installation using the instructions from Chatper 16. As with all walls, laying of the first course should begin at the lowest point of the wall. Setting a stringline aids in keeping the front of the wall aligned (Figure 19–2). Once the base has been compacted and leveled, the first stone can be set. Place the stone and check with a carpenter's level for level side to side and a slight batter toward the back of the wall. A quarter bubble is often used as a standard batter. An alternative to placing a leaning batter on a stone wall is to utilize a step-back batter similar to that used in precast units

Figure 19–2 Stone wall first course installation.

Figure 19–3 Dry-laid stone wall showing set-back batter of 1 inch with each higher course.

Figure 19–4 Placement of dry-laid stone wall material for a convex curve.

(Figure 19–3). While stone is not a manufactured product with safeguards built into the stone, a step-back batter for walls under 2 feet tall should provide enough stability to reduce the potential for failure. Both batters can be combined for increased wall stability.

If stones are not setting correctly, make minor changes by tapping with a rubber mallet. A twisting motion will also adjust the level of stones. Major adjustments require lifting the stone and adding or removing base material as required. After the first stone is set, place the second block next to it and verify that the edges are in contact and the tops are flush. Check the second block for level and batter and adjust as necessary. Repeat this process along the entire length of the wall base. It is imperative that the base course stones be flush and consistently placed since any irregularity will be compounded as later courses are added. Place drain tile behind the base course and leave an opening for the tile through the wall if drain tile is not routed around the ends of the wall. Backfill the void behind the wall with a free-draining, angular material and compact.

Placement, Connection, and Backfilling of Subsequent Courses

Stones for subsequent courses should be placed so that the joints are staggered. Set one course at a time (Figure 19–4). Place backfill and compact. Repeat for each course until the top of the wall is reached. If a set-back batter is used, move the front of each subsequent course back 1 inch from the course below.

In locations where the base of the wall must step up, fill the void beyond the end of the last block with base material and compact. This base must extend under all blocks on upper courses in the same manner as base below the lower course. Level the base material flush with the top of the lower course. Place a block straddling the lower block and the base for the upper course. Check for level and batter, and adjust if necessary. Continue placing stones for the base along the wall alignment.

Square corners should be interlocked between courses (Figure 19–5). If wall is laid using leaning batter, square corners must be butted together rather than overlapped. Concave and convex corners are

typically built using shorter stones and stones that have angled ends that can be turned over to create a tight-fitting radius (Figure 19–6). For concave corners place the shorter face of the stone towards the front of the wall. For convex corners place the longer face of the stone towards the front of the wall. Trim stone as necessary to maintain tight joints. Maintain batter around corners.

Anchoring Course

To cap a dry-laid stone wall, an anchoring course may be installed. At the top of the wall, lay a course of block with the long dimension perpendicular to the face of the wall. Align an end of the stone with the front of the wall and extend the long dimension beyond the back of the wall.

STAIR CONSTRUCTION WITH DRY-LAID STONE WALLS

Stair construction with stone is accomplished by using precut stair pieces or by selecting wider stones that can be used for treads. Both butt stairs and interlocking stairs are suitable for stone. See Chapter 21 for directions on installation of butt and interlocking stairs.

Figure 19–5 Dry-laid stone wall showing ingterlocking courses at corner.

Figure 19–6 Rounded corners with dry-laid limestone walls. Shorter pieces are used to turn tighter corners. (Courtesy of Paul Dykstra, Iowa City Landscaping)

Chapter 20

Gabion Retaining Walls

Gabions are an effective material for landscape walls when stability is paramount. Usually confined to large road and grading projects, the effectiveness of the gabion is ideal for addressing large-scale wall projects. With their wide base and heavy weight, stability is seldom a problem for gabions stacked three or four units high. The open nature of the material also makes hydrostatic pressure less of a problem.

RELATED INFORMATION IN OTHER CHAPTERS

Information provided in this chapter is supplemented by instructions provided elsewhere in this text. Before undertaking activities described in this chapter read the related information in the following chapters:

- Safety in the workplace, Chapter 4
- Materials and installation techniques for retaining walls, Chapter 16

PLANNING THE PROJECT

Preparation for gabion construction requires that the alignment be staked and the rough grading be prepared in the same manner as described in Chapter 16. Grade preparation for gabions can be a shallow trench slightly wider than the gabion cage. Gabions are buried only 6–12 inches rather than an entire first course.

While gabion walls require less craftsmanship than other types of retaining walls, gabions do require special equipment to aid the wall construction. Caution should be used when working with gabions due to their significant weight. While they are typically very stable, slides or shifts can do serious injury to a worker.

GABION WALL CONSTRUCTION

Placement and Leveling of the First Course

Leveling should be accomplished by adjusting the grade below the cages before they are placed. Mark the alignment of the first course and place gabion cages along the alignment. Gabions should be installed with the long dimension parallel to the face of the wall. Note the top of the gabion and orient so that the opening edge faces the side from which material will be installed (Figure 20–1).

Carefully fill each of the cages with a consistent selection of stone that has a diameter larger than the cage openings. Stone selection should be based on availability and aesthetics. Either angular or washed river stone is acceptable. Filling operations for the first course should not dislocate the cages from their alignment. When cages have been filled, the tops should be closed and wired shut (Figure 20–2). Any drainage should be placed behind the cages at this time and the void behind the wall filled and compacted with free draining, angular material.

Placement, Connection, and Backfilling of Subsequent Courses

After the backfill has been placed and compacted for the first course, the cages for the second course can be placed. Subsequent courses should be placed straddling the joint between the two gabions below. A slight set-back batter of 4–12 inches may be used for each

Figure 20–1 Cage placement and filling for gabions.

Figure 20–2 Gabion cages are filled with stone larger than the diameter of the cage openings. Cages are wired shut using 12 gage galvanized wire.

course (Figure 20–3). Fill the cages, close, and wire shut. Backfill behind the wall and compact, then repeat for each course until the top of the wall is reached. Gradual corners can be turned by fanning the gabions along the proposed alignment. Subsequent courses on right-angle corners should be overlapped.

In locations where the base of the wall must step up, fill the void beyond the end of the last gabion in the lower course with base material and compact. Level the base with the top of the lower course and place a gabion cage straddling the lower gabion and the base for the upper course. Fill the cage and continue placing cages along the wall alignment.

STAIR CONSTRUCTION WITH GABION WALLS

Stairs used with gabion walls need to be constructed separately out of an alternative material. While cast-in-place concrete makes a durable step, any of the wall materials described in this section make a suitable step for gabion applications. See Chapter 23 for forming and pouring concrete stairs, and Chapter 21 for butt stair installation.

Figure 20–3 Stacking gabion cages.

Chapter 21

Stairs

When walls are used to create levels on a sloping site, stairs are required to traverse the different levels. Stairs are addressed as part of this section due to their relationship to walls and because many stair projects are incorporated into wall projects. Planning information presented regarding stairs applies not only to stairs as part of a wall project, but also to cast-in-place concrete stairs and wood stairs built as part of a landscape structure. Specific construction techniques for these different types of stairs are covered in chapters describing cast-in-place concrete and wood construction.

The landscape also presents situations where a slope needs to be navigated using stairs where no wall is planned. Such instances require the construction of freestanding stairs. Because the side slopes are so gradual, freestanding stairs are constructed without the benefit of retaining or cheek walls. Stairs in this situation are typically constructed with riser/tread dimensions that match the slope rather than matching the ideal mathematical relationships for steps. This chapter also outlines the methods of planning and installing various types of materials used for freestanding stairs.

RELATED INFORMATION IN OTHER CHAPTERS

Information provided in this chapter is supplemented by instructions provided elsewhere in this text. Before undertaking activities described in this chapter, read the related information in the following chapters:

- Safety in the workplace, Chapter 4
- Basic construction techniques, Chapter 7

PLANNING THE PROJECT

The range of construction difficulty for stairs can be quite wide depending on the materials used and slope on which the stairs are being constructed. Prior to beginning the project, planning should be performed with regard to materials desired, and the number of risers and treads planned.

Local Codes Governing Stairs

Before constructing stairs as part of a landscape, check with local building officials to verify stair requirements. Many communities have regulations on stair **tread** and **riser** dimensions, railing requirements, stair widths, and other stair-related attributes. Access up/down grades may also have to be accompanied by a ramp as well as stairs. Check with building officials for rules or requirements that might supersede techniques described here.

Types of Stairs

Stairs can be constructed as part of wall systems (Figure 21–1), or be independent of walls as freestanding stairs traveling up a gradual slope (Figure 21–2). In certain cases, stairs may be replaced by gradual sloping ramps. The type of construction that is selected depends on the steepness of the slope, the distance covered, and the mobility of the user. Whenever possible, select a ramped walkway over stairs because of the ease of construction and the diversity of audiences that ramps will accommodate. When grades begin to exceed 5% longitudinal (over 5 feet of fall in 100 feet end to end), consider stairs as part of the project. If the slope to be traversed does not require a wall, incorporate freestanding steps to assist the client over the slope.

Figure 21–1 Stairs built as part of a wood tie retaining wall installation.

Figure 21–2 Freestanding concrete stairs.

Figure 21–3 Freestanding concrete steps were formed and poured as part of this entry construction.

Cheek Walls

Stairs constucted as part of a wall require that **cheek walls** be built on either side of the area where stairs are desired. Cheek walls are short walls that run into the hillside the stairs traverse. Cheek walls run perpendicular to the retaining wall, should be the same height as the retaining wall, and should extend into the hillside at least as far as the stairs will extend. Timing of cheek wall construction depends on whether stairs will be interlocked with the cheek wall or built independently and butt into cheek walls.

Material Selection for Stairs Built as Part of a Wall

A preferred choice for stair materials is to utilize the same materials as for wall construction. This provides continuity in design and dimension from the wall into the stairs. In some cases, the same material may not be possible or desirable. When changing materials, it is important that the stair material be integrated in to the construction. One popular choice for stairs is cast-in-place concrete. This material, when properly formed, can be set at any riser and tread dimensions and can be poured to fit snugly with any cheek walls. Forming for stairs is covered in the concrete section of the landscape paving chapter of this book. Freestanding steps can be constructed of any of the wall materials just identified, as well as from cast-in-place or precast concrete.

Material Selection for Freestanding Stairs

Freestanding stairs can be built out of almost any material. To organize the presentation of choices and installation, materials are grouped into four categories as follows:

- Concrete stairs. Concrete can be formed with any riser height and tread length and can be built easily without the benefit of retaining walls alongside (Figure 21–3).
- Wood framing with treads of concrete, brick, or loose material. Stairs along a hillside can be framed on the front and two sides with ties, timbers, or heavy-dimensioned lumber with the tread portion surfaced with a variety of materials. The front portion for each higher step rests on the sides of the lower step.
- Open wood stairs—Steps without sides and with loose fill or turf for a tread. For informal stairs, a tie or timber can be placed across the front of each step with the tread behind left untreated or covered with loose fill (Figure 21–4).
- Stacked slab materials. For informal stairs, large slab-like materials can be placed level in the ground and serve as both a riser and a tread. Each subsequent step rests on the back edge of the lower step. Materials that might be used for this type of step include precast concrete, large, flat stones, tree sections, or other materials

Figure 21–4 Freestanding stairs using railroad ties to create risers with a loose granular material used for the treads.

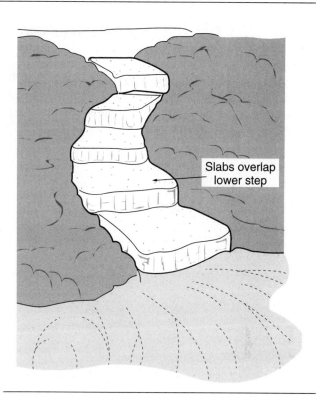

Slabs overlap lower step

Figure 21–5 Freestanding stairs constructed from slab materials.

with thickness and large dimensions (Figure 21–5).

Riser/Tread Calculations for Freestanding Stairs

Unlike stairs that are constructed as part of a deck or wall, freestanding steps are adjusted to match the slope that is being covered. If standard riser/tread dimensions are to be obtained, it will either be a fortunate coincidence that the slope matched the mathematical formula for riser/tread, or landings and/or cheek walls will have to be added to accommodate the mismatch length. To determine the number of risers and treads in a set of freestanding stairs, use the following abbreviated formula:

- Measure the length and height of the slope.
- Convert the measurements to inches.
- Divide the height (in inches) by the thickness (in inches) of the material selected for stairs. The answer will indicate how many risers are needed.
- Divide the length of the slope (in inches) by the number of risers. The answer will indicate the width of each tread. If the tread width is less than 12 inches, the tread width will not accommodate a normal step pattern, creating

potential safety concerns. Consider routing the stairs diagonally across the slope to provide more length for calculating tread width.

When constructing the stairs, use the tread width measurement to determine the placement of the front of each subsequent stair.

Building Freestanding Stairs on Irregular Slopes

When placing freestanding stairs on irregular slopes, adjusting the tread width may be required to avoid significant excavation. If the slope is steep, shorten the tread width to no less than 12 inches. The length cut from these treads can then be added to treads on the flatter portion of a slope. It is best to group at least three steps with short treads together. Avoid alternating a short tread width with a long tread width if possible. If the adjustment of treads is not acceptable, cutting and filling to create a more even slope is an alternative.

INTERLOCKING STAIR INSTALLATION

Interlocking stairs should be constructed as the wall is being erected, with the first course of the stairs installed with the first course of the wall and continuing for each subsequent course. The following steps outline how to install interlocking stairs for ties/timbers (Figure 21–6), segmental units (Figure 21–7), and dry-laid stone (Figure 21–8). These instructions are prepared for stair treads that are approximately 12 inches in depth. For deeper stair treads, additional stair materials are required. If special precast or one-piece stone treads are available, use those materials for stair construction.

STEPS

1. Widen the granular base trench at the stair location 24 inches in back of the entire width of the stair opening. This trench should be the same elevation as the base trench for the wall. Since the first course of the wall will be buried for stability

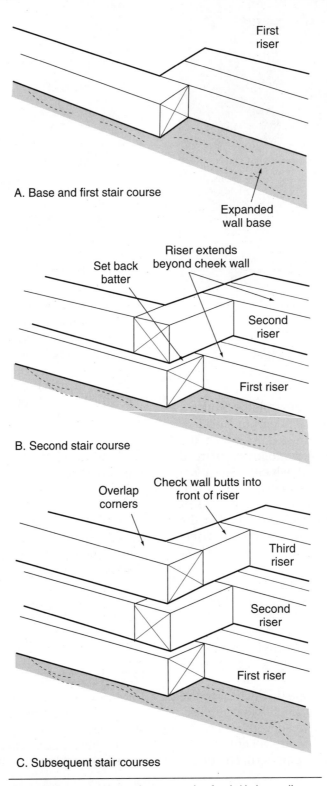

A. Base and first stair course

B. Second stair course

C. Subsequent stair courses

Figure 21–6   Interlocking stair construction for tie/timber walls.

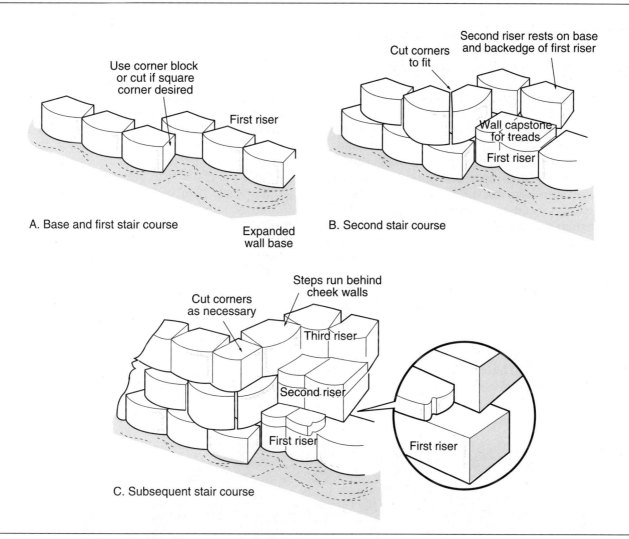

Figure 21–7 Interlocking stair construction with precast segmental units.

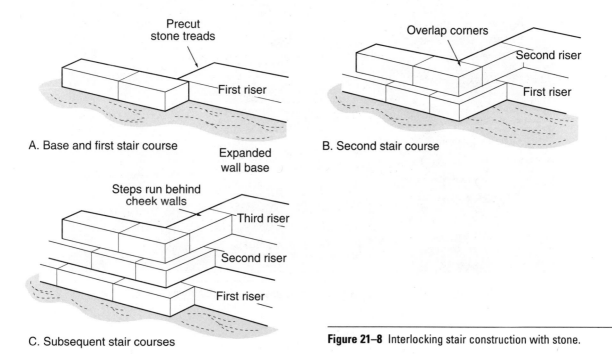

Figure 21–8 Interlocking stair construction with stone.

purposes, this level will serve only as a landing and not as a riser (Figures 21–6A, 21–7A, 21–8A).

2. Route any tile around the back of the widened trench.
3. Fill this widened trench with base material and compact.
4. Place the first course of the wall up to the opening for the stairs, keeping the wall front aligned. Place the first tread behind the first wall course. This tread may be composed of multiple ties, wall units, or stones, or maybe a single-piece tread. The tread should be approximately 18 inches wider than the stair opening, providing enough extra material to allow the tread to extend behind the wall on both sides of the opening (Figures 21–6B, 21–7B, 21–8B).
5. Construct the second course for the wall and stop short of the stair opening on each side of the stair.
6. Excavate a base trench behind the first tread for the second tread. This trench should be 18 inches wide by the length of the tread and 6 inches deep. Fill the trench with base material and compact. If a tile is present, fill over the tile without disturbing

its level. Smooth the base material so that it is flush with the top of the previous tread.

7. Place the second tread across the opening for the stairs. The second tread should be 18 inches wider than the stair opening. The second tread should also overlap the back of the first tread enough to create the desired tread dimension for the first tread. If using tie/timber treads, pin the second tread to the first (Figures 21–6C, 21–7C, 21–8C).
8. Build the cheek wall between the front wall and the tread. Trim wall material as required to fit between the front wall and tread.
9. Backfill and compact behind the wall and tread.
10. Repeat steps 5 through 9 for each subsequent course of the wall until stairs are complete (Figure 21–9).
11. If utilized, place tread coverings for all treads.

BUTT STAIR INSTALLATION

While butt stairs are easier to construct, their stability is not as good as that of interlocking stairs. Because butt stairs are not interlocked into the cheek wall, they

Figure 21–9 Completed interlocking stairs with precast segmental units. (Courtesty of Gary Pribyl, Kings Materials)

may move up or down at a different rate than the walls next to them, causing unevenness or irregularity. Construct butt stairs using the following steps for ties/timbers (Figure 21–10), segmental units (Figure 21–11), and dry-laid stone (Figure 21–12). The instructions are prepared for stair treads that are approximately 12 inches in depth. For deeper stair treads, additional stair materials are required. Precast or one-piece stone treads will simplify tread installation.

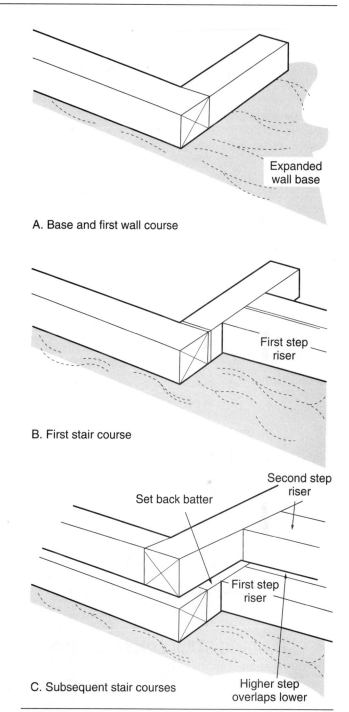

A. Base and first wall course

B. First stair course

C. Subsequent stair courses

Figure 21–10   Butt stair construction for tie/timber walls.

═══════ STEPS ═══════

1. Butt stairs require the construction of wall and cheek walls prior to stair installation. Leave an opening for stairs when constructing the wall and verify that the walls on each side of the opening are level. Adjusting the width of the opening to match tread material dimensions is also desirable, although use of a step-back batter on cheek walls requires cutting tread materials to fit.

2. Widen the granular base trench at the stair location 24 inches in back of the entire width of the stair opening. This trench should be at the same elevation as the base trench for the wall. Since the first course of the wall will be buried for stability purposes, this level will serve only as a landing and not as a riser (Figures 21–10A, 21–11A, 21–12A).

3. Route any tile around the back of the trench where the treads will be placed.

4. Fill this widened trench with base materials and compact.

5. Place tread materials between the cheek walls flush with the first wall course. Level treads side to side and front to back. A slight (1/4 inch or less) fall toward the front is desirable if it does not disrupt the stair construction. Cut and fit partial tread materials along the sides if required (Figures 21–10B, 21–11B, 21–12B).

6. Excavate a base trench behind the first tread. This trench should be 18 inches wide by the length of the tread and 6 inches deep. Fill the trench with base material and compact. If a tile is present, fill over the tile without disturbing its level. Smooth the base material so that it is flush with the top of the previous tread.

7. Place tread materials for the second tread across the opening for the stairs. The second tread should overlap the back of the first tread enough to create the desired tread dimension on the first step. If a wall cap is used for the tread, the overlap should equal the wall unit width minus the cap width (Figures 21–10C, 21–11C, 21–12C).

8. Backfill and compact behind the tread.

9. Repeat steps 6 and 8 for each subsequent course until the wall has been completed (Figure 21–13).

10. If utilized, place cap stones or tread coverings.

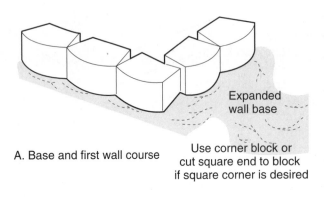

A. Base and first wall course

Expanded wall base

Use corner block or cut square end to block if square corner is desired

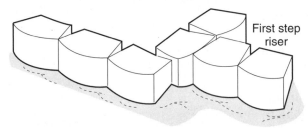

First step riser

B. First stair course

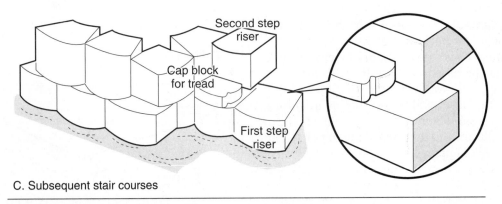

Second step riser

Cap block for tread

First step riser

C. Subsequent stair courses

Figure 21–11 Butt stair construction with precast segmental units.

FREESTANDING STAIR CONSTRUCTION

Site preparation is similar for all of the following types of stairs if they are to be placed on an even slope.

Site Preparation

Mark the location for the stairs along the slope. Cut and remove any vegetative cover and excess soil in the stair right-of-way. Except for concrete stairs, begin placement of stair material with the base step.

Concrete Stair Installation

Concrete stairs can be formed and poured similar to stairs that are built as part of a wall. After excavating the area for the stairs, fill with a small amount of granular base material. Construct the forms and pour the stairs according to the instructions given in Chapter 23, Concrete Paving, for independent forming of stairs.

Wood-Framed Stair Installation

Stairs that have a wood framework on the front and both sides should be installed beginning with the base step. Add a small amount of granular material in the area excavated for the first stair. Cut the front piece the width of the stair. Cut two side pieces, each 1 foot longer than the width of the tread. Place the front and two side pieces on the base material (Figure 21–14).

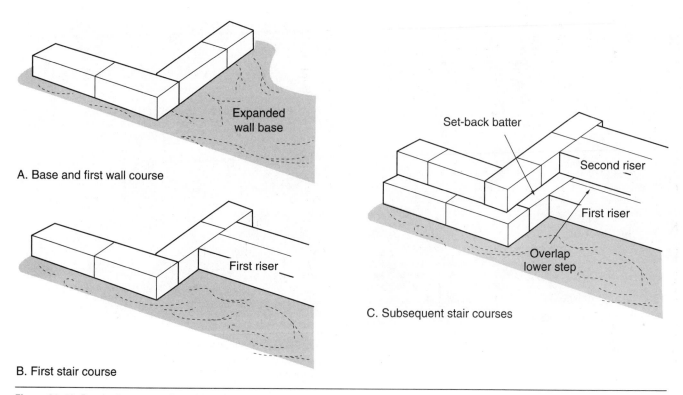

A. Base and first wall course

B. First stair course

C. Subsequent stair courses

Figure 21–12 Butt stair construction with stone.

Figure 21–13 Completed butt stairs with precut stone steps. (Courtesy of Paul Dykstra, Iowa City Landscaping)

Figure 21–14 Building the base step of a freestanding wood timber installation.

Figure 21–15 Subsequent steps of this freestanding stair installation rest on the sides of the lower stair. Treads will be filled with granular material.

Connect the side pieces to the front using galvanized lag screws or spikes. Level all three pieces by adjusting the base material. Fill the voids along the outside edge of the stairs with soil. Fill the tread area between the three pieces with material being used for the tread.

Cut the next set of step boards. Place the front board on the two side pieces for the step below. Anchor the front board for the upper step to the side pieces for the lower step by drilling a 3/8 inch diameter pilot hole through the front board. Drive 12 inch spikes through the pilot holes into the side pieces. Connect the side boards to the front pieces and repeat the filling procedure. Repeat these steps until the top of the stairs has been reached (Figure 21–15). The top stair may require an additional board opposite the front board to hold the tread surfacing material in place.

Wood Riser Stair Installation

Install a small amount of granular base material in the area excavated for the first stair. Cut the first riser piece the width of the stair. Place the riser piece on the base material and level. Fill the void behind the riser with the material selected for the tread. Fill the area along the sides of the tread with soil. Cut a second riser piece. Place the second riser piece on the tread material for the lower stair and at the correct distance from the first riser. Repeat the filling operation and continue with the remaining steps.

Stacked Slab Material Installation

Place a small amount of granular base material in the area excavated for the first slab. Place the first slab on the base and level. Fill around the edges of the slab. Place the second slab with an overlap on the first slab that provides the correct tread length. Fill around the edges of the second slab. Continue this process with the remaining slabs.

onstruction of retaining walls is one of the major aspects of most landscape contracting businesses. Section 5 discussed the preparation for wall projects and installation of tie/timber, segmental walls, dry-laid stone, and gabion retaining walls. Also covered were the installation of stairs as part of a wall project and freestanding stairs built without being incorporated into stairs. Cast-in-place concrete walls are not covered in this text, and freestanding walls are covered in Section 8.

Preparation for a retaining wall project begins with the selection of the proper wall construction material. Many choices are available, with ties/timbers, precast concrete units, and dry-laid stone being the most popular choices. A seldom-used wall material that is easy to use for short walls is the gabion. When choosing wall materials, consideration should be given to aesthetics, ease of installation, and cost. Retaining walls that are cast-in-place with concrete, with or without stone veneers, are also a popular choice for walls, but require the expertise of specialty contractors to construct.

Construction of all wall types requires the preparation of a sound base. Long-term stability of any wall relies on installation of the first course on solid support and with the proper level. First courses for most wall installations are buried to provide extra stability, and proper drainage is provided behind the wall to reduce hydrostatic pressure. This buildup of water is drained away from the wall using a tile placed behind and at the base with a zone of free-draining fill directly behind the wall. Stability of the wall is also maintained by building the wall with a batter, by anchoring the wall, or by doing both. Batter is the backward lean of a wall to counter the effects of gravity and hydrostatic pressure. Anchoring the wall can be accomplished using vertical buttresses, deadmen, or geogrid to secure the wall to the hillside behind. Taller wall installations can be accomplished by terracing (using a series of shorter walls).

Installation techniques vary depending on the type of wall materials selected. Wood products such as ties and timbers can be placed vertically or horizontally in stacked or staggered patterns. Vertical ties/timbers are anchored by placing them in a trench and backfilling. Horizontal ties are placed level with a leaning batter or step-back batter. Stacked ties/timbers are secured using verticals buried in front of the wall at the joints. Staggered ties/timbers walls are secured between courses using nails or pins, and anchored into the hillside using deadmen. Segmental (precast concrete) wall units are placed individually and stacked in courses. Precast units can be placed in a straight line or turn radius or angled corners. Courses are stabilized using blocks with lips, pins, or blocks with special precast and pinned anchors. Most precast unit walls are stabilized using geogrid placed between courses and buried horizontally in the hill behind the wall.

Dry-laid stone walls utilize natural materials cut to standard or random dimensions and stacked without mortar. Stone walls rely on a batter for stability and cannot be reliably anchored with other available technologies. Because of this, the height of stone walls is limited. Gabions are metal cages filled with large aggregate and stacked to create retaining walls. While not common in residential landscape projects, gabions are a viable alternative to other wall materials in locations where aesthetics are not a prime concern.

Walls are usually considered when addressing slopes on a site, and as part of traversing those slopes, stairs are a landscape element commonly built into retaining walls. Stairs constructed as part of a wall project can be built with the tread interlocking into the cheek walls or butting against the cheek walls. A preference for stair material would be the same material as the wall, but stairs may also be formed and poured from concrete as independent units from the wall. Special calculations are required for stair construction to obtain the ideal dimensions for risers and treads, and to determine the number of risers and treads the slope requires. Freestanding stairs can also be constructed without walls from a variety of materials.

Section 6

Landscape Paving

INTRODUCTION

Most landscapes rely on paving to create useable outdoor areas. By providing hard surfacing, we can create drives, walkways, patios, entry areas, and a variety of functional spaces that can be used in all weather conditions. In many projects, the paved area is also a source of aesthetics, introducing color, texture, and materials that enhance the overall design of a project. A paving material will work if it performs the essential function of separating your feet from the mud, but the variety of excellent choices provides aesthetics as well as functionality.

The chapters of this section review material selection and installation techniques for common paving materials used in today's landscape. Emphasis is on selecting a material, preparing the site, and installation steps. Specific paving topics covered in this chapter include concrete paving installation, unit paver installation, dry-laid stone paving installation, mortared stone paving installation, and granular paving installation. Due to the specialty equipment requirements, asphalt paving is described, but its installation is not presented in this text.

Chapter 22

Materials and Site Preparation for Paving

With only minor variations, project preparation is similar for all types of paving. Considerations such as selection of materials, access to the site, and layout of a project should precede the start of a project. A successful paving installation, whatever the surface material, requires a well-prepared, stable **base.** Surfacing is no better than the base on which it rests, and if the base settles or heaves, the paving will usually follow. Identified in this section are methods to prepare base and correct drainage and soil problems that might lead to future problems.

RELATED INFORMATION IN OTHER CHAPTERS

Information provided in this chapter is supplemented by instructions provided elsewhere in this text. Before undertaking activities described in this chapter, read the related information in the following chapters:

- Safety in the workplace, Chapter 4
- Basic construction techniques, Chapter 7

SELECTION OF PAVING MATERIAL

Paving material types are stated on the design documents, or require discussion with clients regarding standard criteria used in determining which material will best suit the project's needs. While selection of paving type is best done during the design phase of a project, there are applicable construction-related criteria that should be reviewed. The following sections identify six choices for paving materials and a description of the pros and cons of each. By using this information and the installation instructions in the chapters that follow, wise decisions can be made about which paving material is best for a project.

Concrete

Concrete is an excellent choice for surfacing outdoor spaces (Figure 22–1). Concrete is moderate in cost and is extremely versatile in finishes and shapes. Because forms are custom built to match the design, almost any shape of surface is possible. There is also a variety of finishes available, from exposed aggregates to colored, impressioned walks that mitigate the aesthetic problem of "plain" concrete. As the complexity of the forming or surface increases, the concrete becomes more expensive. Despite these additions, concrete is still one of the most cost-effective surfacing materials available. Properly installed concrete typically is excellent in durability and safety, and is regularly used in areas with vehicular traffic and in commercial and public use areas. Maintenance is typically performed without problems unless the installation is old or cracks have formed from improper installation. Most concrete pours can be installed with little difficulty if installers are experienced in preparing and completing pours.

Unit Pavers

Unit paver is a collective term that includes clay **bricks, interlocking concrete paving blocks,** adobe pavers, precast concrete units, open-cell grass pavers, and modular resilient pavers. Unit pavers have been among the most desirable and requested types of paving materials because of their aesthetic surface (Figures 22–2 and 22–3). This attractiveness is not without its price, since unit pavers are high in cost and require expertise and effort to install. Strength for

Figure 22–1 Concrete paving is used for this high-traffic entrance.

Figure 22–2 Interlocking concrete paving block is an excellent choice for both pedestrian and vehicular traffic.

Figure 22–3 Brick paving provides rich colors and patterns.

Figure 22–4 Dry-laid stone paving provides texture and natural feel.

properly installed unit paver surfaces is very good. Failure of base/subgrade under unit pavers can create potential safety problems and definite maintenance problems. In these same failures, precast concrete units often break. All unit pavers have numerous joints between the pavers, and even with a perfect installation, these joints can hinder snow removal. Concrete paving block with a heavy-duty base is often used in areas where vehicular traffic is expected or for commercial and public use areas. Brick can be used in these circumstances if installed on a proper base.

Open-cell pavements are typically used in turf locations where compaction from foot or vehicle traffic is a problem. Open-cell pavers can be constructed of precast concrete or plastic rings attached to a mesh. Resilient pavers are manufactured from recycled materials and the installation techniques are similar to those for precast concrete pavers. Resilient pavers have applications where softer surfacing is required, such as playgrounds. With a cost of nearly twice that of other unit paving choices, placement of resilient pavers is limited to specialty uses.

Stone

Patios of limestone, bluestone, granite, or slate are an aesthetically pleasing paving surface (Figure 22–4). The intricacy of shape, form, and pattern suggests a high level of craftsmanship. Concurrent with increased aesthetics for stone paving are higher costs and higher skill requirements for installation. Material availability has an impact on price, with lower costs in areas where stone is commercially quarried and available. Balancing the highs just mentioned are moderate levels of strength and concerns about safety, particularly for stone paving in public areas. Maintenance and safety issues are similar to those for unit pavers. Stone paving has joints that can be irregular, creating the potential for tripping and making snow removal difficult. Any base failure or water below the paving can force the stone out of position, amplifying the problems. Only under certain circumstances (some granite units, and cut stone placed on permanent base) is stone paving recommended for vehicular traffic or paving in commercial projects and public areas.

While the material for stepping stones can be selected from any paving choice, **flagstone** is utilized

Figure 22–5 Mortared stone provides a rich, stable stone paving.

most often for informal walkway applications. Unmortared stepping stones placed on earthen or thin, granular bases are inexpensive, but are suitable only for private uses.

Mortared Paving

Mortared paving utilizes a concrete-like mixture called mortar to space and support the paving materials (Figure 22–5). Mortar is also placed in the joints between materials to provide even spacing and prevent water penetration. Paving materials that can be installed as a mortared surface include tile, brick (usually half the thickness of nonmortared brick), and stone (particularly random-shaped stone). Mortared paving can be very durable if properly installed and maintained. If water is allowed to penetrate the joints of the surface, it will eventually damage the paved surface. This problem is far more damaging in areas of the country that experience freezing temperatures. Aesthetics of mortared surfaces are very good, but installation is difficult and costly when compared with other paved surfaces. With proper maintenance, the strength of a mortared surface is acceptable. Surfaces that are improperly installed, poorly maintained, or utilize improper base materials may suffer from durability problems. A maintained surface is typically safe, but surfaces that are not maintained can present tripping problems.

Granular Paving

Granular surfacing is typically found in drives, walkways, trails, and, occasionally, outdoor living areas (Figure 22–6). Granular paving includes such materials as crushed stone, crushed brick, pea gravel, and other permanent materials available as small pieces. Selection of such materials requires a careful review of use and placement to be effective. Durability of thin applications of granular paving limits it to only outdoor living areas and private walkways. Aesthetics are neutral for granulars, as they provide consistent texture and color, but do not suggest a high level of craftsmanship. Installation is easy and costs are typically low for granulars, which provide an impetus for their use in residential settings. Granulars with permanent **edgings** such as stone, brick, or concrete balance cost and aesthetics for walkways and outdoor living areas. Safety problems such as washouts, poor traction, and possibility of sharp edges exist for some materials. Maintenance is high for granulars, with the constant refurbishing of surface materials, edging upkeep, leveling, and cleanup after materials that track inside. Granulars are typically not applicable where erosion or public safety are concerns.

Asphalt

Hot mix **asphalt** paving is used for roadways, drives, walks, and trails in the landscape. Versatile installation

Figure 22–6 Granular paving provides functional walkways.
(Courtesy of Margaret Sauter)

also makes asphalt useful as a base material for unit pavers. Asphalt is composed of bituminous materials mixed with aggregate and sand, then heated to high temperatures to improve workability and bonding. Installed while hot, using special spreading and compaction equipment, asphalt provides a functional wearing surface for foot and vehicular traffic. When properly installed over a stable base, asphalt can be relatively free from maintenance. Due to the limited aesthetics of the surface, asphalt is seldom used when a high level of craftsmanship is desired. Cost for asphalt is dependent on labor and material prices. Installation requires the use of highly specialized equipment and crews with experience working with materials and equipment. A variation of asphalt—cold mix—has a similar composition but is used only for patching potholes in paved surfaces. Asphalt installation techniques are not covered in this text.

ACCESS TO SITE AND DELIVERY OF MATERIALS

Once the project has begun, workers need to have access to the site with base, paving, and edging mate-

rials. Ideally, this access route will not cross over the prepared base of paving surface. Plan a route that provides for access to the site by equipment as large as a concrete truck as well as wheelbarrows and foot traffic. Delivery of unit pavers typically requires access to the site by a large truck with a loading crane. Delivery of granular paving requires access to the site by a dump truck or by a skid-steer loader moving material from a stockpile. Planning the installation so the pavers can be delivered, and extra pavers picked up and removed without conflicting with completed installations, will save labor and time. A staging area where several pieces of stone can be laid out will speed the selection process required when laying stone. A stockpile of granular material can be delivered and extra picked up and removed without conflicting with completed installations, which will save labor and time. It may be beneficial to locate an area where additional granular material can be stockpiled so that the client can refurbish the surface in the future.

PROJECT LAYOUT

Measure and place project location stakes and stringlines to guide the initial excavation. Painting or "chalking" (pouring a line of base path chalk) the limits is often easier than fighting with strings. Mark and excavate an area that goes 1–2 feet beyond the actual paved area to create an apron that can be used for construction and later used to gradually match the new paving to the existing grade.

Once paving has begun, the highest quality and efficiency is achieved by moving through the project without pause. This requires that materials be delivered and prepared for installation. Any equipment that will be necessary for site preparation or cutting should be present at the proper time.

STRIP SOD

Remove sod from the entire construction area and recycle to bare areas of the yard, or compost. Once the sod is stripped, remark the project edges and identify project elevations, typically with grade stakes.

MATCHING EXISTING CONCRETE AND ASPHALT SLABS

When new paving abuts old cast-in-place concrete slabs or asphalt paving, it is beneficial to prepare a straight edge against which new paving is placed. To prepare the edge of existing pavement, mark and cut a straight line using a cutoff saw or mobile concrete saw.

Figure 22-7 Using a skid-steer to excavate the base for an expanding paved area.

A dry diamond or carborundum blade is required to cut most paving materials. The cut should be vertical and through the pavement. Chip away rough edges with a chisel. Excavate away old pavement and prepare the edge for new pavement. Concrete paving may also require joining old paving to new paving (see Chapter 23). If asphalt is cut, a spray tack coat may be applied to improve adhesion of new pavement.

EXCAVATION

Using the most efficient means available, dig out excess soil until a uniform depth is achieved over the entire paved area (Figure 22-7). For paving that utilizes an edging, the base excavation should extend 1 foot beyond the finished edge to support the edging. Check the depth often by running a string between grade stakes and measuring down with a tape measure. Over-excavation should be minimized to avoid creating a less-stable fill area. Excess topsoil should be stockpiled for use in backfilling edges.

Actual depth of the excavation depends on which surface material is used, the subgrade soil conditions, and the weight of the traffic on the surface. Table 22-1

suggests appropriate base depths based on the type of paving material and the type of traffic on the surface. To determine excavation depth, add the paver dimension, the thickness of any sand or mortar setting bed, and the thickness of the base. If the paving material is designed to be seated, or vibrated into the **setting bed,** as is the case with some unit pavers, subtract the manufacturer's recommended dimension (typically 1/4 inch) from the excavation depth. Verify base requirements with a design professional or manufacturer for the situation. Poor soils require additional depth or alternative base treatments.

Rough Grade Preparation

Since paving creates relatively flat surfaces with slight slopes introduced for drainage, creating the proper grade is important. Attempting to accommodate slopes that are too steep leads to difficult installation and surface slippage. Slopes that are too flat result in future problems with water that does not adequately drain from the surface. Grade also directs surface runoff away from structures. A slope of 2% (1/4 inch fall for every 1 foot of horizontal distance)

Table 22–1 General Base Thickness Chart for Various Paving Materials

Paving Type	Pedestrian Traffic	Occasional Vehicles	Auto Traffic	Heavy Traffic
4 inch concrete slab	3 inches	4 inches	6 inches	8 inches
Interlocking paving block with 1 inch sand setting bed	4 inches	8 inches	8 inches	16 inches
Clay paving brick with 1 inch sand setting bed	4 inches	8 inches	8 inches	Not recommended
2 inch unmortared stone with 1 inch sand setting bed	4 inches	Not recommended for this traffic type		
2 inch mortared stone with 1 inch mortar base	4 inches	Not recommended for this traffic type		
3 inch granular paving	3 inches	4 inches	Not recommended	

The depths are general recommendations from manufacturers. Verify actual base depths required for a project with an engineer.

is recommended for most paving. Flat cross-slopes (even sloping the entire surface one direction) are the easiest to construct and the best for drainage, but occasionally it may be necessary to introduce variable cross-slopes, or warping, in the pavement to accommodate drainage and match existing elevations. Warping makes installation difficult, but is preferable to poor drainage.

Subsurface Utility Work

Because paving is not easily changed, it is important to ensure that other work that might affect an installation is completed before paving. All underground utilities in the area should be in place and trenches below the paving backfilled and compacted. Empty conduit can be placed if future utilities are anticipated. See Chapter 15 for a discussion of related utility work.

Correcting Subgrade Problems

The best subgrade is undisturbed, sandy-clay soil. Placing base on topsoil or unconsolidated soils can expose paving to subgrade movement. Therefore, consider removing unsuitable materials and backfilling with base material. A soils engineer should be consulted when there is uncertainty about the condition of the subgrade.

Soil Problems. Occasionally, subgrade soil problems must be corrected. Identify problem soils by looking for areas that are wet, spongy, or sink under foot traffic. Also look for soils that have leaf and twig debris or a foul smell. If present, these soils should be excavated and replaced with base material. If the depth exceeds 1 foot, special subgrade restoration will need to be designed by the soils engineer.

Drainage Problems. Subgrade drainage problems can be identified by the presence of standing water or water seeping into the base excavation from any

direction. During dry seasons, there may be no indication of water, but mucky, gray soils that have an odor of sewage are an indicator of wet conditions. Water that percolates through the paving surface is usually minimal and is expected to be absorbed into subgrade soils. Sloping the subgrade the same direction as the surface assists in removing water. If it is determined that subsurface drainage problems exist, the most effective solution is to install subgrade drainage tile in the wet area. Place a 4 inch perforated plastic drain tile wrapped in a geotextile sock in 1 foot deep trenches spaced 5 feet apart (Figure 22–8). Run the trenches the length of the paved area from the high side to the low side. Connect each tile to a collector tile at the low end of the paved area and outlet this tile to a drain, storm sewer, or swale. Backfill and compact each trench with base material.

Grade Preparation and Subgrade/Base Compaction

Undisturbed earth is preferable for subgrade. If a paving installation is constructed in an area where the subgrade has been disturbed, proper compaction is necessary. If walls are required to create a flat surface for paving, the walls should be constructed in advance and backfill compacted to meet base specifications. It is also helpful to ensure that all heavy traffic (deliveries of building materials, etc.) through the area is complete, or that alternative routes exist that protect the paving from traffic for which it was not intended.

Compacting subgrade in most small landscape operations is typically done using a vibratory plate compactor. Alternatives to a plate compactor include a rammer plate, vibratory roller, or hand-operated compaction plates. For best compaction results, select a plate compactor that has a compactive force of 5,000 pounds at 75 to 90 cycles per second. Operate the vibratory plate compactor much like a self-propelled

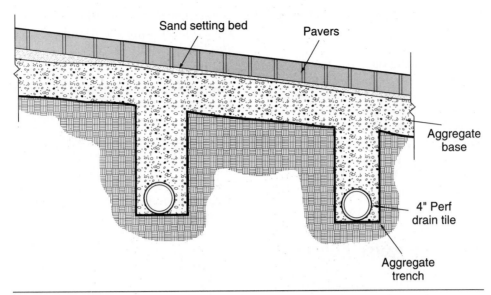

Figure 22–8 Cross section of tile placement to drain paved area base.

lawn mower. Start the engine and steer using the handles. To work tight corners, the handles may be lifted up vertically and the plate compactor twisted to change direction. If the plate compactor being operating has multiple operation settings, use the high-force/low-frequency setting for subgrade and base compaction. Always run the plate compactor at full throttle. Utility trenches under paved areas should be compacted in layers, or lifts, as the trench is backfilled. Small areas that cannot be accessed by a plate compactor should be hand tamped.

Testing the compaction of subgrade and base is recommended to verify that the foundation for paving is stable. Engineers utilize measuring standards for determining if a material has been compacted to its maximum density, and the most commonly used standard is modified Proctor density. Projects performed under contract may require that soil and base samples be taken to a testing lab to verify proper compaction. For projects without construction documents, obtaining Proctor test results may be cost-prohibitive. An informal test of base for firmness is to observe if heavy equipment operated over the surface leaves any rutting or depressions. Such conditions indicate incomplete or improper compaction, or an unstable subgrade.

GEOTEXTILE INSTALLATION

Unit pavement, dry-laid stone, and granular paving projects may be designed for placement of a geotextile below the base to assist in maintaining separation between the base and subgrade soils. Select appropriate geotextiles with woven construction. Do not use

weed barrier or geogrid to substitute for geotextiles. Place the geotextile on the subgrade and extend the material up the sides of the excavation and any structures around the entire perimeter of the project. Geotextile may be stapled to a structure to keep it out of the way during construction. Excess can be trimmed after the pavement is placed. If the area requires more than one piece, overlap the joints by at least 18 inches.

BASE MATERIAL INSTALLATION

Choice of base material varies from geographic region to region, but in most areas, a base of granular crushed stone ranging in size from dust to pieces that pass a 3/4 inch sieve is preferred. In some areas of the country, this material is called 3/4 inch roadstone. A more formal description is National Stone Association Class 2 aggregate. When compacted, this blend usually provides adequate support for residential paving projects. Base material should have enough moisture that a handful holds together when squeezed. Dry material requires wetting during installation to compact properly.

Spread base material in 2 inch lifts over the paving area and level with a rake or **screed**. Remove all clumps of soil or other debris and compact with a plate compactor. Make two compaction passes with the plate compactor at different angles. Best compaction results are obtained by working from the outside edge to the center, from a lower side to a higher side, or by working up and down slopes.

Check the elevation of the base after each lift is installed. The slope should not have deviations of more than 3/8 inch from desired grade and should reflect the slope desired for the surface. To level the base for an evenly sloped area, first check along a structure or existing edge for proper elevation. From that edge, check the sides of the base installation for proper grade. Between the sides, a stringline or straight board may be used to check for high or low spots across the center. The base can also be checked by rolling a pipe over the surface and looking for bumps or depressions. Mark any unleveled areas with paint. Add a thin layer of base if low, or scratch out excess base with a rake if high. Recompact and verify for proper base grade. Repeat the process until the base is complete.

EDGE RESTRAINTS FOR PAVED SURFACES

Depending on the choice of paving, some edging materials should be placed after the base has been installed and before the setting bed is installed. Many of these edgings require special construction techniques, and will seriously disrupt the pavers that have been installed. Other edgings work better when placed after the majority of the pavement has been placed. These edgings work best when installed before the partial pavers have been cut and installed. The common types of edgings in landscape pavement and their installation techniques are identified in the following sections (Figure 22–9).

Stone (Pre- or Postpaving Placement)

Wall stone makes an acceptable edging for dry-laid stone and clay brick. If stone is used elsewhere in the design, it carries the material into the paved areas. Unless a stable burial depth is used, stone edging is not recommended for surfaces where the pavers must be seated. To install stone, select a heavy wall stone material, typically 4 inches to 8 inches wide. Longer stones make a more stable installation, with shorter stones used to turn corners. Excavate a trench around the area to be paved and set the stone level with the desired finish grade for the patio. If desired, a strip of landscape fabric can be placed under the stone to reduce weed growth between blocks. After placement, backfill and compact around the outside edge and build paving layers up on the inside (Figure 22–9A).

Concrete Curbing (Prepaving Installation)

Concrete curbing makes a durable and long-lasting edging for a paved area. To install curbing, begin by excavating a trench along the perimeter where the curb is to be placed. Make sure the depth is correct for the thickness of the curbing selected. If too deep, fill with a granular backfill. Construct forms (see con-

crete, Chapter 23) with the top of the forms set at the finish elevation of the paved area. Fill with concrete and finish (Figure 22–9B).

Precast Concrete (Prepaving Installation)

Precast concrete edging units can be used to secure most unit pavements. Install prior to installation of the setting bed and paving. Mark the alignment of the edging and set the blocks in place. If openings for spikes are available, secure into the base using 10 inch spikes. Fill any voids between paver and edging with base material. Backfill and compact the area outside the paved area prior to placing the pavers (Figure 22–9C).

Plastic (Pre- or Postpaving Installation)

The current standard for concrete paving block is the use of flexible plastic **edge restraint** specifically designed for this application. This edging is notched so that it can be bent to very tight radii, is installed with metal stakes or spikes, and is not visible after installation. Install this edging either before placing the setting bed or after completing installation of full pavers and before edge pavers are cut and placed.

Prepaving installation requires marking the alignment of the pavement edge. Place the edging firmly on the base and secure using stakes or 10 inch spikes. Trim edging to fit. Joints should be as tight as possible, with gaps covered with geotextile to prevent setting bed sand from seeping out. Use this edging as a screeding guide for setting bed placement.

Postpaving placement should begin by using a steel trowel to cut a vertical edge down from the pavers, through the setting bed sand, to the base. Scrape this sand away. Beginning at a structure or a corner, take a length of the edging and, with minimal disturbance to the paved surface, tuck the notched side of the edging under the setting bed. Placing the edging directly under the setting bed and on top of the base will assure that enough of the paver is covered to hold it in place. Slowly work the edging under and tightly against the pavers along the length of the side (Figure 22–10). If necessary, tap edging lightly with a hammer to force it under the pavers. Trim off extra and verify that at least half the length and thickness of each paver is covered with edging. Place and drive metal edging stakes or 10 inch spikes through the openings and into the pavement base at 1 foot increments along the edge. Driving the stake or spike at an angle toward the pavement base will draw the edging in tighter against the pavement. Fill any voids between the edging and paver with base material. Backfill and compact outside the edging before seating pavers (Figure 22–9D).

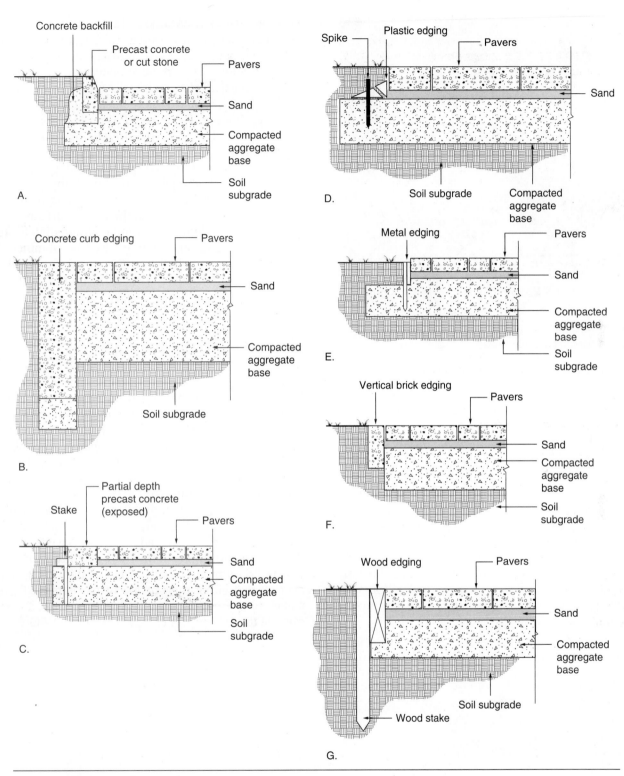

Figure 22–9 Pavement edge restraints.

Figure 22–10 Installation of plastic edging after placement of pavers.

Metal (Postpaving Installation)

Metal edging can be used to hold the pavers in place. Install metal edging after all pavers are cut and placed. Measure and precut pieces of edging for each side of the paved area so that the entire perimeter is covered. Using a steel trowel, cut the setting bed sand away in a vertical edging along the pavers to the base. Starting at a corner, hold the metal edging 1 inch below the top of, and against, the sides of the pavers. Through the notches in the edging, drive edging stakes into the base at 1 foot increments along the edge. Backfill outside the edging and compact immediately after placement (Figure 22–9E).

Vertical Bricks (Postpaving Installation)

This method provides limited stability and works best with brick pavement. Vertical brick edging should be installed after all bricks have been placed, including half and partial brick. To install vertical brick edging, carefully excavate a 9 inch deep trench along the entire perimeter of the paved area. The trench must form a vertical edge going straight down along the outside of the pavement. Place a small amount of granular material in the trench and place brick vertically against the pavement. By adding or removing granular material, adjust the height of the edging so that it is flush with the paved surface. Backfill outside the edging and compact immediately after placement. Additional stability can be obtained by backfilling with a small amount of mortar in place of soil. When hardened, cover the mortar with soil (Figure 22–9F).

Wood (Pre- or Postpaving Installation)

A wood edging will provide restraint, but should be considered a short-term option due to problems with warping, decay, and heaving from frost action. Cutting and filling of partial pavers should be completed after this edging is installed. To install wood edging, select a decay-resistant, dimensioned lumber and cut to length for each side of the paved area. Larger-dimensioned lumber, such as 4 × 4 or 6 × 6, reduces problems with warping, but a 2 × 6 may also be used for short-term installations. Retreat cut ends with a wood preservative. Excavate a vertical trench 6 inches deep along the outside edge of the paved surface. Place the wood edging in the trench tightly against the pavers. To improve stability, drive treated, 2 × 2 stakes every 2 feet along the outside of the edging. Fasten stakes to the edging using galvanized nails. Backfill outside the edging and compact (Figure 22–9G).

Chapter 23

Concrete Paving

oncrete can be placed in any shape or form, can be colored or stamped to look like another material, is cost-effective, and is as durable as a paving surface. For these many reasons, the use of concrete is a staple in the world of landscaping. Countering the versatility of concrete is the challenge of installation. While competing paving materials also have their challenges, the preliminary work required to assure a competent concrete pour is unmatched by that of other paving types. This chapter describes the forming, pouring, and finishing of a typical concrete installation. Included in this chapter are instructions for forming both flatwork and stairs, along with information about typical finishes available for concrete.

RELATED INFORMATION IN OTHER CHAPTERS

Information provided in this chapter is supplemented by instructions provided elsewhere in this text. Before undertaking activities described in this chapter, read the related information in the following chapters:

- Safety in the workplace, Chapter 4
- Basic construction techniques, Chapter 7
- Site preparation for paving, Chapter 22

PLANNING THE PROJECT

Preparation of base for concrete projects should be completed according to the instructions provided in Chapter 22.

Concrete Finishes

Several choices exist for concrete textures, and their procedures and timing differ depending on the choice. The following list explains five common choices for finishes. Instructions for creating each finish are listed in the finishing section of this chapter.

- **Float** finish. Using the float as a finish produces a slightly rough surface that provides adequate traction.
- Broom finish. A stiff-bristled shop broom texture provides an excellent nonslip surface. When combined with the bordered segment approach described for a float finish, it creates an attractive appearance (Figure 23–1).
- Colored, impressioned finish. A recent trend in finishing is to use concrete that can be colored and stamped with special forms to create a surface that looks like stone, granite, or brick rather than plain concrete.
- Seeded exposed aggregate. Exposed aggregate surfacing can be installed by sprinkling pea gravel or other fancy stone on a smoothed, wet concrete surface and using a trowel to embed stone in the top of the slab (Figure 23–2). After the concrete begins to dry, it is "scrubbed" with a hose and broom to remove the top film of concrete to "expose" the aggregate that was embedded earlier. One recurring maintenance problem with an exposed aggregate surface is that moisture can work under the aggregate and break it away from the surface.

Figure 23–1 Broom finish for concrete walkway.

- Integral exposed aggregate. A variation on the previously described aggregate surface is an integral exposed aggregate. The problem of surface deterioration is minimized by mixing the special aggregate right into the concrete at the plant. Instead of being a weak addition to the surface, it is mixed completely into the full slab. The finished product of this surfacing is obtained by scrubbing the slab before set or sandblasting after the slab is dry.

Planning Concrete Delivery and Pouring

Concrete pours require planning. While preparing for a pour, consider how the mixer will get into the site, where the pour should begin and end, and how access will be gained to finish the surface.

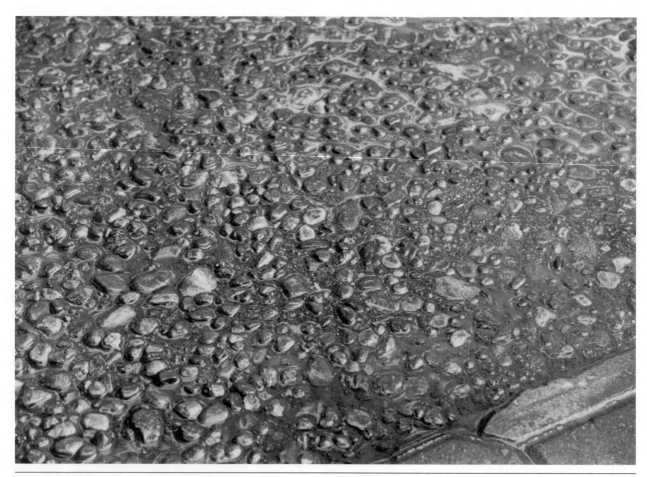

Figure 23–2 Exposed aggregate concrete finish.

It is imperative that a route for the mixer be found that will not cause it to contact overhead utility lines or damage underground installations or valuable plant material.

Table 23–1 Water/Cement Ratio Effect on Compressive Strength of Concrete

Days after pour	Water: Cement Ratio by Weight			
	0.4	0.5	0.6	0.7
	Compressive Strength in PSI (pounds per square inch)			
28	5,800	4,700	3,700	3,000
7	4,000	3,100	2,400	1,700
3	2,500	1,900	1,500	1,000
1	1,000	700	480	250

Notes:
1. Compressive strengths shown are plus/minus 200 pounds per square inch (PSI).
2. Data shown is for air-entrained Type I Portland cement with moist curing at 70 degrees F.
3. Flexural strength is approximately 79% of compressive strength.
4. Ratio example: 0.4 indicates approximately 1 pound of water (approximately 1 pint) for every 2.5 pounds of cement in a concrete mixture. This ratio will vary with cement and aggregate type.

Source: Adapted from *Design and Control of Concrete Mixtures,* 12th edition, Portland Cement Association.

It is best if the mixer can access the entire pour, eliminating the need to transport concrete by wheelbarrow. At times, it may be beneficial to leave a section of forms out and have the mixer drive across the base to access a remote corner. The many extensions that a mixer carries and the experience of a driver will save a great deal of labor. If access is not possible, a wheelbarrow can carry 2–3 cubic feet of concrete at a time to the necessary locations. Plan the pour so that there is open area at the end of the pour to allow screeding beyond the forms. It is also helpful to have open area along both sides of a pour to provide space for floating, finishing, and edging. Working against a wall and in corners creates problems finishing the slab from the edge. Verify that the **forms** and base are prepared before the concrete arrives.

It is best to order the concrete ahead of time. Even in small communities, it may be necessary to call a day or two ahead to reserve concrete. Do not expect that small orders will take priority over larger jobs that the concrete supplier must supply, especially if calling at the last minute. Inform the concrete supplier of any special colors, aggregates, fibremesh, or other additives that need to be in the concrete mix well in advance of the pour. Some of these items may not be stocked by the supplier and require ordering time.

For most landscaping pours, order 4,000 **PSI** concrete with a 3 inch slump. Slump is a measure of the amount of water in concrete. Water directly affects the strength of concrete—the more water in the mixture, the weaker the slab. Table 23–1 shows the relationship between water/cement ratios and strength for concrete. When ordering concrete, it should be ordered drier than actually desired. Water can be added at the site before placement. Check the consistency when the mixer arrives to see if more water is needed (consistency checks are described in Placement of Concrete later in this chapter).

The occasion may arise when concrete must be mixed in small batches by hand. This operation requires a mixing tub or wheelbarrow that will accommodate approximately 3 cubic feet of material. Mix one part cement (one-half bag, approximately 47 pounds), two parts sand, and three parts aggregate together with approximately 3–4 gallons of water. Work the materials back and forth with a mortar hoe or shovel until thoroughly mixed. A chopping motion works well when attempting to incorporate moisture into dry materials. If additional water is necessary, add in small amounts and remix until the proper consistency is obtained.

Curing of Concrete

Curing is a process whereby moisture loss and temperature are controlled to improve the strength and durability of concrete. Curing is most critical in the first few days following a pour. Methods to cure concrete include covering with plastic sheets or wet burlap, or spraying with a membrane compound. Ideal duration for curing is to keep the surface covered for 3–7 days following the pour.

JOINTING AND REINFORCING

While forming, prepare any joints and reinforcing that can be done before pouring. This work includes installation of any **expansion joints** next to fixed objects, preparation of **construction joints,** and installation of wire mesh for reinforcement. **Contraction, or control, jointing** is performed after the pour is complete and is covered later in this chapter.

Reinforcement

Reinforcement material is a subject of much discussion in concrete circles. Some say wire mesh does little to help strengthen concrete, especially if it is improperly installed. Others still insist on the value of **welded wire mesh** (WWM) to reduce the effects of contractive cracking and to strengthen the slab. Welded wire mesh is a 10 gauge wire that is welded in 6 inch squares. If utilizing WWM, roll it out flat across the entire area of the pour before the concrete arrives (Figure 23–3). Overlapping the joints slightly provides seamless reinforcement. Hold the mesh back a few inches from the edges in case it moves when the pour begins. Use a fencing tool to cut WWM when required.

> **CAUTION**
>
> WWM is typically supplied in rolls that must be unrolled and flattened before use. When unrolling WWM, hold the end in place to prevent recoiling and injury.

Cut holes around objects that are in the middle of the slab. Keeping the mesh flat requires significant bending and shaping, but is necessary since bows will cause the WWM to protrude above the slab.

Fibermesh is an alternative to WWM that requires no preparation, but can be a problem in the finishing stage. Fibermesh, which is tiny strands of fiberglass, is added to the concrete while it is being mixed. The strands bond to the concrete and form an integral reinforcement. Finishing is slower since the strands of fiberglass drag on the equipment. The finished look of the slab with fibermesh has a "fuzzy" appearance until the strands wear. The fine fiberglass strands of this reinforcement appear to add a great deal of strength to the slab and may serve as an adequate strength replacement for WWM.

Another reinforcing material used in concrete work is deformed steel reinforcing rod. Reinforcing rod, also termed **rerod,** is a rigid bar used to strengthen concrete in pours. The surface of rerod has tiny ridges, or deformations, that create additional friction when the

Figure 23–3 Welded wire mesh reinforcing cut and placed before pour.

concrete hardens around them. Cut rerod using a hacksaw, heavy-duty bolt cutter, or cutting torch. Rerod can be placed in a grid network inside thick slabs or placed along the edge of a slab to add strength. Rerod is purchased in lengths and is specified by fractional diameter. This diameter is typically stated in eighths—for example, a #3 rerod is 3/8 inch in diameter. Diameters are available up to 2 inches, but most landscape work requires #3 or #4 rerod.

Expansion Joints

Expansion joints absorb the expansion of concrete when temperatures warm the material. **Expansion joint material** should be placed between a slab/stair and any permanent object in or adjacent to a pour. This list typically includes buildings, curbs, deck posts, light posts, hydrants, foundations, and any other object that can be moved as the slab expands.

Materials used for expansion joint include a pre-manufactured 1/2 inch thick plastic that comes precut in 4 inch and 6 inch widths in varying length rolls, or a 1/2 inch thick by 4 inch wide asphalt-impregnated fiberboard that comes in 8 foot lengths. The plastic material is very flexible and fits most installations, while the fiberboard is rigid and works best in straight installations. To attach expansion joint material to structures, mark the top of the slab with a chalk line and use concrete nails to fasten the joint material along that line. If the object does not accept nails, try wrapping the joint material around the object and taping it in place, or attach it using a construction adhesive (Figure 23–4).

Figure 23–4 Preformed plastic expansion joint placed between slab and building.

Installing Construction Joints and Anchoring Adjacent Pours

If plans are to complete the pour in more than one session, a construction joint, or a clean vertical joint from which future work can be started, will need to be formed (Figure 23–5). This joint is created by installing a 2 × 4 form staked into place where the pour stopped. If using wood joints in the pour, any of the wood headers can also serve as a construction joint for the project.

When pouring slabs at different times, one of several joining methods should be used to hold both slabs in the same position. An existing concrete slab may be undercut to help join slabs. Underexcavate the existing slab 4 inches deeper and back 4 inches from the edge. When pouring the new slab, be certain to tamp concrete into this void. Another method for joining slabs is to drill 1/2 inch diameter holes 4 inches into the edge of first slab and insert 1/2 inch (#4) rerod into the holes. The rerod should extend into the new slab area 12 inches (Figure 23–6). Space these rerod connectors every 2 feet along the length of a slab's edge. A third way to joint slabs is to utilize a keyway joint. A **keyway** is a metal channel that is nailed to the pour side of the board forming the construction joint (Figure 23–7). Keyway channels form an indentation into the slab, creating a lip into which concrete from the next pour will flow and interlock the slabs.

Attempting to join two 4 inch slabs with rerod or a keyway joint results in only 1–1-1/2 inches of concrete above the rerod/keyway that is weak and prone to cracking. If the keyway or rerod joining is desired, the slab should be thickened to 8 inches at construction joints to provide more concrete above and below the rerod/keyway.

FORM CONSTRUCTION FOR SLAB POURS

Concrete pours depend on reliable forms. Properly installed forms should be anchored to withstand the force of several dozen pounds of wet material being dumped into them at high rates of speed. Forms should also be stable enough to maintain the proper grade at the perimeter of the project, and strong enough to withstand the **screeding** and finishing work that will be performed upon them.

Selection of materials for forms depends on the thickness and shape of the pour, with 2 × 4s providing the most common method of pouring a 4 inch thick, straight-edged slab (a 2 × 4 produces a slab that is 3-1/2 inches thick). Thicker pours may require dimensioned lumber that is wider (2 × 6s or 2 × 8s) or rental of steel forms. Curved sections of a pour can be constructed with forms of 1/4 inch masonite siding cut into strips that match the thickness of the slab. Stakes

Construction joint showing
rerod which will extend
into next slab

Figure 23–5 Construction joint set at a breaking point in pour.

Figure 23–6 Rerod placement along edge of pour to anchor new slab to existing slab.

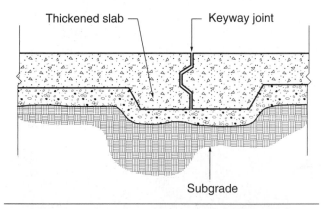

Thickened slab — | — Keyway joint

Subgrade

Figure 23–7 Keyway joints and thickened slabs.

should be 2×2 or 1×4 material and 8 inches longer than the slab is thick (i.e., 12 inch stakes for a 4 inch slab). **Form nails** are the best choice for nailing dimensioned lumber forms since the double heads make them easy to remove, but using deck screws to connect forms reduces the disruption of forms due to hammering.

Begin form installation by selecting a starting point at the highest corner of the project or next to a structure. Set form material on edge along the outside of the project edge (Figure 23–8). Forming should match any existing pavement grades that abut the new slab (Figure 23–9). At the high end of the form, drive one of

Figure 23–8 Installing and leveling forms for a concrete pour.

Figure 23–9 Installing form flush with top of adjacent slab.

the stakes along the outside edge of the form material. Drive a form nail through the stake and into the form material, holding the top of the form at the desired elevation for the top of the slab. Placing the head of a sledge on the inside of the form opposite the nail will make driving the nails easier. Repeat this process at the other end of the form and then adjust the elevation by tapping the stakes down or prying the form up. Adjust until the entire form follows the required grade along the outside edge of the project. Add additional stakes along the form every 2 feet, driving and nailing carefully so that the elevation is not disturbed.

Once in place, move to the next form and repeat the process. Be sure the forms butt together tightly. Add an extra 1×4 stake behind the joint for stability. Forms work best if the sections are at least 2 feet long, but use whatever length is required to make a continuous connection. Continue forming around the entire project. Curved forms should be set up in a similar manner. Nail from the inside of the form into the stake since the nail will project into the concrete if nailed from the stake side. If using masonite forms, use nursery nails to fasten forms to the stake. Stake curved forms every 1 foot for additional support.

At locations where posts or other improvements will penetrate the surface of the concrete pour, an opening, or box-out, must be formed. A box-out will frame this opening and provide identification of an area where concrete should not be placed. Frame the opening with form material that matches the slab thickness. Position the box in the correct location and drive stakes into the base on the inside of the box-out. The correct elevation of the box-out forms can be determined by using a string level resting on side forms. Place the string level on the side forms passing over the box-out. Adjust the box-out until the correct elevation is obtained and fasten to the stakes.

Use of Wood Headers

An alternative to constructing forms that are to be removed following a pour is to construct forms that will remain and serve as expansion joints and decorative edging to the concrete surface (Figure 23–10). This type of construction must be done with a decay-resistant lumber. Kiln-dried redwood or dried-treated lumber (wet-treated lumber may shrink) are available choices. Construct the forms in the same manner as that used for the previously described temporary forms. In each location where a contraction joint or header is planned, cut and connect forming material that is set on edge and placed flush with the top of the forms. Every foot along the inside of the forms, drive

Figure 23–10 Wood header between paved slabs.

galvanized 16d nails halfway into the header. These nails will project into the concrete and, when the concrete sets, will anchor the header to the slab. When all forming is complete, place wide masking tape over the tops of all headers. This will protect the lumber from scuffing during concrete placement and finishing, and can be removed when the pour is complete.

Checking Forms

After forming is complete, check the grade around the perimeter to be certain the elevations are correct. Cross-slopes can be checked by placing a 2×4 on edge from the form on one side to the form on the other. A level placed on top should show a slight drop between the sides. Backfill to the top of the outside of the forms and add base material to just below the bottom on the inside. Any gaps or openings of 1/2 inch or less can be filled using expanding spray foam insulation from an aerosol can. Spray the foam in the opening and let the foam expand and harden. Trim any excess to match the surface of the forms. Carefully saw off the tops of any of the stakes that stick above the forms so they are not in the way for screeding and finishing.

Leveling Base

Make a **base screed** by cutting a 2×4 (or the dimension of lumber selected for forms) 10 inches shorter than the width between the forms. To serve as a handle, at each end of this screed nail a 1×4 laid flat, on top, that extends 8 inches beyond the end of the screed. The flat handles of this screed can now be set on top of the forms and, using a sawing motion, moved from one end of the pour to the other to check the depth of the formed area. Add additional base material and screed inside the forms to ensure a uniform slab thickness (Figure 23–11). Sprinkle a fine granular base material (stone screenings that pass a 3/8 inch sieve) into the areas that look low, and rake material away from any area that appears high. This screed should move easily from one end to the other with the compacted base just below the bottom of the screed. Compact the base using a vibratory plate compactor.

FORM CONSTRUCTION FOR STAIR POURS

One of the most difficult forming and pouring jobs is the placement of stairs in landscapes. Not only must the setup hold concrete horizontally between forms, but it must also hold concrete vertically. This situation is often compounded by the existence of walls, structures, or other landscape elements that do not provide an open work area. Limitations for stairs constructed

Figure 23–11 Screeding 3/8s minus granular base to a consistent depth. Base will be compacted after screeding.

by a landscape contractor are in the range 6–10 feet wide and five risers. Longer or wider stairs may require special construction techniques beyond what can be expected of the landscape contractor. Form construction for three types of stair pours are: (1) independent forming, or pours that allow forming the stairs before other improvements are installed; (2) suspended forming, or pours that require forming after improvements such as wall structures are installed, and that require hanging forms between the improvements; and (3) a method in which forms are wedged into place.

Independent Forming of Stair Pours

Pouring stairs before any of the surrounding landscape improvements have been installed often provides the opportunity to independently form a set of stairs. This allows the contractor to form all parts of the stair, including sides and risers. Be certain to verify the location and elevation of stairs before forming. Independent pouring can be completed in four steps, as illustrated in Figure 23–12.

Begin the form construction by preparing forms for the sides of the pour. Utilize a 2 × 12 or 3/4 inch plywood for side forms. Mark the alignment of risers and treads for one side of the entire stair run on the lumber. Repeat this marking for the other side of the stairs. Side forms do not need to be cut.

Excavate the pour site and place each side form, verifying that the treads are at the correct location and level. Drive stakes behind the forms, and anchor them in a manner similar to that used in anchoring forms for flat work. Prior to constructing and installing riser forms, place any reinforcing along the ground in the pour area. If using WWM, bend the mesh to conform

closely to the stair contour. An alternative to using WWM is to use fibermesh for stair pours.

Using 2 inch dimensioned lumber that is wider than the calculated riser height, measure and cut risers to match the width of the stair opening. Any gap larger than 1/16 inch at either end will allow concrete to seep through when pouring. If the stairs are not a constant width, mark and measure each stair in its location to verify the correct length. After trimming the length, cut the lumber the long dimension (rip cut) to the correct height for the riser. Rip cutting at a 45 degree bevel will provide room at the bottom of the riser form for smoothing the tread using a trowel (Figure 23–13). The riser for the bottom stair should be 4 inches deeper than the others to provide the depth necessary for paving that will abut the stairs. The bottom riser form does not need the bevel cut.

Beginning with the bottom riser, nail the riser form in position using form nails driven through the outside of the side forms. Check for level side-to-side across the top of the riser form and plumb down the front, then secure using deck screws placed from the outside of the side forms into the end of the riser form. Repeat this process for each of the risers. Before completing connections for each subsequent riser, check for a slight fall between the bottom of the upper riser form to the top of the lower riser form. Properly calculated and constructed stairs should have a tread that tilts 1/4 inch per foot toward the front of the stair to prevent standing water.

Cleats should be added in the corner formed by the back of the riser form and the side form to brace the form. Forms wider than 3 feet may require additional bracing to prevent bowing after concrete placement. Near the center of the stairs, screw a 1 × 4 to the tops of all the riser forms.

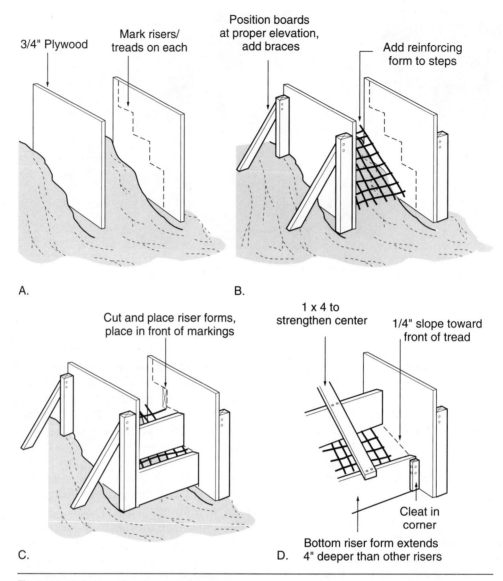

Figure 23–12 Independent forming of concrete stairs.

Suspended Forming of Stair Pours

Occasionally, the landscape contractor is required to construct forms for poured concrete stairs that must fit between existing walls and structures. Use of walls or structures as the "sides" to the forms saves a step in construction, but it also eliminates the support for the riser forms. To accommodate this lack of support, the riser forms are cut and hung, or suspended, from beams that run over the top of the pour. This type of pour should be completed before any paving is done above or below the stairs to allow anchoring support beams.

Before constructing this type of form, place any required expansion material against structural objects against which the stairs will come in contact. Concrete stairs expand and contract, creating pressure against a structure foundation. If expansion joint material is not installed, damage to the structure may occur. Also place any reinforcing along the ground in the pour area. If using WWM, bend the mesh to conform closely to the stair contour. An alternative to using WWM is to use fibermesh for stair pours.

Installing an anchor between the stairs and a building foundation or concrete wall may also be required. Foundations and concrete walls have surfaces that allow stairs to rise or settle over time. To anchor the stair to the sides, drill 3/8 inch diameter holes 2 inches into the wall. Cut 6 inch long pieces of 3/8 inch (#3) steel reinforcing rods and drive them into the holes. Leave the 4 inch stub projecting into the stair pour area.

Constructing suspended forms begins by marking the riser and tread locations on the sides of the pour. Select a piece of 2 inch dimensioned lumber that is

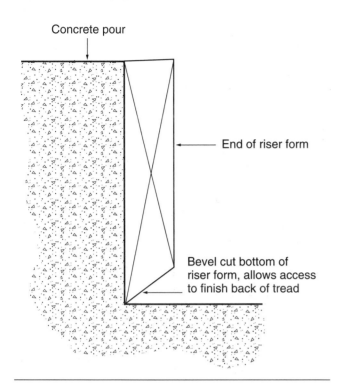

Figure 23–13 Bevel cutting the bottom of rise forms.

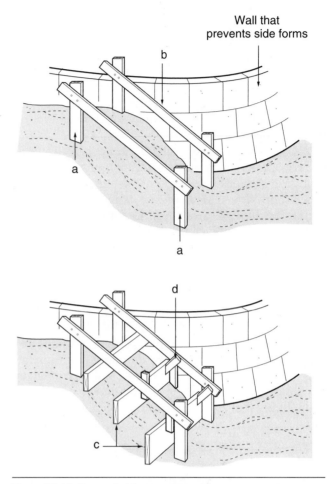

Figure 23–14 Suspended form construction.

wider than the height of the riser. Measure the length for the first riser and cut the lumber to the correct length. Rip cut the lumber to the riser height using the 45 degree angle described in the independent pours section. Repeat for each riser. The riser for the bottom stair should be 4 inches deeper than the others and does not need the bevel cut.

Beams mounted on stakes running from the top to the bottom of the pour will be used for support of the riser forms. Support beams must be spaced close enough and braced well enough to hold the weight of the riser forms, the concrete, and the laborers who will lean on them when finishing. Place a beam 1 foot in from each side of the stairs, with an additional beam in the center, and one for every 3 feet of width for the stairs. At the top and bottom of the stair location, and in alignment with each of the beam locations, drive 4 foot long 2 × 4 stakes 18 inches into the ground. Between the pair of stakes at each side of the pour, attach a 2 × 4 that sets just above the height of the planned risers. If a stair run has more than five risers, these beams should be constructed of 2 × 6s. Once installed, the stakes may need to be braced to prevent movement. The two outside beams will be used to initially hang the riser forms, with the center beams added after the riser forms are secured and leveled (Figure 23–14).

Beginning with the bottom riser, attach 18 inch long 2 × 4 extensions to the back of the riser form at loca-

tions under where the beams will pass. Using the riser markings on the side walls as guidelines, place the precut riser form in the correct position under the support beams and tack the 2 × 4 extensions onto the beam. Verify that the riser form is level along its length and plumb down the face, then securely connect the riser form extensions to the beam at both ends. To further brace the riser form, connect angled braces from the beam to the base of the riser form. This will prevent the concrete from pushing out at the bottom of the riser. Repeat this process for each riser form to the top of the stairs. Before completing connections for each riser, check for a 1/4 inch per foot downward slope between the bottom of the upper riser form and the top of the lower riser form.

When the risers are secured to the beams at each edge of the stairs, center beams and riser connections may be installed. Use caution not to disturb the level previously established for the risers. Additional bracing may be added to secure the forms, but leave access to the inside of the forms for placing and finishing the concrete. Placement of concrete by hand reduces the chance of form failure.

Wedge Forming

When a stair pour is located between two flat, solid, well-anchored objects such as building walls, stairs may be formed by wedging a riser form between the two objects (Figure 23–15). This method saves time-consuming construction of form supports, but also has the risk of form failure if the riser form cannot be tightly wedged between walls.

Place any reinforcing material and expansion joint material before installing the form. On the surfaces at each side of the stairs, mark the location for the stair treads and risers. Measure the length between the marks for the first riser form. Cut a piece of 2 inch dimensioned lumber 1/8 inch shorter than this measurement. Rip cut the length of the lumber to match the height of the riser. Rip with a beveled bottom similar to the riser forms for independent and suspended pours.

Position the riser form and drive a shim between the form and wall to wedge the form tightly into position. If necessary, adjust the location using a hammer. Fill gaps of 1/4 inch or less with expanding foam insulation. Let the foam expand and harden, then trim away any excess. Repeat for each riser. When pouring concrete, hand shovel concrete into the stair behind the wedge forms to reduce the chance of form failure.

PREPARATION BEFORE POURING

Temporary Screeding Rails

Screeding of concrete is an important step in the pouring process. Screeding utilizes a long, straight piece of lumber or metal to smooth the rough concrete into a surface that can be finished. Occasionally there are pours where the slab is wider than available screeds or where there is no form on one side, such as next to a building. If this is the case, consider a temporary screeding rail as a solution. A temporary screeding rail is a rigid metal conduit placed parallel to forms and supported on short stakes. This rail supports the screed and is slid to a new position as the screeding progresses. As the screed is moved, concrete is shoveled into the void left by the rail and smoothed by following the steps of the concrete finishing process. The stakes that supported the screed are left in the slab. Failing to slide the pipe every 2 feet or so will make reaching the void more difficult. At the end of the pour, the rail is removed.

Stakes that support the temporary screeding rails are made of 1 × 4s, driven into the base to within 2 inches of the finish grade. Two nails spaced the width of the conduit are driven into the top of the 1 × 4. Leave only 1 inch of the length of the nails sticking up so

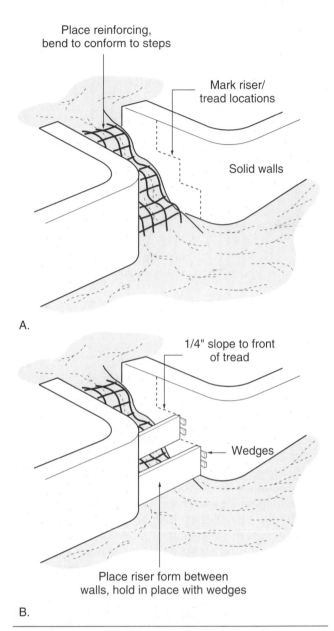

Figure 23–15 Wedge forming stair risers. (A) Marking rise/tread locations and placing reinforcement. (B) Installing and securing riser forms.

they will not project above the concrete surface when finished (Figure 23–16). Stake height must be adjusted to place the top of the temporary screed rail at the desired height (Figure 23–17). A typical screed rail is a 2 inch diameter, galvanized, schedule 40 electrical conduit. An alternative is to set up temporary wood forms that are removed, and fill the void after screeding.

Form Break

Spray the inside of the forms with a form break compound or light kerosene. These chemicals reduce the

Figure 23–16 Support stakes for temporary screeding bars. These stakes will be buried in the slab; verify that nails do not project above screed bars.

bonding of concrete to the forms and makes removal of forms an easier chore with less concrete damage.

Final Check

Go through the checklist again before the concrete arrives. Included in the checklist are some last-minute items to do in preparation:

- Sprinkle the base with a light shower of water before pouring to slow curing on hot days.
- Make sure all necessary tools are easily accessible.
- Prepare ahead of time for bad weather. Have a cover ready for the surface.
- Make sure any route for a wheelbarrow is free of debris to avoid accidentally dumping a load.
- Have preselected locations for excess concrete and mixer cleanout.

PLACEMENT OF CONCRETE

Begin the pour by checking the slump of the concrete. Correctly checking the slump requires filling a 12 tapered cone and tamping the concrete inside. Invert the cone on a solid surface, remove the cone and measure the amount the concrete "slumps" below the top of the cone. A 3 inch slump is standard for slabs. Lacking methods to make a formal slump check, other methods can be used to determine concrete preparation. In a

Figure 23–17 Temporary screeding rails used for screeding wide pours. The top of the rail should be set at the desired thickenss of the slab.

visual check, concrete with proper consistency should look like a thick stew. If the concrete fails to move down a wetted chute, it is probably too dry; if it pushes water out along the sides of the chute, it is probably too wet. Another check for concrete is to chop through a pile of concrete and see if ridges remain. If ridges crumble, the concrete is too dry; if ridges slump, the concrete is too wet. It is better to have concrete delivered too dry and add water than to have the other extreme. If the concrete pour is on a slope, a low slump will be necessary to keep it in place when working from the top of the slope to the bottom.

> **CAUTION**
>
> Concrete contains caustic materials that can cause burns after prolonged contact with the skin. Avoid contact between bare skin and wet concrete and wash thoroughly after working with concrete.

Placement and Screeding

Once the concrete is prepared for pouring, position workers to make the work efficient. Placing one worker in rubber boots in the middle of the pour to push concrete and help pull the screed with a rake is an effective use of labor. Position two workers on either side of the pour, one to operate the end of the screed and the other to shovel concrete to the locations required.

When the proper consistency is obtained, place the concrete in piles, working the chute from one side of the forms to the other. Using rakes and shovels, spread the concrete in front of the screed. If using WWM, use the rakes to hook the mesh and pull up gently to seat it in the center of the slab. If the WWM rises above the surface during the pour and cannot be forced back down, lift the mesh up, cut out the offending section with a fencing tool, and bend the edges back down into the slab.

When approximately 2 linear feet of the pour has been leveled with rakes and shovels, begin screeding. Using a 2 × 4 cut 2 feet longer than the width of the pour, rest the screed on the forms. With a side-to-side sawing motion, gradually pull the screed forward until the concrete is leveled and the aggregate is pushed down into slab (Figure 23–18). Be certain to keep the screed resting on the forms to ensure the proper thickness of the slab. If the screed is difficult to move, try dragging the screed forward 1 to 2 feet without the sawing motion to rough level the concrete, then go back and rescreed using the sawing motion. Tilting the top of the screed forward can also make the screeding process easier. It may also be easier to pull the screed

Figure 23–18 Placing and screeding concrete.

rather than to push it. If large pours are planned, the work may benefit from renting a power screed. The power screed is a flat, horizontal plate mounted on a framework. A motor vibrates the screed and workers push the machine forward across the forms.

In corners or areas where both ends of the screed cannot rest on a form or temporary screeding rail, a technique termed **wet screeding** can be utilized. Wet screeding is performed by resting one end of the screed on a form, holding the unsupported end of the screed at the approximate finish elevation, and performing the sawing action of screeding. While this technique is not recommended for large areas, small spaces or short runs can be successfully leveled using the previously screeded area as a guide for the elevation of the area being wet screeded.

If low spots or honeycombing voids appear after screeding, workers with shovels should sprinkle concrete on the spot and repeat the screeding. Screeding should be completed in as few passes as possible. When finished, the surface should be level, without holes, and should resemble a sheet of rough-sawn plywood.

Bull Floating

Almost immediately after the screed passes, workers can begin floating the surface with the bull float. The

Figure 23–19 Bull floating a concrete slab.

bull float, a wide, flat metal blade with a long handle, is pushed and pulled over the surface several times. This process further smoothes the surface and embeds the aggregate in the slab, and may not be practical in tight areas or on small pours. Most bull float handles are adjustable to allow proper operation on any reasonable width of slab by a person of any height. Because the handle is long, watch out for power lines and windows behind and above the path of the float handle.

> **CAUTION**
> Review the area before bull floating. Contact of the handle with power lines can cause electrocution. Damage can also be caused to structures and individuals if struck with the bull float handle.

The long-handled bull float is set on the near side of the slab and slowly pushed across the slab, keeping the trailing edge of the float low by holding the handle low. When the float reaches the other side, lift the han-

dle up and slowly pull back along the same path, keeping the trailing edge low by holding the handle high (Figure 23–19). If the float sinks into the concrete surface, remove the float and wait a few minutes before continuing. If the float digs into the concrete, the front edge is being held too low. Using a repetitive jerking motion may help the float move easier on the first pass. Float the surface as many times as necessary to create a smooth, glasslike surface without holes or voids, then pick up the float and move it down one width, and continue. Tiny ridges will exist between sections where the float was moved, but will be trowelled out later. If the float becomes coated with chunks of concrete or debris, stop and rinse off the surface.

Darby Floating

A **darby float** may be used in areas that are too tight to reach using a bull float. The darby is a narrow, 4 foot long float made of wood or metal, with handles on top. Operate the darby by lifting the leading edge when moving the float across the surface. Make several arcing passes with the darby, with slight pressure on the trailing edge. This pressure will bring moisture and the fine particles of concrete to the surface to fill any small voids or holes.

Finish Floating and Trowelling

Trowelling is the laborious step of working the surface with flat-bladed tools that push the aggregate farther down into the pour while causing water and fine particles to move up to the surface. Two trowelling steps, finish floating and steel trowelling, are recommended for concrete slabs. Exterior slabs may need only the first step, as is the case when planning a rough, broom texture for the surface. Smoother surfaces and interior slabs may require both steps to prepare the surface properly for finishing. Trowelling works best starting at the point where the pour began, working in the same direction that the pour progressed. This finishes the surface in approximately the same order in which it will harden. Watch for portions of the pour that are in direct sunlight, or areas where the concrete may harden faster. These areas may need to be trowelled sooner.

Finish floating can be done with wood or **magnesium floats,** also know as mag floats. Begin once the surface has begun to lose the wet sheen that is left from bull floating. Test the surface before finishing. If the tool sinks easily into the surface when pushed straight down, wait a bit longer. Working quickly, move the float in a sweeping, arcing motion across the concrete surface to remove any ridges, holes, or other irregularities. Pushing down hard with the back edge of the float while holding the leading edge above the surface makes floating easier and more productive (Figure 23–20). Floating is done while the concrete is just beginning to dry, so the majority of this step is completed with the worker standing to the edge of the slab. Reach as far inward as possible, then move to the other side of the slab or utilize kneeboards as described in the next paragraph.

If the slab is too wide to reach from outside the pour, utilize a pair of 24 inch × 24 inch sheets of 1/2 inch plywood, termed kneeboards, to kneel on while reaching the center. Gently position the first kneeboard and step onto it, then place the second kneeboard within a step from the first. If the board sinks completely into the surface from your weight, delay until the concrete hardens a bit more. If the board does not sink at all, move quickly. Kneeboards should be spaced close enough to step from the first board to the second. If unable to reach the center of the pour, then reach back and pick up the first kneeboard and move it closer to the center. Repeat this process as many times as necessary. When positioned, float/trowel the area around and up to the kneeboard.

Figure 23–20 Floating a concrete slab from the edge.

Step back to the kneeboard closer to the edge (Figure 23–21). Peel up the vacated board and reverse the placement process used to get to the center. Weight on the board will cause it to stick to the concrete surface, requiring it to be "peeled" from the concrete. Once removed, any indentation can then be floated/troweled out. If the concrete is hardening quickly, several passes over the surface while exerting pressure on the float will bring moisture back up to the surface. Extra concrete may be placed by the shovelful to fill low areas or repair serious damage. Properly floated/trowelled surfaces appear smooth, wet, and without any holes, ridges, or irregularities. The surface should look prepared enough that it could be left in that condition if no other finishing were to take place.

Steel trowelling should be done after finish floating if a smooth surface finish is planned or the surface is still rough after floating. Begin trowelling after the sheen has disappeared from the surface, and utilize the same process outlined in the finish floating description. When done properly, steel trowelling cre-

Figure 23–21 Using kneeboards to reach center of a concrete slab for floating. Concrete must begin to set before boards can be placed on slab.

Figure 23–22 Broom finishing a concrete slab.

ates a very smooth surface that is slick in exterior applications. Consider renting a power trowel to save a lot of labor on surfaces that are intended to be to left in a smooth condition. The power trowel is a large piece of walk-behind equipment that uses a motor to drive a fan of trowels. Used primarily for interior slab finishing, the power trowel has limited applications in exterior use.

SURFACE TEXTURING AND FINISHING

When the concrete has begun to set, surfacing and texturing can begin. Instructions for obtaining the finishes described in the chapter introduction are described here. Proper timing of each operation is also identified.

- Float finish. Floating for a finish should be done after the surface has begun to harden, since the intention is different than that of the first floating operation done. Use a back-and-forth motion that eliminates a pattern and provides uniform results. It may be necessary to apply extra pressure on the float in order to get a rough surface. Jointing the slab into squares, rectangles, or other geometric segments creates

an interesting surface out of what could otherwise be a very dull finish. Using the edger around each segment creates a border that highlights the finish inside.

- Broom finish. Brooming should be done when the surface is firm to the touch but not dry. It is best if the ridged pattern created by the broom is perpendicular to the direction of traffic, but parallel to the direction of drainage off the slab. Extend the broom to the far edge of the slab or to a location where a joint will be placed, set the broom down lightly, and pull with a single, smooth motion to the near edge of the slab (Figure 23–22). Repeat this motion the full length of the slab, using a slight overlap from one pull to the next. If the broom becomes plugged with wet concrete or begins to drop small "kernels" of concrete, set the broom head on end to the side of the slab and tap it a few times to dislodge the debris. Small hand brooms can be used to surface tight areas.

- Colored, impressioned finish. (Some versions of this process may be restricted to use by contractors who hold licenses from manufacturers.) This process involves two steps: coloring the concrete and stamping the surface. The first step is accomplished by adding a powdered color to the concrete when mixed or sprinkling the powder on the wet, screeded surface and floating it into the concrete. Manufacturer's instruction's will recommend the best method for adding color. It is important that the powder be mixed thoroughly and evenly into the surface of the concrete. Finishing proceeds in the same manner as that of plain concrete until after trowelling. When the sheen from trowelling has disappeared, stamps are carefully placed on the slab surface and are either pounded down with a rubber mallet or forced down by the weight of a person standing on them. If the stamps sink into the surface, postpone the operation until the concrete has hardened a bit more. When one area has been successfully impressioned, the stamp is pulled up and placed on the slab adjacent to the first area stamped. This process is repeated until the entire surface has been impressioned. Small areas and irregular spaces that will not allow stamping can be hand finished by pushing the blade of a brick set into the wet concrete. Follow the same pattern made by the stamp in the areas that are hand finished. Work quickly with the stamping operation to avoid the concrete hardening before the impressioning work is complete. This type of surface is not edged or jointed as described for the float and broom finishes.
- Seeded exposed aggregate. Broadcast the special aggregate over the surface after bull floating and embed it into the top of the slab using a trowel. After the concrete begins to harden—in approximately 2 hours—lightly "scrub" with a hose and broom to remove the top film of concrete and "expose" the aggregate that was embedded earlier. If the scrubbing operation gouges the surface, stop immediately and repair the holes, then wait for the slab to harden further. Six to seven days later, wash the surface with a mixture of nine parts water and one part muriatic acid or trisodium phosphate (TSP).
- Integral exposed aggregate. Pouring and finishing proceeds in the same manner as that of plain concrete until just before the slab is nearly hardened. When the slab supports the weight of a person walking—in approximately

CAUTION

Muriatic acid and trisodium phosphate (TSP) can cause severe burns. Follow manufacturer's instructions for mixing and application of these chemicals. Wear protective clothing and avoid contact with skin or inhalation of fumes.

2 hours—begin a light "scrubbing" of the surface with water and a broom. Scrub only enough to wash the top film of concrete off the surface and expose the aggregate near the surface. An alternative to scrubbing the surface is to lightly sandblast the concrete 2–3 days after the pour to expose the aggregate. This approach should also be done carefully so that gouging of the surface does not occur. Jointing and edging of this type of surface should be done carefully so the aggregate is not dislodged.

Contraction Jointing

Before the concrete has lost all workability, edging and contraction joints done with a jointer should be completed. While these two operations use tools and methods that are similar, the purpose for each is quite different. Edging rounds and smoothes the edge corners of a slab to reduce chipping of the edge and to help separate the slab from the forms. Edging also provides an aesthetic function by providing a contrasting surface texture when compared to the finish of the interior of the slab. Jointing controls cracking caused when the slab dries. As the concrete project dries, the slab will contract. This contraction shows up as cracks in the surface of a slab, and contraction (or control) joints are placed in the surface to create weak points where that cracking will occur. Without these joints, the cracking would occur at random weak points in the slab, and seldom in aesthetically pleasing patterns. Jointing may be done with a jointing tool while the slab is still wet, or can be done 1–2 days after the pour with a saw. Commercial installers typically saw, but the equipment and blades are expensive. Like edging, jointing also provides a contrasting surface texture and helps define squares or other geometric forms created in the slab.

Planning joint locations requires the application of the mathematical formula presented in Chapter 5. In addition to spacing, it is also important to avoid joints that create acute angles. Acute angles leave weak points that will crack without a joint. It is better to space joints wider to avoid an acute angle. Consult an engineer or design professional with questions regarding joint placement.

Jointing with a hand tool requires a guide to keep the joint straight across the slab. This is usually provided by laying a piece of dimensioned lumber on the forms from one side to the other, parallel to where the joint is desired (Figure 23–23). A 2 × 8 should be adequate for this. Gently lay the 2 × 8 in place and step halfway across the slab. Begin at the center and press the **jointer** into the concrete. Work the jointer forward and backward, always holding the tool tight against the 2 × 8. The best results are obtained by lifting the leading edge of the tool while pressing down on the back edge. If the slab is drying quickly, pieces of aggregate may be worked up by this operation. Use the flat part of the tool to push them back into the slab, and then work back and forth over the aggregate with the back edge of the tool until smooth. Going back and forth several times over the surface will usually work up enough moisture that the surface can be reworked. Continue the jointing until the edge of the slab is reached.

Move to the other side of the slab and repeat this process from where the previous joint began. Carefully lift the guide board and move it to the next joint location. If necessary, retouch the surface texture when the board is removed. Running joints down the center of a long slab is a difficult process that requires setting the guide in the correct location and accessing the guide using kneeboards. When completed, jointing will provide a 3/8 inch deep joint that creates the weak area in which contraction cracking will occur.

Jointing after the concrete has dried can be accomplished with a cutoff saw equipped with a carborundum or concrete blade. After the concrete has hardened enough to walk on (2–3 days), use a chalk line to snap the location of all joints. Start the saw and slowly run the blade along each mark. Use caution to make the cuts only 3/8 inch deep and straight. Use of a 2 × 4 guide may help keep the sawn joints straight.

> **CAUTION**
> When cutting concrete, always wear long-legged pants, hard hat, ear protection, safety glasses, dust mask, and gloves to protect from dust, noise, and debris from the saw.

Edging

Edging should be done following the completion of surface texturing (colored impressioned concrete is not edged) and may require additional work with the tools since the slab may be nearly dry (Figure 23–24). Dip the edging tool in a bucket of water. Use the **edger** in the same manner as the jointer, pushing the edger along the form with the leading edge lifted and putting downward pressure on the trailing edge. Work around all edges that need to be finished. If it is difficult working corners, start the edger in the corner and work outwards, then work back toward the corner to cover any marks left in the surface.

COMPLETING THE POUR

Curing

After all finishing operations are complete, cover the work with a plastic sheet or wet burlap, or spray on a curing compound. Covering retains heat and moisture as the slab cures, improving the strength and reducing surface cracking by slowing the drying process. The cover should lie flat on the surface to avoid discoloring. A cover also provides some protection from rain and sunlight, and identifies the project site as a construction zone. With concerns about traffic passing over a

Figure 23–23 Jointing a concrete slab using a jointing tool run alongside a 2 × 4 guide.

Figure 23–24 Edging a concrete slab.

new slab, it would be wise to erect a snowfence around the project to further protect the work. Leave the cover in place a minimum of 24 hours. Coverage for 3–7 days is ideal. Remove when moisture no longer condenses on the underside of the plastic cover.

Removing Forms

Approximately 2–4 days after the pour is completed, forms can be removed. Uncover the slab and use a claw hammer to remove the form nails from the stakes. Using a careful, prying motion with a shovel or pry bar, gently work the stakes up and away from the forms. When all stakes are removed, gently work the shovel or pry bar between the slab and the forms and pry outward. Use great care during this step so that the edges of the slab do not chip. A finished slab will still be "green" (the color is actually a dark gray that will dry to a light gray over a few weeks), but it is ready for backfilling the edges. Keep foot traffic off the slab for 2–3 days, and any vehicle traffic off the slab for 1–2 weeks.

Chapter 24

Unit Pavers

This paving classification, which includes man-made paving material that is laid by hand, provides opportunities for color, texture, and pattern unique to the landscape industry. This chapter covers the installation basics used for unit paving materials such as brick, interlocking concrete pavers, asphalt block, and adobe. Included in the information are descriptions of paving patterns and specific installation techniques. Open cell paving installation is described at the end of the chapter.

RELATED INFORMATION IN OTHER CHAPTERS

Information provided in this chapter is supplemented by instructions provided elsewhere in this text. Before undertaking activities described in this chapter, read the related information in the following chapters:

- Safety in the workplace, Chapter 4
- Basic construction techniques, Chapter 7
- Construction staking, Chapter 8
- Site preparation for paving, Chapter 22

PLANNING THE PROJECT

Construction of unit paver surfaces can be a time-consuming project. To avoid problems, planning of the installation should precede the work. Following are some considerations for planning a unit paver installation.

Paving Patterns

Examples of common paving patterns and the pavers that work best in those patterns are identified in Fig-

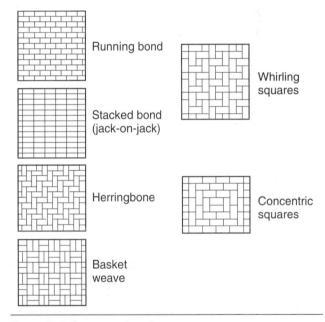

Figure 24–1 Common unit paving patterns.

ure 24–1. Certain paving materials may allow only one pattern due to the special shapes in which they are manufactured. Use caution when purchasing brick pavers. If the units are not modular (length twice as long as width), patterns such as herringbone and basket weave will not work.

- Running bond. Running bond is a pattern in which pavers are placed in staggered, horizontal rows. Pavers are placed end to end in one row, and in the next row they are set end to end but are offset by one-half paver. Pavers that work in this pattern include most interlocking concrete pavers, all brick, and any paver that is not modular. Old roadway bricks can often be used only in this pattern.

- Stacked bond (also called jack-on-jack). Stacked bond is a pattern in which pavers are placed side by side in even horizontal and vertical rows. Pavers that work in this pattern include many interlocking concrete pavers (all sizes), most brick, adobe, and any paver that is modular.
- Herringbone. Herringbone is a pattern that places horizontal and vertical blocks in a diagonal pattern across the paved area. The finished pattern leaves a zigzag, or herringbone, appearance. Halves are required for starting this pattern. Any modular paver can be used in this pattern.
- Basket weave. Basket weave places two pavers vertically, and then two pavers horizontally. This alternating pattern is repeated across the entire paved area. Any modular paver can be used in this pattern.
- Whirling squares. Whirling squares place four pavers in a square pattern around a half-brick center. This pattern can be offset to create an even more interesting pattern. Halves are required for this pattern. Any modular paver can be used in this pattern.
- Modified basket weave. Variations of the basket weave are possible. A common variation includes placing an extra vertical between the vertical and horizontal pairs in every other row. Any modular paver can be used in this pattern.

Interlocking concrete pavers are also available in a variety of shapes that create patterns when installed in stacked or running bond. Some common shapes of interlocking pavers include (Figure 24–2):

- Rectangle. Modular rectangular paver.
- Double score rectangle. Modular rectangular paver with a score line in the center to create an impression of two halves.
- Octagonal. Paver with a octagonal body and interlocking tab.
- Dentated. Paver with an interlocking angled edge.
- Cobble. Combination of modular rectangular and square stones.
- H block, or I paver, or dogbone. Paver with an interlocking H shape.

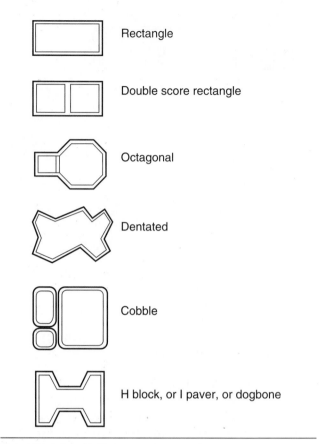

Figure 24–2 Interlocking concrete paver shapes.

BASE MATERIAL

Base preparation for most unit paving installations is similar to the process described in Chapter 22. However, alternatives do exist to use of an aggregate base with a setting bed for some pavement types. Installation over a previously paved surface or pavement that must carry heavy traffic loads may require an alternative to a crushed stone pavement base. Regardless of material, the base under the entire paved area should be consistent. There is a potential for differential settlement in areas where the unit pavers move from one type of base onto another. Valleys and ridges typically develop over time where base materials change. Some alternative treatments are:

- Concrete or asphalt base. Placement of brick or concrete pavers on a sand setting bed over concrete or asphalt is possible if the base is level and without cracks or serious surface disruptions. If the proposed base surface has broken joints or other disruptions, paving is not recommended. The installation may suffer long-term deterioration if placed on a base that does not drain properly. If pavers are placed on an impervious base that has no drainage, holes should be drilled at the low edges of the subpaving to allow for moisture that has passed through the pavers to drain through the base.

- Placement on mortar base. There are applications where brick or stone paving is placed on a mortar base. See Chapter 26 to review the situations and methods for this type of paving.

EDGE RESTRAINTS FOR PAVED SURFACES

All unit pavers require some sort of edge restraint to prevent the outer courses from wandering. The methods and timing of the placement vary depending on the design of the project, but several choices are described in Chapter 22.

ADHERING PAVERS TO STEPS

Installations that require that pavers be placed on step treads or edges of stoops require setting pavers in a thin bed of mortar or using an adhesive to bond the paver to the material below. Mortared installations should be completed according to the instructions provided in Chapter 21. Pavers bonded to stairs should be cut with joints of 1/16 inch or less. Use a one-part urethane or similar type of adhesive to bond the pavers to the outer edge of the stoops.

PLACEMENT OF SETTING BED

A thin layer of material on which the pavers will be set should be placed over the compacted base. The setting bed should be screeded to a consistent thickness between 1 inch to 1-1/2 inches thick across the compacted base. Because the pavers are resting directly on this material, it is also important that the setting bed be sloped in the proper direction and be set at the proper elevation. Clean, coarse, concrete sand is the preferred material for use in a setting bed. Stone dust, limestone screenings, and other materials provide inconsistent compaction after pavers are set. Setting bed sand should contain enough moisture that it forms a ball when squeezed in the palm of the hand. Sand that is too dry will need to be wetted slightly prior to screeding.

Precise screeding of the setting bed can be accomplished using a board notched to fit adjacent pavement or preplaced edge restraint. If using a paved edge for screeding, be certain the base screed is

Figure 24–3 Screeding setting bed for unit pavers.

notched to accommodate the thickness of the paver minus 1/4 inch (Figure 24–3). Screeding can also be accomplished by working off 1 inch diameter screed rails that are set directly on the base. Screed only the area of the setting bed that will be paved immediately. Screed perpendicular to the rails when possible. Small sections may need to be screeded in areas where the pavement cross-slopes, or warps, to match existing grades. Screed the setting bed and lift out the temporary screed rails, filling the voids left by the rails with additional sand and smoothing with a steel trowel.

Headers

Brick paved surfaces can be separated into smaller paved areas with the use of headers similar to those discussed in Chapter 23 for concrete paving. Typically,

Figure 24–4 Soldier course at edge of pavement.

decay-resistant wood headers are installed prior to the placement and screeding of the setting bed. The headers are then used as a screed support for leveling the bedding sand.

UNIT PAVER PLACEMENT

Patterns and Edge Course
Before beginning a project, lay a test area large enough to see how the pavers will fit together. For each of the patterns, it is recommended that a soldier course of full pavers be placed around the entire perimeter (Figure 24–4). This course will eliminate placement of cut and partial pavers adjacent to the edge restraint, and will reduce breaking and movement problems.

Beginning Placement
To obtain a pattern that is tight and uniform across the paved area, always work from a line that is perpendicular to the edge from which paving begins. Beginning in the wrong place or starting on two sides may lead to a "pinching" or "spreading" of the pattern near the center of the paved area. This error may require additional cutting or relaying of pavement. One method

commonly used for paver placement is the T method. The T method begins paver placement in a line perpendicular to one of the project's 1straight edges. Often, the straight edge used for alignment is a structure or existing paved surface that the paved area abuts. Near the center of this straight edge, a line is laid out at a 90 degree angle to the straight edge that runs through the center of the paved area. Snapping a chalk line on the screeded bedding sand along this alignment will help maintain the pattern. The pattern is extended along this line for four or five courses and then the area between the straight and perpendicular lines is filled (Figure 24–5). This method works well for rectangular-shaped areas where minor variations from perfect horizontal and vertical lines will be noticed. Most patterns can be laid with this method.

Another method for placement is to begin at a structure or wall and work out one row at a time. This method works well for curved walkways and irregular shaped areas where minor variations from perfect vertical and horizontal alignment will not be noticed. This method is also easier for placing herringbone patterns. Without a structure or paved area against which to align paver placement, snap a chalkline across the screeded bedding sand for alignment.

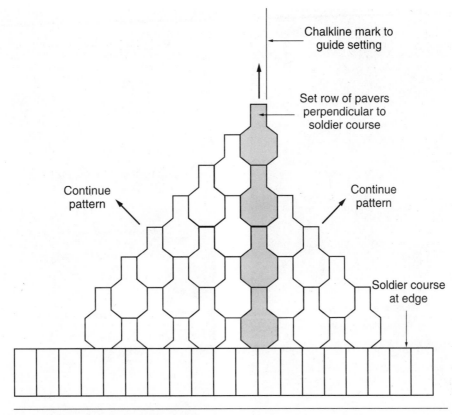

Figure 24–5 T pattern for placement of unit pavers.

Placement of pavers using the herringbone pattern may benefit from beginning in the corner of a project rather than using the T or horizontal methods. A corner beginning point allows the establishment of this complex pattern and the continuation of paver placement in one direction across the project. Problems will be encountered with either method if the paved area is between walls that are not at right angles. If this situation is encountered, expect to do a great deal of cutting along one or both walls.

To speed the work, stack several piles of pavers near where the paving placement will begin. Become familiar with the paving units, since interlocking concrete paving block and precast pavers have a top side with a beveled edge. Other pavers may have patterns or textures on the upper side. Pavers placed upside-down will need to be removed and reinstalled with the wearing surface facing up. When paving large areas, several pallets of material may be required. To avoid problems with minor color variations between pallets, select pavers randomly from each of the pallets provided.

Paver Laying

Begin laying pavers according to one of the placement methods and patterns previously described. When placing the pavers, set them straight down onto the setting bed. Avoid dropping, angling, or twisting the paver. Allow the weight of the paver to set it onto the sand bed. The paver should be placed with joints of approximately 1/16 inch (Figure 24–6). Some paver varieties have spacers cast into the sides of the paver, which hold the pavers the correct distance apart. Clay brick and other pavers should be set with sides and ends touching.

If a paver pushes too far down into the setting bed, it may leave a small ridge of sand around the bottom edge. This ridge will prevent tight placement of the next pavers. Lift the paver out and lightly smooth the setting bed with a steel trowel. Use caution not to compact or vary the surface level when resmoothing the setting bed.

Paver alignment should be checked often as the laying process moves along. Verify that the joints are straight and that the pattern is correct. Paver alignment can be checked every ten courses by placing a stringline along joint lines of the pavers. Adjust pavers using a pry bar. Avoid stepping near or adjusting pavers within 1 foot of an unrestrained edge. Continue laying pavers up to the edging that is in place or to the edge marking. Paving operations that are not completed in a single session should be temporarily restrained using a piece of edging. Before continuing paving operations, remove the edging and repair the surface.

Figure 24–6 Beginning placement of interlocking concrete pavers.

Figure 24–7 Setting cut pavers.

CUTTING AND FINISHING EDGES

Few unit paving jobs can be completed without cutting and placing partial pavers. Planning and selection of certain pavement patterns may reduce the number of cuts that have to be made, but anytime an irregular shape or structure is encountered, cutting can be expected. Pavement should be planned so that full pavers are placed in areas where visibility and traffic is high. This will improve aesthetics and help to avoid breaking and movement. The basic construction techniques in Chapter 7 describe and explain several techniques for cutting pavers.

Placing Cut Pavers

Placing cut pavers into position may require tapping the paver with a rubber mallet or shifting of pavers with a masons' trowel. Pavers should fit precisely into the opening (Figure 24–7). If the fit is too snug, trim a bit more off the paver. If the paver can be removed without contacting any adjacent pavers, the cut was too extreme and should be redone using a new paver.

SEATING AND FINISHING THE SURFACE

Following placement of all pavers and edging, the final steps can be applied to a paved surface. This process may vary depending on manufacturers' requirements, but most require that that the pavers be "set" into the setting bed and that sand be swept on the surface.

Seating Pavers

Concrete paving blocks require a mechanical seating of the paving units (Figure 24–8) using a vibratory plate compactor. The seating of pavers vibrates them into the setting bed and forces the setting bed sand up into the joints between pavers. When seating clay brick pavers, utilize a plate compactor with a rubber mat on the base to avoid chipping the units. Sweep all debris from the paved surface. Place the plate compactor on the surface, start the equipment, and work the surface from the outside edge into the center. Two passes should be made in this initial seating operation. If the plate compactor has multiple settings, use

Figure 24–8 Seating pavers using plate compactor.

Figure 24–9 Sweeping dry sand into paver joints.

the high-frequency/low-force setting. Maintain the plate compactor at least 3 feet away from unrestrained edges. This mechanical seating will push blocks down into the sand leveling course as much as 1/4 inch. When working against existing paving or a preset edging, run the plate compactor on both surfaces at the same time. Placing the plate compactor only on the new pavers may cause them to settle below the existing surface. Examine the surface for any damaged paving units. Pavers may be removed at this stage using a paving puller or two screwdrivers to lift the unit out of the surface. Replace damaged or broken units with new pavers.

Sweeping Joints
Filling the joints between the pavers helps create interlock and contributes to the waterproofing and stabilization of the surface. This is accomplished by shovel-

ing dry, coarse sand onto the surface of the pavers and sweeping it into the joints. Sweeping diagonally to the joints in two different directions is most effective in working the sand down. (Figure 24–9).

Repeating Compacting and Sweeping Operations
Seating with the plate compactor and sweeping of joints should be repeated until joints are full.

SEALING UNIT PAVER SURFACES
Protection of some unit paved surfaces from stains and soiling can be accomplished by sealing the surface. Pavement sealing is a specialized operation that involves washing and application of formulated sealers. Paving manufacturers or distributors should be contacted for recommendations on sealing procedures for unit pavers.

INSTALLATION OF OPEN CELLULAR PAVEMENT

To reduce the problems associated with compaction of turf by pedestrian and vehicular traffic, PVC rings attached to a fiber mesh or precast unit pavers with an open-cell design can be placed directly on a prepared subgrade. (Figure 24–10). Excavate the area where the paving is to be placed. If the paved area is intended to support vehicular traffic, install a compacted granular base as described in the section introduction. Install cellular pavers level and flush with adjacent pavement. Backfill with topsoil to within 1/2 inch of the top of the pavers and apply turf seed over the surface.

Figure 24–10 Open-cell pavers with grass growing in cells.

Chapter 25

Dry-Laid Stone Paving

Stone paving combines craftsmanship with technique in creating one of the most aesthetic paved surfaces. This chapter covers the techniques of dry-laid stone paving using limestone, slate, and other regionally available flagstone materials. Techniques for base preparation and laying also support the installation of modular paving materials such as granite and cut stone. In addition to dry-laid base materials, stone paving can be effectively laid in a mortar base. Mortar base stone paving is covered in Chapter 26, while this chapter covers stone paving laid on a granular base. Also included at the end of this chapter are techniques for placing dry-laid stones informally as stepping stones.

RELATED INFORMATION IN OTHER CHAPTERS

Information provided in this chapter is supplemented by instructions provided elsewhere in this text. Before undertaking activities described in this chapter, read the related information in the following chapters:

- Safety in the workplace, Chapter 4
- Basic construction techniques, Chapter 7
- Construction staking, Chapter 8
- Site preparation for paving, Chapter 22

PLANNING THE PROJECT

Paving Patterns

Selection of a pavement pattern for stone pavement is based on the type of stone used for construction. Pavement patterns are generally identified as either modular patterns, for stone that is cut into square and/or rectangular shapes, or random patterns, for stone that is left in natural, uncut shapes. Examples of common paving patterns and the stone that works best in those patterns are as follows (Figure 25–1):

- Random-irregular. Pattern using irregular sized stone. Stone should be placed with small and large pieces evenly intermixed.
- Random-fitted. Pattern using irregular sized stone that is placed creating horizontal and vertical joints. Stone sizes are intermixed but joint arrangement is maintained.

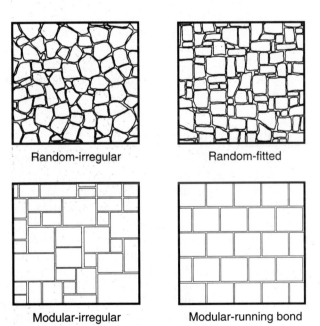

Random-irregular   Random-fitted

Modular-irregular   Modular-running bond

Figure 25–1 Common stone paving patterns.

- Modular-irregular. Pattern using stone of varying sizes with straight-cut edges. Sizes are typically in multiples of 2, 3, or 4 inches so stone can be fit with horizontal and vertical joints.
- Modular-running bond. Pattern using stone of same size with straight-cut edges. Stone is placed in staggered pattern similar to that found in brick installations.

BASE MATERIAL

Base preparation for most dry-laid stone paving installations is similar to the process described in Chapter 22. Alternatives to a granular base with setting bed are:

- Undisturbed soil. There is always a temptation to place stone directly on a soil base. This can be done if the base is undisturbed and the expectations for a long-term, level surface are not high. The amount of work in preparation and placement is somewhat less than that for a proper base preparation, but the time and labor required for long-term maintenance offset the short-term time savings. It is not recommended in any area where public liability or heavy traffic will be an issue.
- Placement on mortar base. In addition to being dry-laid, flagstone may be set on a bed of mortar. See Chapter 26 to review the situations and methods for mortared stone paving.
- Placement over concrete or asphalt. Concrete or asphalt will substitute for granular base material if the surface is level and stable. Surfaces with deep depressions or ridges should be removed and replaced with a new granular base. If paving over an existing slab, place a setting bed of 2 inches to accommodate base variations.

PLACEMENT OF SETTING BED

Because stone can be variable in thickness, it requires a thin setting bed of fine granular material to maintain a level surface. Depending on the region, this may be accomplished using concrete sand or stone dust. Each is placed over the base and screeded smooth. Setting beds should not be less than 1 inch or more than 2 inches in thickness. Maintaining the 1 inch course provides the most stability for an installation. Because the stones are resting directly on this material, it is also important that the setting bed be sloped in the proper direction and be set at the exact elevation below finish grade required. The proper elevation for the top of a setting bed is the finish elevation minus the average thickness of the stone.

Screed the setting bed using preinstalled edge restraint or temporary screeding rails. If using a paved edge for screeding, be certain the base screed is adjusted to accommodate the thickness of the stone. In installations where it is only important to maintain a positive slope, screed from stringlines stretched between grade stakes. This will create a surface with some minor variations but is a much faster way to prepare for paving.

Small amounts of setting bed may be screeded as the work progresses, or the entire area may be screeded before any stone is laid. If an edger has not been preplaced, use a stringline or paint line to mark the edge of the paved area.

If the surface of the setting bed has dried out before starting, wet the surface with a light spray from a hose attachment. Do not soak the surface, and use caution that the wetting process does not disturb the setting bed.

EDGE RESTRAINT

If the installation requires a permanent edge restraint such as a concrete curb, walk, or cut stone, this edge should be installed before placement and screeding of the setting bed. This minimizes disturbance of the paving after it has been placed and assures that the finish grade of the stone matches the edging material. Edge restraints are described in Chapter 22.

STONE PLACEMENT

Laying stone requires a great deal of placement and rearrangement of materials, particularly with random patterns. Lay a test pattern large enough to see how the stone fits together. Regardless of the pattern selected for the body of the paved surface, it is recommended that full stone be placed around the entire perimeter. This placement reduces stability problems that occur when small stone is placed adjacent to the edge.

Beginning Placement

The objective of laying stone is to maintain a tight and uniform pattern across the paved area. Stone should have joints of less than 1/2 inch, and adjacent stones should touch. Stone that is placed in random patterns is best laid in a fan shape. Begin in one corner of the paved area and work along the edges, then fill the space between (Figure 25–2). It is advisable to use a structure or corner as the straight edge(s) against which the first fan of stone is aligned. Outer edges may be moved slightly, or the stones trimmed to create a tight fit.

Figure 25–2 Proper placement of dry-laid stone. Note that surfaces are flush with surrounding stone and each stone makes contact with adjacent material.

Another method for placement is to begin at a structure or wall and work out one row at a time. Select pieces with straight edges to use in placement next to structures or other restraints. This method works well for irregular-shaped areas where minor variations from perfect vertical and horizontal alignment will not be noticed. Snapping a chalkline on the setting bed will aid in maintaining alignment for patterns with straight joints.

Stone Laying

Begin laying stone according to one of the placement methods previously described. Have a mag float handy to smooth out any variations in the setting bed. Tapping the stone with a rubber mallet will also help fit it into tight locations. Before placing a stone, examine the edges for bevel. Bevels should be placed down when setting stone, reducing open joints in the surface (Figure 25–3). When placing the stone, set it

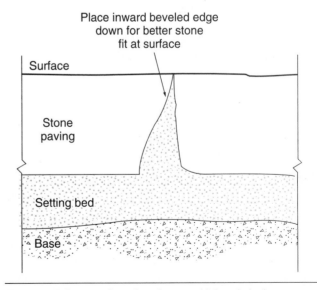

Figure 25–3 Proper orientation of stone with beveled edge.

Figure 25–4 Placement of dry-laid stone in pattern.

straight down onto the setting bed and twist it slightly to seat. Modular stone must fit tightly against any adjacent stones already in place.

Irregular-shaped stone should make contact with adjacent stones at a minimum of three points to provide stability. Random patterns should have minimal joints. While some joint space is expected, any stone that has fewer than three contact points or joints over 1 inch should be set aside, and a new stone should be placed (Figure 25–4). A flush surface must be maintained, requiring that stone height be adjusted as necessary by adding or removing setting bed material below the stone.

Pattern alignment should be checked often as the laying process moves along. For modular stone, verify that the alignment is straight and that the pattern is correct. Corrections are made by relaying stone or by gently tapping the sides with the rubber mallet. Continue laying stone up to the edger that is in place or to the stringline or paint marker at the edge. Modular stone can be cut to complete edges, while random

stone should be fit and/or trimmed to fill all voids up to the edging.

CUTTING AND FINISHING EDGES

Planning and selection of certain pavement patterns may reduce the number of cuts that have to be made, but anytime an irregular shape stone or structure is involved, a certain amount of cutting can be expected. To the greatest extent possible, avoid the use of small stone pieces along the pavement edges. Methods for marking and cutting stone for the project are described in Chapter 7.

FINISHING THE SURFACE

Following placement of all stone and edgers, the finishing steps can be applied to a paved surface. This process varies depending on the type of stone, but most require that material be swept into the joints of the surface.

Sweeping Joints

Filling joints between the stones helps hold them in position. This is accomplished by shoveling concrete sand onto the surface of the pavers and sweeping it into the joints. Sweeping in several different directions is most effective in working the sand into joints. Sand can be mixed at the rate of two parts sand to one part mortar mix to create a weak mortar for flagstone paving. Sweep the mixture into joints and lightly wash the surface with water. This mix hardens after the surface is washed and performs the role of a spacing material. Expect this weak mortar joint treatment to crack soon after placement, since it is not a full-strength mortar placed on a stable base. Repeat the sweeping operation twice more within a week on all types of surfaces.

SETTING STEPPING STONES

Placement of paving materials for informal paths can be accomplished by placing the material directly on an earth surface. Place stepping stones in locations where the path is desired. Arrange as necessary to accommodate a comfortable stepping pattern.

For a more stable installation, place the paving material and sprinkle basepath chalk around the edge of the paving material. Lift the paving material and excavate a shallow, level area at the location of the markings. The excavation should not be deeper than the thickness of the paving material and should extend slightly outside the markings. Replace the paving material and check for stability. If paving material rocks, lift and place a small amount of sand in low areas. Twist back into the excavation and recheck for level. Backfill around the material with soil to complete the installation.

Chapter 26

Mortared Paving

Mortared paving is a surface in which bricks, stone, tile, or other paving materials are set in a bed of mortar with mortar-filled joints. When improperly installed, this type of paving can create durability problems in areas of the country where temperatures fluctuate above and below freezing. With proper installation, applications are appropriate in many situations. Choice of materials for mortared pavement is almost as great as the choice for dry-laid pavement and unit pavement. Stone pavement, brick, and tile are common choices. This chapter provides the basic techniques required to set appropriate pavement types in a mortar base. Included are instructions on base preparation, mortar preparation and placement, placing, and finishing surfaces.

RELATED INFORMATION IN OTHER CHAPTERS

Information provided in this chapter is supplemented by instructions provided elsewhere in this text. Before undertaking activities described in this chapter, read the related information in the following chapters:

- Safety in the workplace, Chapter 4
- Basic construction techniques, Chapter 7
- Site preparation for paving, Chapter 22
- Concrete paving, Chapter 23

PLANNING THE PROJECT

Base Material

Base preparation for mortared pavement typically requires that a solid base be available on which to place a mortar setting bed. This base is typically a concrete slab because of the rigid nature of concrete. In some random stone applications, a compacted base similar to that described in Chapter 22 is prepared, and mortar is applied to that base. The possibilities of cracks developing in that type of surface are very high.

Base Preparation

If a concrete base is being used, prepare the surface in the manner described in Chapter 23, stopping after the screeding operation. The surface needs to be level and slightly rough so that the mortar will adhere to the base. Be certain when pouring a base for a slab or stairs that will have a mortared surface that the elevation is lowered to leave an allowance for the mortar bed thickness and pavement material thickness. See Chapter 22 if a compacted granular base is selected.

Paving Patterns

Any of the pavement patterns available for unit pavers can be accomplished as a mortared pattern. Examples of common paving patterns and the materials that work best in those patterns are located in Chapters 24 and 25.

Use of Half Brick

Stair locations may require application of half brick on mortar to maintain proper riser dimensions. Half bricks are paving bricks with full length and width dimensions, but half the thickness. This narrow paver can set on a mortar bed with minimal increase in stair riser dimensions.

Figure 26–1 Preparing for placement of mortared limestone.

PREPARATION FOR PAVING

Mortared paving requires preparation to avoid delays and problems when the actual paving begins. Practicing the patterns before performing an on-the-job installation speeds the work. Working small areas limits the amount of mortar that must be mixed. Observing weather conditions is also important. Unlike other unit paved surfaces, rain destroys an incomplete mortared project if the mortar has not set. A staging area for mixing mortar and preparing paving materials that is close to the project also speeds the construction process (Figure 26–1).

Test Layout of Paving

Test patterns by laying a test area large enough to see how the paving material fits together. On small projects and stairs, it is advisable to lay out the entire paved area and precut the material. This test layout will save time during placement of pavement and will avoid placement of cut materials in locations that are unstable or highly visible. When placing unit pavers, such as brick, in mortar, a soldier course of full pavers is recommended around the edge of the project. The

soldier course will reduce small pieces of paving material adjacent to the edge and improve durability of the surface.

CUTTING PAVEMENT MATERIALS

Methods for marking and cutting are discussed in Chapter 7. Stone and unit pavers that are set on mortar should be cut using a hydraulic splitter or wet masonry saw. Cleaving with a set may slow the process and leave uneven edges. Hand shaping of stone with a mason's hammer may be necessary to improve fit of irregularly shaped pieces.

MIXING AND PLACEMENT OF MORTAR

Mortar is used as the setting bed, which holds the paving material at the proper elevation. Mortar is also used between the paving materials to hold them in position and help create a surface impermeable to water. Mortar types vary from project to project, and may be custom mixed or purchased in premixed bags. When purchasing premixed mortar, type M or spec mix is typically used for flat, exterior applications. Table 26–1 identifies common mortar types and suggested mix ratios. When custom mixing, use a ratio of three parts Portland cement, one part hydrated lime, and nine parts sand. Small mortar quantities can be mixed in a clean wheelbarrow, in a wood box, or even on a sheet of plywood, but large jobs may require rental of a portable mixer. Blend the dry ingredients and mound in a pile using a shovel or mortar hoe. Make a depression in the top of the pile similar to the crater of a volcano. Slowly add water in the depression and mix, using a pushing and pulling motion at first, then a chopping motion while pulling the mortar forward and backward (Figure 26–2). Shovels of the mortar may be picked up and slapped down to break up clumps.

Mix the mortar in batches that are workable. When working with a small crew, several small batches are preferable to a large batch that will harden before it is used. Deliver the mortar to the work area on a rigid material. 18 × 18 inch squares of 3/4 inch plywood can be placed in convenient locations around the pavement surface and used for storing delivered mortar. If mortar begins to harden before it can be used, it can be tempered, or rewetted. Sprinkle a small amount of water over the mortar and remix with a trowel. Use a chopping motion, occasionally picking up small amounts and slapping them back on the board.

Laying mortared pavement requires working from a kneeling position on the base. Begin work in a corner that will allow exit without walking over newly laid material. Lay the pavement working backward, away

Table 26–1 Mortar Mixes

United States–ASTM C270

	Parts by Volume					
Mortar Type	**Portland Cement or Blended Cement**	**Masonry Cement Type M**	**S**	**N**	**Hydrated Lime or Lime Putty**	**Aggregate***
	1	—	—	1	—	4½ to 6
M	—	1	—	—	—	2 ¼ to 3
	1	—	—	—	¼	2 13/16 to 3¼
	½	—	—	1	—	3⅜ to 4½
S	—	—	1	—	—	2¼ to 3
	1	—	—	—	Over ¼ to ½	·
	—	—	—	1	—	2¼ to 3
N	1	—	—	—	Over ½ to 1¼	·
	—	—	—	1	—	2¼ to 3
O	1	—	—	—	Over 1/14 to 2½	·

Canada–CSA A179M

	Parts by Volume				
Mortar Type	**Portland Cement**	**Masonry Cement Type S**	**N**	**Hydrated Lime or Lime Putty**	**Aggregate***
S	—	—	1	—	2¼ to 3
	½	—	1	—	3½ to 4½
	1	—	—	½	3½ to 4½
N	—	—	1	—	2¼ to 3
	1	—	—	1	4½ to 6

*The total aggregate shall be equal to not less than 2 1/4 and not more than 3 times the sum of the volumes of the cement and lime used.

Notes: 1. Under both ASTM C270, Standard Specification for Mortar for Unit Masonry, and CSA A179, Mortar and Grout for Unit Masonry, aggregate is measured in a damp, loose condition and 1 cu ft. of masonry sand by damp, loose volume is considered equal to 80 lb. of dry sand (in SI units 1 cu m of damp, loose sand is considered equal to 1,280 kg of dry sand).

2. Mortar should not contain more than one air-entraining material.

Courtesy Portland Cement Association and the Canadian Standards Association. Canadian material presented with the permission of the Canadian Standards Association; material is reproduced from CSA Standard A179–94 (Mortar and Grout for Unit Masonry), which is copyrighted by CSA, 178 Rexdale Blvd., Etobicoke, Ontario M9W 1R3. Although use of this material has been authorized, CSA shall not be responsible for the manner in which the information is presented, nor for any interpretations thereof. This material may not be updated to reflect amendments made to the original content. For up-to-date information, contact CSA.

from the beginning point. Shovel or trowel the mortar onto the base in a layer over about 2 square feet of a corner of the work area (Figure 26–3). Spread the mortar in a rough layer over this area with a trowel or rake. This layer should be approximately 1/2 inch thick for brick and 3/4–1 inch thick for stone.

PAVEMENT PLACEMENT

Placement of paving material requires proper alignment and elevation for all materials. For materials that are placed in a modular pattern, it may be beneficial to utilize a stringline to maintain alignment. Random patterns typically have enough flexibility that aides are not required to maintain the pattern.

Lacking a structure or paved area to begin placement of pavement, lay a stringline across the area and begin the first row by placing stone lightly against the string. Setting a stringline slightly above the finish grade will help make certain the paving material is following the desired alignment and assist in maintaining the proper elevation (Figure 26–4). Problems may be encountered with modular patterns if the paved area covers a corner between walls that are not at right

Figure 26–2 Mixing mortar with mortar hoe.

Figure 26–3 Mortar placement.

angles. If this situation is encountered, cutting of pavers along the wall may be required.

Placement of pavement is similar for all materials. Placement requires that the paving units be set into the mortar bed and the joint filled separately. Before placing stone, examine the edges for bevel. Bevels should be placed down when setting stone, reducing the irregularity of the surface. Stone should be placed with joints of 1/2 inch–1 inch. Tile and brick should have uniform edges and a typical joint spacing of 1/2 inch. Push stone or tile into the mortar bed with a twisting motion and use the handle of the trowel to tap the material into alignment. Check for level and correct as necessary (Figure 26–5). Material that is significantly out of level or at an incorrect elevation may require removing it and adding or removing mortar. Replace the material and recheck for level and elevation.

Pattern alignment should be checked often as the laying process moves along. Verify that the alignment is straight and that the pattern is correct. Corrections

are made by relaying stone or by gently tapping the sides with the trowel.

JOINTING AND FINISHING THE SURFACE

After a few paving units have been set, the joints may be filled. Pick up a small amount of mortar on a trowel and, holding the side of the trowel at an angle next to the joint, push the mortar into the joint using a pointing tool. Joints can also be filled using a mortar bag filled with mortar that is squeezed into joints for finishing. Fill the joint and scrape away the excess with the trowel. Using a jointer or small trowel, press down on the mortar that was placed in the joints between paving units. This pressing will smooth the joint surface, press the mortar into voids in the joint, and compact the mortar surface (Figure 26–6). Refill with mortar and smooth all voids between edges of stone. Flat or concave joint surfaces are the best for horizontal paving.

Figure 26–4 Placement of stone using stringline to maintain proper elevation.

Figure 26–5 Leveling individual stones when placed.

Cleaning Surfaces

> **CAUTION**
>
> When cleaning pavement using acid and caustic cleaners, follow all safety precautions identified on the product label. Wear proper safety clothing. Contact and exposure to acids and caustic cleaners can cause personal injury and can harm plants.

After the mortar has hardened, the paved surface should be cleaned of any excess mortar and scrubbed with a solution of nine parts water and one part muriatic acid or trisodium phosphate (TSP). Surfaces should be rinsed thoroughly. Difficult stains or deposits can be removed using a wire brush or trowel.

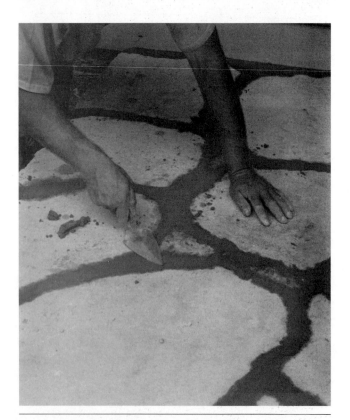

Figure 26–6 Finishing joints using mortar trowel.

Chapter 27

Granular Paving

Granular paving materials are typically not thought of as being an important paving surface, but there are several applications for this type of surfacing in today's landscape. Private patios, historical paving surfaces, and playground surfaces are common uses of loose, granular materials. Each of these paving surfaces requires consideration of project criteria to assure that the installation is durable and in accordance with safety and design goals. The following sections provide information specific to the applications of granular paving surfaces to the landscape.

RELATED INFORMATION IN OTHER CHAPTERS

Information provided in this chapter is supplemented by instructions provided elsewhere in this text. Before undertaking activities described in this chapter, read the related information in the following chapters:

- Safety in the workplace, Chapter 4
- Basic construction techniques, Chapter 7
- Site preparation for paving, Chapter 22

PLANNING THE PROJECT

As with all previous paving projects, advance planning improves the effectiveness and efficiency of the work operation.

Base Material

Base preparation for most granular paving installations is similar to the process described in the section introduction. Because the wearing course is resting directly on the base material, it is critical that the base course be sloped in the proper direction and be set at the exact elevation below the finish grade required.

Some alternative base treatments include:

- Undisturbed soil. There is a temptation to place granulars directly on a soil base. This can be done if the base is undisturbed and the expectations for a long-term level surface are not high. The amount of work in preparation and placement is somewhat less than that for a proper base preparation, but the instability of the surface and ongoing maintenance, especially refreshing of the surface, will offset the initial time savings. If using a very fine, granular wearing surface, there is the potential that the material will be "pumped" into the base by traffic. Pumping occurs in wet conditions when traffic pushes the granular material down and forces moisture and subsoil up. Eventually, the granular material will be completely embedded in the subsoil. Granular materials work their way into a soil base much faster than into a prepared base, requiring refurbishing of the surface sooner than if a prepared base is installed. Placement directly over soil is not recommended in any area where public liability or heavy traffic will be an issue.
- Concrete or asphalt base. Granular paving should *not* be placed over existing concrete or asphalt bases. Granulars will not bond with a solid paving base and will very quickly be eroded or worn away over such bases. Slipping of the granular material will compromise safety. The best solution is to remove existing pavement below where granulars are to be placed.

Landscape Fabric or Geotextile

Landscape fabric or geotextile may be placed on top of the base material and under the granular paving if desired. Installation of landscape fabric or geotextile reduces the loss of granular paving by segregating the base from the granular material. There may also be an initial reduction in weed growth with the use of a landscape fabric, but this benefit will lessen over time. Landscape fabric can create problems if areas of high wear expose the fabric. Once exposed, the fabric can create a tripping hazard and be pulled up, disrupting the surface of the paved area. If fabric is used, lay the fabric or geotextile flat over the entire base, trimming the material at the edge restraint (Figure 27–1). Overlap any joints by 18 inches.

EDGE RESTRAINT

All granular paving installations require edge restraint to prevent migration and settling at the outer edge of the paved area. The methods used to edge vary depending on the design of the project, but some choices are described in Chapter 22. Any edging selected must cover the entire perimeter of the project to prevent granular paving from washing through open joints between edging pieces.

Preplacement of Edging

Edging material should be preplaced to provide a screeding guide for wearing course material. Install all edging at the finish elevations desired around the entire area to be paved. Edging may be temporarily left out of an area where access to the site is necessary and replaced as the paving reaches that segment of the project.

PLACEMENT OF GRANULAR WEARING COURSE

Installation of a granular paving surface is much like the placement of concrete. Granulars are dry and moveable with most common landscape equipment and tools. A small crew may place a great deal of this material in a short time if the correct equipment is available.

Beginning Placement

Using the most efficient means available, deliver the granular material to the area to be paved. Provide an access point that will not disturb any of the preplaced edging. Begin placement at a point opposite the access point, and fill hard-to-reach corners first (Figure 27–2). Spread the granular material in a layer that is slightly thicker than the desired grade.

Figure 27–2 Placement of granular paving material.

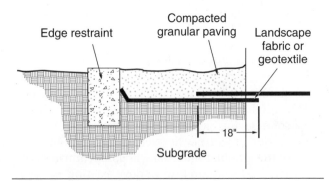

Figure 27–1 Proper placement of landscape fabric below granular paving.

Leveling the Wearing Course

To develop a level draining surface, the placed material should be screeded. Select a long, straight 2 × 4, and nail pieces of lath to the bottom of the screed at locations where it will rest on edging. This will leave the screeded surface slightly higher than the surrounding paved areas to allow for compaction. Screed the paved area utilizing the preplaced edging material as a screed support. To provide precise screeding of the open areas in the center, utilize temporary screed rails, as described in Chapter 23. Unlike concrete applications, stakes used to support the temporary screeding rails should be removed and the course around the disturbed area repaired by raking material into the opening.

FINISHING AND COMPACTING THE SURFACE

Following placement and screeding of all granular material, the finishing steps can be applied to a paved surface. This process varies depending on the type of stone, but most require that the material be raked and the surface compacted.

Raking

Adjust the surface elevation of the granular material by lightly raking with a garden rake. Most granular materials require continual maintenance of this type to maintain worn areas. Loose granulars should be raked along the adjacent existing pavement to assure that the two surfaces are flush.

Compacting

Fine granular materials should be compacted using a sod roller to create a tight, drainable surface (Figure 27–3). The surface should be dry to prevent material sticking to the roller drum while compacting. Work from the low end to the high end, rolling and adjusting the level as required. If a high area is encountered, use the teeth of a garden rake to loosen the surface, and the top of the rake to drag material out. Low areas should be loosened with a rake and material sprinkled over the depression. Recompact all adjusted areas. Most loose granular materials benefit from a light rolling; however, some materials may well up under the roller and prevent even compaction. The materials that do not accept rolling should be hand tamped.

Figure 27–3 Leveling and finishing granular paving installation.

Specialty paving of outdoor use areas is a basic skill required of most landscape contractors. Section 6 covered the selection of materials and the preparation of base for most paving types. Specific instructions were provided for installation of concrete paving, including the forming for flat work and stairs, unit paving, dry-laid stone paving, mortared paving, and granular paving.

When choosing a paving material for a project, consideration should be given to the durability, safety, maintenance, aesthetics, cost, and installation characteristics of each material available. With the variety of paving materials to choose from, a selection that meets the demands of the project should be possible. One of the more common paving materials for landscape projects is concrete. Whether left "plain" or treated with one of several surface textures and colors, concrete is a durable choice. Unit pavers are also widely used for landscape applications. Interlocking concrete paving block, asphalt pavers, brick, cut stone, adobe, and pavers from recycled materials create a look of craftsmanship. Compacted turf areas can be treated using open-cell pavers, which have openings filled with soil and seed. An aesthetically pleasing surface can also be obtained using a variety of locally available stone paving materials. Many of the unit materials and stone can be given added stability by placing the material in a bed of mortar with mortar joints. Stone and other paving materials can be used for stepping stones. Granular paving is an inexpensive way to cover flat walkways and outdoor use areas.

Preparing for paving requires that a stable subgrade be developed. In most situations, this will involve excavating to the proper depth and examining the subgrade for any moisture or soil problems. If any unstable areas or problems with drainage are encountered, these must be addressed before paving base can be installed. Any compaction and/or utility work required below the paved area should also be completed before base installation proceeds. Undisturbed earth provides the most stable subgrade, and fill areas require special compaction techniques to avoid settling problems. Base material is then placed over the subgrade and compacted in preparation for the various types of paving. If edge restraint is required for a paving material, verify whether the restraint should be placed before or after paving installation.

Concrete paving installation requires extensive preparation prior to the actual paving activities. Preparation of forms, leveling of base material, installation of reinforcement, and planning joints must all be completed before concrete arrives. Form materials are typically dimensioned lumber for straight sections and flexible masonite for curved edges. Forms must be securely anchored with staking and backfill to prevent failure during the pour. Stairs can be formed in open areas by building forms for sides and risers, and in enclosed areas by suspending riser forms from beams running over the top of the stairs. For all forms, check their stability carefully before concrete arrives to reduce the chance for failure.

Concrete pours begin with the placement of concrete directly from the truck or transported by equipment to remote areas. The material is leveled initially using a hand or mechanical screed. Smoothing of the surface with a bull float and hand float follows. In locations where a very smooth surface is desired, an additional floating with a steel trowel follows. Each of these steps is intended to embed the aggregate into the slab and bring the fine materials to the surface. Before the slab hardens, special surface finishes are applied and the surface is edged and jointed. The new pour is sprayed with a curing compound, or covered for 3 to 7 days for curing. Forms are removed after 2 to 4 days. If control joints are to be cut with a saw, it is done at this time.

Installation of unit pavers requires the installation of a setting bed over the base course. Composed of sand, this course is placed approximately 1 inch thick and screeded to obtain a level surface that mirrors the desired finish surface. One of the variety of patterns is selected for the project and pavers are placed beginning with a T or horizontal pattern, working out from a wall or straight edge. Pavers are individually placed across the entire area to be paved and, if not already in place, edge restraint is installed. Working with unit pavers requires special edging to hold the pavers in place. Many choices are available, based on aesthetics and requirements of the paving installation. Unit pavers require cutting and placing of partial pavers to fill voids along edges and around corners. Pavers are seated into the setting bed using a plate compactor and the joints swept with dry sand to lock the pavers into place.

Dry-laid stone paving is installed in a manner similar to that of unit pavers, with the exception that some stone choices have no formal pattern to guide placement. A setting bed is installed and screeded, and individual stones are placed beginning in a corner or along a straight edge. Random stone must be carefully selected for fit, while cut stone may be placed in patterns similar to those of unit pavers, or special patterns developed by the designer. After placement, edge restraint is placed and joints are swept with fine granular material. Stepping stones are marked and the base excavated and leveled. Steppers are then replaced and leveled.

Stone and most unit pavers can be installed on a mortar bed with mortared joints if the design requires. For a more durable installation, a concrete base is recommended in place of a granular base. Patterns for installation are the same as those used with unit pavers and dry-laid stone, with the exception that allowance must be made for mortar joints between paving units. After the base has been installed, mortar is mixed and placed in a thick layer over the base. Paving materials are set in the mortar bed and adjusted to the correct elevation using a trowel. Placement continues with the installer backing across the base while laying pavement. Edges and corners require cutting of materials, but edge restraint is not required. When the setting bed has begun to harden, the joints are filled using a masons' bag or a trowel and pointing tool. Joints are smoothed with a joining tool. When all mortar is set, the surface can be cleaned with a weak acid mixture if necessary.

Granular pavement can be installed with or without base material. Consideration may be given to placing landscape fabric under the granular material to reduce weeds and prevent the paving material from being pumped into the base. Edge restraint is required and must be placed before granular material is installed. Place the paving material over the base, and screed or level with a rake to the appropriate depth across the entire paved area. Compacting the surface is required for most materials.

Section 7

Wood Landscape Structures

INTRODUCTION

Wood structures provide practical, useable space in the landscape. Whether the structure is a multi-level deck or a simple trellis, the presence of the structure adds warmth and interest to the design. When structures are incorporated into a landscape concept, they instantly change the dynamics of the design. Compared to lawns and spaces defined by plant material, structures are very precise in their definition of space. Because of their "instant" nature, they tend to dominate the vistas of the landscape until plant material matures. Structures are also labor- and cost-intensive aspects of the design.

Construction of wood elements for the landscape requires a level of craftsmanship not typically associated with the landscape industry. Because of the dominant characteristics of a structure, the quality of the craftsmanship is obvious to the client and visitor. Many think of outdoor construction as being rough and approximate, yet the building of decks, gazebos, or other landscape elements requires as precise a level of worksmanship as any structure.

This section approaches wood structures and carpentry by describing the construction techniques for the basic components of typical structures rather than describing how to build each specific structure. The information in this section addresses the construction of foundations and platforms for decks, adding railings, stairs, seating, and skirting to a structure, and building structure roofs. This approach allows the carpenter to mix and match components required to build a variety of structures. This section also introduces the basic types of structures, and describes the variety of materials used in exterior carpentry. The text does not attempt to cover every aspect of carpentry nor every possible design available to the landscape contractor. The approach uses use basic construction techniques to help the contractor become familiar with building with wood and to provide a foundation for exploring advanced construction methods. Enclosed porches with footings that are integral with a building are not covered in this text because of the structural and building code implications related to their construction.

TYPES OF STRUCTURES

Landscape structures can be loosely classified as deck structures, roofed structures, or freestanding structures. While this may be an oversimplification of structure classification, most wood landscape elements include one or more of these characteristics. An expanded definition of each follows.

Decks

A deck is a surfaced, structural platform that can serve as a landing for an entryway, an outdoor use area, or walkways between points in a design. Many decks are constructed to accommodate a severe change in grade between use points, but some are constructed simply because the client prefers the use of wood for outdoor use areas. (Figure SI–2). Decks consist of a foundation that is supported on an existing building or by a structural system anchored in the earth, a framework of structural lumber, and surfacing on which one can safely walk. Further components often associated with decks include railings for safety, stairs for access, and possible seating for convenience. Decks can be combined easily with various types of roofing and fencing to create even more useful and interesting outdoor use areas.

Arbors, Shade Structures, and Trellises

Arbors are open-roofed structures that are typically used to accent entry points in the landscape. Composed of posts that support a framework of open roofing materials, arbors are one of the simplest structural ways to introduce overhead enclosure into the landscape. The roof structure is not intended to be weatherproof and can be composed of any number of creative patterns or materials (Figure SI–3).The designer may consider adding vining plant material as a natural enhancement of the arbor landscape structure. Designs may have the arbor arcing over a walkway or enclosed on one or both sides along a walkway.

Shade structures are similar to arbors in construction, but are typically used to reduce light intensity over patios, decks, or other outdoor use areas. Clearance under shade structures (also called sunscreens) and placement for maximum shading potential may alter the construction slightly from that of arbors, but the use of creative patterns and plant material as part of the structure is the same. A shade structure most often has two or three sides open to the landscape, but partial enclosure of the sides is sometimes introduced to restrict further access by sun. Trellises are wall-mounted or overhead structures with an open framework (Figure SI–4). Many landscapes use a trellis to support vining plant growth.

Gazebos and Freestanding Structures

Gazebos and other freestanding structures are closed-roof structures typically used as an outdoor use facility and focal point (Figure SI–5). Components of the typical gazebo include flooring created by a deck or concrete platform, railings and/or sidewalls, support posts for the roof, and an enclosed, weatherproof roof. Gazebos are built in a variety of sizes and shapes and

Figure SI–2 Wood decks create functional outdoor spaces.

Figure SI–3 An open-roofed landscape arbor placed in a prairie setting.

Figure SI–5 Gazebos add form and interest to a landscape.

Figure SI–4 Landscape trellis supporting grapevines.

can have enclosed sides to further weatherproof the structure. Use of gazebos to articulate historical landscape themes has led to the introduction of intricate details and ornamental embellishments of gazebos.

PRODUCTIVITY SUGGESTIONS

Suggestions for making work more productive include:

- Double-check measurements to avoid cutting twice or wasting lumber.
- Square the ends of boards before measuring. Factory cuts are often not square.
- Mark dimensions with a V, where the point of the V is the desired dimension.
- To mark the location of pilot holes for bolts or connectors that are already in place, hold the piece of lumber in correct position and tap slightly with a hammer opposite the bolt location. This leaves a small indentation at the exact location for pilot holes.
- To mark the center of a square or rectangular piece of lumber, draw diagonal lines from opposite corners. The lines cross at the center of the square or rectangle.

- Tie the ends of power cords in a loose knot to prevent annoying unplugging while moving around a construction site. Cords can then be moved by "whipping" them over piles rather than hand moving.
- Rent a power auger if the project requires digging several holes.
- Make sure framing is square before attempting to build any roof. A small error from square will amplify in the roof structure.
- To keep rafters level, install both end rafters and run a stringline from one to the other that rests on the top side of the rafters. New rafters should not push the stringline up or have a gap between the rafter and string. Use stringline at both the top and bottom.
- To determine the trim angle for stair posts, place a board on top of two or more stairs next to a post. Place a speed square between the board and the post and set the angle.

- Establish a consistent procedure for marking that takes into account the width of the saw blade.
- The elevation of interior floors can be located by making measurements down from windows and doors. Establish elevations on the outside of a building by measuring down on the exterior side of the same windows or doors.
- Go to the basement and check the perimeter of the structure in the area where work will be taking place. This may aid in locating rim joists and spotting potential problems such as utilities running around the perimeter of the interior.
- Premanufacture sections of railing on a flat surface, then position in place and fasten.
- In locations where holding the lumber to start a nail or screw is difficult, start the nail or screw on a flat surface and hold into position to connect. Marking the position of the lumber before connecting also makes work easier.

Chapter 28

Materials for Exterior Carpentry

Many material choices are required when performing landscape carpentry tasks. Selection choices include wood products, connectors, fasteners and hardware types, finish choices, and a wide variety of premanufactured materials made from wood, metal, and vinyls. With the rapidly changing nature of carpentry products and construction methods, it is important to visit the lumberyard often to explore for new products.

WOOD PRODUCTS

Wood Types for Exterior Use

Selecting the proper wood for a project requires consideration of the characteristics of wood products that are available and the proper application of the product. Primary choices for wood products used in exterior carpentry include treated or naturally decay-resistant hardwoods and softwoods milled into boards or dimensioned lumber. In certain applications, plywoods and other engineered panels may also be used.

The terms "hardwood" and "softwood" refer to the source tree type used in the milling and manufacture of lumber. Softwoods are milled from coniferous, or cone-bearing, trees such as pine, spruce, and fir, and are the dominant wood used in the framing and structural segments of construction. Hardwoods are milled from deciduous, or broad-leafed, plants such as oak, maple, and walnut, and are predominantly used in cabinetry and finish carpentry. Engineered panels such as plywood and sheathing can be manufactured from either hardwoods or softwoods.

Treated Woods. Treated woods include a variety of southern yellow pine, firs, and spruces that are treated by one of several methods to resist decay and insects.

See the information on lumber treatment later in this chapter.

> **CAUTION**
>
> Avoid prolonged skin contact with treated lumbers. Working with treated lumbers requires wearing gloves and long-sleeve shirts to prevent exposure to chemicals used in treatment. Wear a dust mask when cutting treated lumbers. Wash hands thoroughly after contact with treated lumbers. Wash clothes worn when handling treated lumbers separately from other clothes. Do not let treated lumbers come in contact with food. Do not burn treated lumbers.

Treated lumber is an acceptable choice for structural members and any part of the construction that will not be visible. Most treated woods are durable and strong enough to use for posts, beams, joists, and decking materials. Use of treated lumbers for finished components is acceptable, provided potential contact between lumber and skin is limited and the client does not mind the colored cast left from many treatment methods. This discoloration can remain for several years, but eventually most treated lumber will weather to a light gray.

Western Red Cedar. Cedar heartwood is a soft, naturally decay-resistant lumber that makes excellent surfacing and trim material. Cedar is usually *not* cut in structural dimensions (2×8s, 2×10s . . .), but is available as $1 \times$ dimensioned lumber and deck surfacing in a form called 5/4s, which has a nominal, or actual, thickness of 1 inch and various widths and lengths. The edges of 5/4s cedar are eased, or rounded, to reduce splintering. Cedar deck surfacing is soft enough that it can be installed with limited predrilling. Cedar can be left to weather out to a medium gray or stained as desired. Rough-sawn cedars are often used as trims for exterior work.

Redwood. Redwood heartwood is a soft, naturally decay-resistant lumber that is excellent for decking and trim material. Available in limited quantities, all heart redwood can also be used for structural members in exterior conditions. Redwood deck surfacing is soft enough to work with without drilling pilot holes. Without finishing, redwood weathers to a warm, dark gray.

Plywood and Wood Panels. Plywood is an engineered lumber product that utilizes thin layers, or plies, of wood glued together to create a sheet of wood. Plywood has applications in exterior carpentry such as sheathing for roofs and siding for structures. Typically, plywood is available in sheets with 4 foot × 8 foot dimensions, and can be cut and nailed like boards. Most plywoods require staining or painting to improve aesthetics and protect and preserve the material. Also included in the wood panel classification is a variety of exterior siding treatments that are manufactured as surfaced and textured panels.

Selection of Appropriate Materials

When a project is performed under the direction of construction documents, the material choices will be made by the design professional. When the decision is the responsibility of the contractor, factors that influence the material choices range from how the product will be used to ecological considerations. Primary considerations in selection include a wood's strength and resistance to decay and insects. Additional considerations include workability, availability, and aesthetics. The following sections and Table 28–1 discuss several of the common selection criteria.

Strength and Related Structural Properties. Selecting lumber types and dimensions that will support the loads expected on a structure requires an understanding of the structural properties of lumber. Natural properties make most lumbers highly resistant to breaking when exposed to weight pushing down on the member, or compression forces. Working opposite the compression forces are tension forces, or forces attempting to pull the wood apart. The resistance to failure from these forces is a wood's strength. Woods such as southern yellow pine and fir are high in strength, while cedar and redwood are considered low in strength. Strength is typically measured in pounds of weight that can be placed on every square inch of

the board. Wood also has a structural quality of stiffness, or elasticity, which plays a part in how much "bounce" a deck platform may exhibit. Southern yellow pine and fir are moderate in elasticity, while cedar and redwood are very flexible.

Each of these structural properties can be calculated if factors such as type of lumber, orientation of the wood fibers, thickness of the member, and spacing between supports are provided. Other factors that enter into strength calculations are live and dead loading. Live load is the weight a piece of lumber will have placed on top of it, while dead load is the weight of the piece of lumber itself. Both live and dead loads need to be determined when calculating the strength of lumber. The complexity of these variables underscores the importance of having a qualified design professional determine the types and sizes of wood members used in construction.

> **CAUTION**
> To prevent structural failure, calculations for wood members should be prepared by a registered design professional.

Lumber Grades. Grading is a process utilized by wood processors to classify the strength, source, moisture content, and related aspects of lumber. In the United States, associations such as West Coast Lumber Association, Western Wood Products, Redwood Inspection Service, and Southern Pine Inspection, grade the products harvested in their geographic region.

Grading criteria typically include an indication of the following wood characteristics:

- Grade indicating bending strength ratio. Higher grades are rated for more stress and heavier loads. Terms such as "select" and "No. 1" indicate higher grades. Larger numbers, such as "No. 2," indicate lower grades. To accurately compare stress rating nomenclature, each association's standards need to be reviewed.
- Moisture content. Lumber that has been dried to less than 19% moisture content is stamped "Dry." Surfacing of the lumber may take place before the lumber has dried, indicated by an "S-Grn" label, or after the lumber has been dried, indicated by an "S-Dry" label. Moisture contents below 19% are important in reducing problems with insects and lumber warping.
- Mill. The location of milling is indicated.
- Grading association. Initials or name of grading association is included on the stamp.

Table 28–1 Softwoods

Kind	Color	Grain	Hardness	Strength	Workability	Elasticity	Decay Resistance	Uses	Other
Red cedar	Dark reddish brown	Close medium	Soft	Low	Easy	Poor	Very high	Exterior	Cedar odor
Cypress	Orange tan	Close medium	Soft to medium	Medium	Medium	Medium	Very high	Exterior	
Fir	Yellow to orange brown	Close coarse	Medium to hard	High	Hard	Medium	Medium	Framing, millwork, plywood	
Ponderosa pine	White with brown grain	Close coarse	Medium	Medium	Medium	Poor	Low	Millwork, trim	Pine odor
Sugar pine	Creamy white	Close fine	Soft	Low	Easy	Poor	Low	Pattern-making, millwork	Large clear pieces
Western white pine	Brownish white	Close medium	Soft to medium	Low	Medium	Poor	Low	Millwork, trim	
Southern yellow pine	Yellow brown	Close coarse	Soft to hard	High	Hard	Medium	Medium	Framing, plywood	Much pitch
Redwood	Reddish Brown	Close medium	Soft	Low	Easy	Poor	Very high	Exterior	Light sapwood
Spruce	Cream to tan	Close medium	Medium	Medium	Medium	Poor	Low	Siding, subflooring	Spruce odor

- Species. If not inferred by the association name, the species may be indicated.
- Method of rating. Machine-rated lumber may also have the strength stamped on the board.

Grades are typically stamped on the lumber piece. Typical grades for exterior carpentry include southern pine No. 2 structural grade for structural work, and southern pine appearance or finish grade for visible surfaces such as trim. Construction heart redwood is used for structural work and clear heart is used for visible surfaces.

Plywood Grades. Plywood used in exposure to moisture should be classified as *Exterior* grade by the American Plywood Association (APA). Strength of panels is identified by the span rating stamped on the sheet. Typical plywood panels used in exterior carpentry include sheathing-rated plywood and composite sheets.

Resistance to Decay and Insects. A vast majority of the landscape contractor's work is exposed to exterior conditions and moisture. This necessitates the use of materials that are either naturally resistant to decay and insects or are treated to obtain the same properties of resistance. Woods such as cedar and redwood are naturally resistant, and as long as the heartwood from these species is used, there is no need to further treat these lumbers. Most remaining softwoods are susceptible to decay and insects if not treated by one of several methods and materials. Each treatment method is utilized to reduce potential damage from insects and decay, and most treatments are registered pesticides, restricting their use and application. Look for **treated woods** indicating "Ground Contact" to assure use in exterior structural projects. See the cautions listed under the Treated Wood section earlier in the chapter before using chemical treatments or treated lumber. Treatment methods for decay resistance include:

- Pentachloraphenol, also termed Penta. Penta is an oil-borne registered pesticide product. It is injected under pressure into poles and structural members.

- ACQ®. ACQ is a mixture of alkaline copper and quaternary. ACQ is less toxic to the environment than other chemical preservation techniques.
- Inorganic arsenicals. Arsenicals are water-borne, registered pesticide treatments derived from different copper and zinc compounds. Inorganic arsenic treatments are used on deck and structural components as well as certain plywoods.
- Creosote. Creosotes are oil-borne, registered pesticides used to pressure-treat woods. Creosotes are tar-based compounds used to treat posts and foundational wood members.

Most of the these treatment materials can be applied to lumber using the following methods:

- Pressure treatment. Pressure treatment utilizes special chambers to place lumber under high atmospheric pressures. This forces preservatives deeper into the wood than other methods of treatment achieve.
- Nonpressure treatments. Nonpressure treatments expose the surfaces of the wood to preservatives under normal atmospheric pressures. Nonpressure treatments cover the surface of the lumber, but the penetration into the wood is limited. Methods included in this category include brushing, dipping, and soaking. Other processes, such as thermal treatment and vacuum processes, are combined to improve the effectiveness of nonpressure methods.

When treated wood is cut for use in a project, retreatment of the cut ends is necessary. Selection of a hand-applicable or dippable treatment method is the most efficient method for retreating.

Aesthetic Appearance. The aesthetic appearance of wood products can be referred to, in part, by the characteristic of color. Generally, color can be further described as the initial color of the lumber when installed and the weathered color of the lumber after extended exposure to the elements. Exterior woods have distinctive initial colors, with most weathering to various shades of gray if left unfinished. Cedar is initially a light tan and redwood is a reddish brown. Treated woods begin with either a light green or brown cast.

Plywood appearance is graded using the letters A, B, C, and D, corresponding to its surface veneer quality, with A being the best and D being the poorest. Table 28–2 describes the letter grades.

Table 28–2 Plywood Veneer Grades

Veneer Grade	Description of Surface Quality
A	Smooth, paintable. Not more than eighteen neatly made repairs.
B	Solid surface. Repairs and tight knots of up to 1 inch across. Some minor splits permitted.
C	Splits of up to 1/8 inch, defects limited to 1/4 inch × 1/2 inch. Plugged. Broken grain and wood/synthetic repairs permitted.
C	Tight knots up to 1-1/2 inches with limits. Defects that do not impair strength.
D	Knots and knotholes up to 2-1/2 inches. Limited exposure.

Cost and Availability. Selection of a wood product is often dictated by which material is available at a reasonable cost. Influences such as markets, transportation costs, and production schedules are beyond the control of the contractor and may necessitate use of whatever acceptable material is available. The factors of cost and availability are typically interrelated. The more available a product, the lower the cost. Products that are in short supply tend to be more expensive.

Many of the woods that are treated, such as southern yellow pine and fir, are harvested from large stands and are typically available in greater quantities than woods such as redwood and cedar. Dwindling supplies of redwood have also interjected the ecological issue into material selection, causing some design professionals to select alternatives, particularly for structural components, in order to reduce the demand for harvesting redwood.

Quality of Wood Products

Quality of the products being used has a direct relationship to the quality of the product produced by the carpenter. Awareness of common wood defects improves selection and installation of wood products. Many of these defects are cause for rejection of a piece of lumber. Some, such as cupping, can be overcome by reorienting the lumber if the defect is not severe. Others require trimming the defective portion from the lumber to make it useable. See Figure 28–1 for illustrations of common wood defects.

Warping. Twisting and deforming of the board shape caused by drying too quickly and improper storage (among other things) is referred to as warping. Several types of warps can be present in lumber, including:
- Bow, or curving of the ends of the lumber.

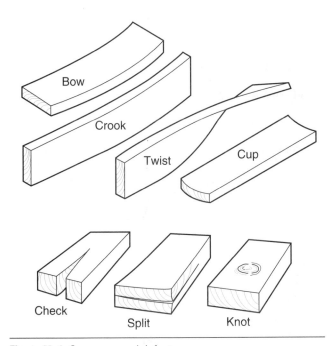

Figure 28–1 Common wood defects.

- Crook, a curving of the edge of the lumber along its length.
- Twist (or wind), an end-to-end twisting of the lumber.
- Cup, or curving of the edges of the lumber from side to side.

Checks. Splits in the end of the lumber.

Shakes. Cracks between and parallel to annual rings.

Knots. Branch cross sections. Knots are a problem if the defect is large and/or loose.

Lumber Dimensioning

One source of confusion when working with wood is the use of nominal terminology for lumber dimensions. When a 2 × 4 is first cut, its nominal dimensions are actually 2 inches thick by 4 inches wide. After drying and planing, boards are reduced to their actual thickness and width. Using the **nominal dimensions** rather than the actual dimensions when measuring and cutting will cause errors. To avoid this problem, memorize and use the dimensions listed in Table 28–3. Length is not affected by this same issue, however it is recommended to verify the length of any board before installing.

CONNECTORS, FASTENERS, AND HARDWARE

Connectors and construction hardware present an almost endless variety of ways to attach and support

Table 28–3 Nominal and Actual Dimensions for Stick Lumber

Nominal Dimension	Actual Dimension
1 inch	3/4 inch
2 inches	1-1/2 inches
4 inches	3-1/2 inches
6 inches	5-1/2 inches
8 inches	7-1/4 inches
10 inches	9-1/4 inches
12 inches	11-1/4 inches

Read Example: 2 × 12 actual dimensions are 1-1/2 inches × 11-1/4 inches.

landscape projects. Specific selection depends on the application situation that a project presents.

Connectors, fasteners, and related hardware for exterior projects should be made of materials that resist rust naturally or are treated to resist rust. **Galvanized** materials are made of steel that has been dipped in a zinc coating to reduce the chance of rusting in an application exposed to moisture. Galvanized materials are suitable for use with any type of treated lumber, but may stain if the galvanized coating is damaged. Polymer-coated fasteners are made of steel that has been coated with a thin polymer formulated to resist rust. Polymer-coated fasteners resist rust and ease installation due to their slick coating. Select stainless steel connectors when working with cedar or redwood for a finish free of stains.

Fasteners

Fasteners include the variety of nails, screws, and bolts used to fasten construction components together. Terms used when discussing fasteners include the head, or the flattened portion used to drive the connector into the wood, and the shank, or the long shaft of the fastener. Nails have either a smooth shank or a deformed shank, and screws and bolts have a threaded shaft.

Nails. Fastening pieces of lumber together can be done using a variety of connectors. The most common fastener used in exterior applications is the nail. Nails are pieces of hardened wire of varying diameters that are cut to length and flattened on one end to facilitate driving into the wood. Nail sizes are determined using a measurement called "penny" and symbolized using a number and the small letter "d" (a sixteen-penny nail is labeled 16d). Several types of nails are available for use in specialty situations (Figure 28–2). Common nails are used for general purpose applications. Truss nails, or **joist hanger** nails, are short, stubby nails used

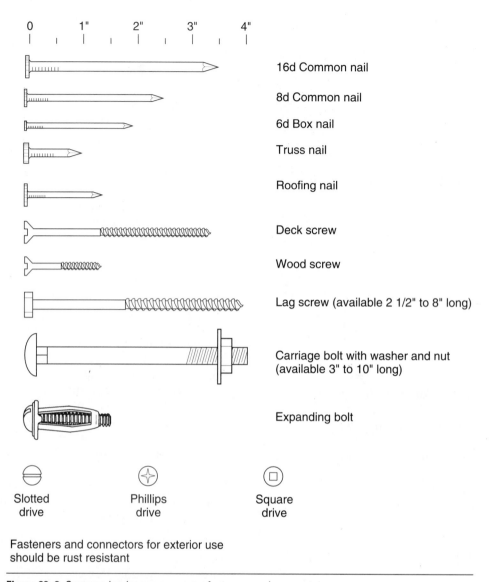

Figure 28–2 Common landscape carpentry fasteners and connectors.

to connect galvanized hangers. Box nails are thin nails with flat heads used for nailing trim. Roofing nails have large diameter heads to reduce tearing of roofing materials.

Wood Screws. Screws are similar in length to nails but the shanks are tapered and threaded at one end so they can be twisted into the materials to form a stronger hold. Recommended for fastening deck surfacing and most trim, the screw provides stronger holding characteristics when compared to the nail (Figure 28–2). Most common among the screws used in exterior applications are the deck screw, which is similar in length to the 16d nail. Deck screws can be driven with hand or power screwdrivers using **slotted, phillips,** or **square-headed** drives (Figure 28–2). Pilot holes the diameter of the shank should be used for installing screws into most woods. Screws should

always be torqued, or twisted, into the wood and never driven using a hammer.

Lag Screws. The lag screw is a heavier and longer screw used for anchoring structural members. Lag screws come in a variety of shank diameters and are available in lengths from 2 inches to 8 inches and larger (Figure 28–2). Lag screws require pilot holes that are drilled to the same diameter of the shank. To determine the proper diameter of a lag screw pilot hole, hold a drill bit in front of the lag screw. The proper diameter drill bit will cover the solid portion of the shank but the threads will be visible. Lag screws are always torqued in place using wrenches and never driven using a hammer.

Carriage Bolts. Carriage bolts are large-diameter bolts with a rounded head used for connecting structural pieces. The strength of the carriage bolt far exceeds the strength of the other connectors discussed previously. Carriage bolts come in diameters from 1/4 inch to 3/4 inch and in lengths of 3 inches to 10 inches and larger (Figure 28–2). Carriage bolts require a nut and washer at the end opposite the head to complete the connection. A washer is not required under the head. Carriage bolts require a hole drilled through all pieces of lumber being connected the same diameter as the bolt, and must be connected using a box or open-end wrench.

Specialty Bolts. Many types of specialty bolts are available for applications in appropriate situations. One common bolt used in landscape applications is the **expanding bolt** (Figure 28–2). When connecting to brick, block, concrete, or other hard material, expanding bolts are inserted into a predrilled pilot hole. Expanding bolts have a flange that enlarges when the bolt is tightened. The flange pushes against the side of the pilot hole and holds the bolt in place.

Specialty Connectors and Hardware

To aid in the connection of posts to footings, posts to beams, and other structural connections, a wide selection of galvanized hardware is available (Figure 28–3). Post supports can be set into a concrete footing to support the post for a structure. Post supports hold the end of the post above the footing to reduce the potential for contact with moisture. Joist hangers are U-shaped metal pieces that aid in connecting joists to **ledgers** or end joists. These hangers improve the strength of joist/beam connections and make the work easier. Joist hangers can be purchased at standard angles or bent to fit around corners and at edges of structures. Saddle connectors are available for joining 4 × 4 lumber.

Numerous galvanized metal fasteners are available to make or supplement wood connections. Straps, Ts, stair supports, angle braces, and post caps are all examples of metal connectors that can be used to make connections that are difficult to nail, screw, or bolt. Most connectors can be installed using galvanized nails or rust-resistant screws whose length is two-thirds the thickness of the wood piece being connected. When nailing a connector, hold the piece in proper position and, with the same hand, hold a nail that can be driven to "hang" the piece. Drive nails into the remaining opening to stabilize the connector. When using screws, hold the connector in position and mark the openings. Remove the pieces and drill pilot holes for each screw location. Replace the piece and install the screws to stabilize the connector. Specialty connectors achieve rated strength only when all fasteners have been installed. Nailing in only half the

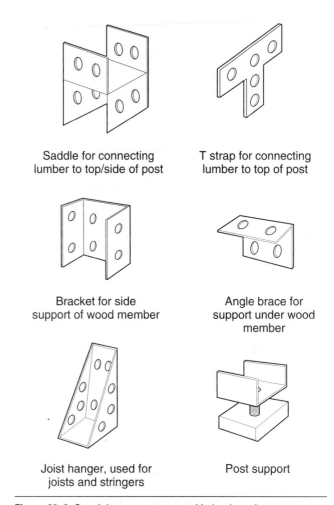

Saddle for connecting lumber to top/side of post

T strap for connecting lumber to top of post

Bracket for side support of wood member

Angle brace for support under wood member

Joist hanger, used for joists and stringers

Post support

Figure 28–3 Specialty connectors used in landscaping.

openings provided or improper installation will reduce the strength of the connection.

EXTERIOR FINISHES FOR WOOD

Wood used in exterior conditions faces a variety of weathering conditions that will age the materials, change their color, and possibly begin the process of decay. A choice should be made during the design phase of the project if the wood selected for construction is to be treated to deter the effects of weathering or if the lumber will be allowed to age naturally. A common misconception is that woods suitable for exterior use must be treated to avoid decay. If a wood is properly selected and prepared for exterior use, decay should not be a problem. The only reason to finish-treat such woods is to maintain the new look of the lumber or to change the wood's natural color. Once finish treating is selected for exterior woods, it should continue for the life of the project. Wood finished and then allowed to weather will become unattractive as the natural, untreated finish blends with the remaining treated finish.

All exterior use woods can be allowed to weather to their natural color without doing structural damage to the wood. Pressure-treated southern yellow pine begins with a light green color and takes 1 to 2 years to weather out to a natural gray. Western red cedar begins as a white color and takes 1 to 2 years to weather to a mottled gray in an unprotected location and a consistent gray in protected locations. All heart redwood begins as a rich, red color and takes 2 to 3 years to weather to a consistent medium to dark gray color. To extend the natural color of a wood, a water repellent with an ultraviolet light inhibitor must be applied to slow the weathering process. Water repellents eventually wear off and must be reapplied periodically to maintain the protection of the wood.

To change the natural color of the wood, a surface must be **stained** or painted. Stains enter the surface of the wood and temporarily bond a new color to the surface. Two different types of stains commonly used with exterior applications include opaque stains, which provide a painted coating over the wood, and transparent stains, which provide coloring to the surface while allowing the grain of the underlying wood to show through. Solid stains should not be used on exterior surfaces. As with water repellents, stains must be reapplied periodically to maintain a consistent surface color.

Application of all surface treatments should be completed according to the manufacturer's instructions. Provide protection for all existing improvements, including structures, paving, amenities and plant material, before applying surface treatments.

PREFABRICATED MATERIALS

To speed the construction of landscape structures, many pieces can be purchased prefabricated. Available in a variety of natural and artificial materials, contractors can find **lattice,** stair **stringers,** rail posts, and many other commonly used parts, cut and ready for installation. Following are exterior building components that are commonly available.

Screening materials. **Screening** materials are valuable for insect barriers on decks, screened porches, and gazebos. Screening is available in sheets or rolls that are 36 inches or 48 inches wide. Contractors can choose from inexpensive polyester or the more durable aluminum screening materials.

Lattice. Lattice is thin strips of wood fastened together at right (or more acute) angles in a evenly spaced gridwork pattern. Precut and nailed lattice is available in 4×8 foot sheets. Lattice is available in treated and naturally decay-resistant woods in both **lath** or 1×2 lumber. Diagonal and gridded lattice patterns can be found in some locations. Paintable vinyl lattice is also available in many locations.

Stair carriages. Stair carriages are the notched, dimensioned lumber that supports the stairs. Pressure-treated stair carriages are available precut to standard riser/tread dimensions.

Columns and posts. Precut and routed columns and railing posts are available in a variety of materials. 4×4 posts in treated and naturally decay-resistant woods have a variety of designs. Stainable porch columns are available in wood and vinyl.

Post caps. Protective and decorative caps for posts are available in metal, vinyl, and lathed wood. Metal and vinyl are manufactured with integral color, and wood can be stained or painted.

Railing and fence Sections. Preassembled railing and fence sections are available in a variety of lumber types and designs. Fencing panels are also available in vinyl materials that are ready for assembly.

NON-WOOD PRODUCTS

New materials are regularly introduced to the carpentry industry as the availability of natural resources decreases and new technologies emerge. These alternative materials offer expanded opportunities to exterior carpenters in the form of decking materials of recycled plastics, premanufactured vinyl products, and similar wood substitutes. When considering an alternative material, review with the supplier the material's performance using the same criteria outlined in this chapter. Materials may change, but the conditions in which they are expected to perform seldom do.

Chapter 29

Wood Decks and Platforms

The concept of a deck is usually seen as an open-air space, often with railings and stairs. While this is one version of a deck, another is the platform from which many of the landscape structures are derived. Stoops at the entry to a building are actually small decks with stairs and railings attached. Gazebos are decks with roofs and side railings. The deck is the basic structure that, with a few additions or modifications, supports the diversity of landscape structures available.

This chapter covers site preparation, construction of the foundation, substructure, surfacing, trimming, and finishing for a deck. Construction of railings, stairs, benches, and roofs for a deck or other structures is covered in Chapters 30 and 31. The integral nature of wood construction may require the review of all chapters in this section to gain an understanding of how the components relate to each other and fit together. While one piece may be constructed as a stand-alone, many wood projects require the combination of all three components. Methods and techniques described in this chapter would be superceded by any construction documents prepared for a project.

RELATED INFORMATION IN OTHER CHAPTERS

Information provided in this chapter is supplemented by instructions provided elsewhere in this text. Before undertaking activities described in this chapter, read the related information in the following chapters:

- Safety in the workplace, Chapter 4
- Basic construction techniques, Chapter 7
- Construction staking, Chapter 8

- Materials for exterior carpentry, Chapter 28
- Wood steps, railings, seating, and skirting, Chapter 30
- Wood overhead structures, Chapter 31

SITE PREPARATION FOR WOOD STRUCTURES

Structure Layout

Before building any structure, it is advisable to lay out the foundation and post locations. Proper placement of these elements will save time removing and relocating misplaced elements. Preparation of a plan that precisely calculates the location of footings and posts will aid in proper placement. Measuring these elements to the center line will also avoid mistakes common with adding and subtracting lumber dimensions. If the site has a slope, it may be difficult to accurately place elements that are above or below each other. Place tall stakes just beyond the edges of the structure and then run level stringlines at the finish elevation of the structure to obtain an accurate picture of how the structure will be positioned. When the stringline network is set square and at the proper dimensions, use a plumb bob to locate the proper center position of post holes and foundation locations. Batterboards (Chapter 8) provide a means of locating improvements throughout a project.

Ground Preparation Below Structures

Decks, gazebos, and other structures are often positioned slightly above finish grades. Structures that are positioned with a crawl space underneath require preparation of the ground below to prevent erosion and weed growth. One common treatment of this area includes removal of all vegetative growth within the perimeter of the structure, positioning of landscape fabric over the exposed soil, and covering with a non-decomposing material such as washed river rock, pea gravel, or field stone. This treatment will hold the soil in place and reduce the weed growth in an area that is difficult to access.

FOUNDATIONS FOR WOOD LANDSCAPE STRUCTURES

All landscape structures require some form of anchoring, either to a building or to the ground, for stability. Most structures also require that this anchoring be prepared in a manner that provides adequate protection from frost heaving. Anchoring can be accomplished by attaching structures to an existing building through the use of a ledger plate; supporting structures on existing slabs, footings or foundations; or supporting structures on posts anchored in the ground. Another function of most types of support is the provision of lateral support to prevent the structure from leaning or twisting.

CAUTION Obtain the assistance of a design professional for designing wood structural systems. Serious injury or death can occur with structural failure of any wood structures described in Chapters 29, 30, and 31.

Attachment to an Existing Building

Use of a ledger plate provides an adequate anchor for structures that can be attached to a building. Attachment to an existing building also protects the landscape structure from differential movement. In the unlikely event the building goes up or down, so does the structure. Lateral support is provided only for structural members attached to the ledger. Additional support methods are required for portions of the structure away from the building. Bracing may also be required to prevent lateral movement of the structure.

Using a ledger plate requires preparation of the building where the ledger is to be installed. The building is marked at the location where the landscape structure will be attached. Mark the building on the exterior surfacing with a pencil or chalkline. Be sure to include the dimensions of any surfacing, railing, and trim that will attach to the building in the markings. Verify that the markings are level and match the elevations of any doorways that the structure must serve.

If the building is new construction and siding has not yet been installed, proceed directly to the installation of the ledger. If the building is an existing wood structure, the exterior surfacing must be removed until sheathing or the foundation is exposed. A circular saw set at the depth of the exterior surfacing material can be used to cut along the markings. A pry bar should be used to carefully remove the surfacing from the building. Use caution not to damage remaining surfacing when removing exterior surfacing from a building. When the surfacing has been removed, remark this exposed surface with the location of the ledger, again verifying for level and proper elevation.

Holding the ledger in the correct place, **tack** the ledger to the building using galvanized 16d nails. Use the nailing pattern on the sheathing as a guide to locate building joists and studs behind the sheathing. If possible, avoid connecting to the ledger studs and building joists to avoid future problems in the event the building is remodeled. Connecting to the building **rim joist** is preferred. At the location where the rim joist is positioned behind the sheathing, drill 1/4 inch diameter pilot holes every 16 inches through the ledger and into the sheathing and building rim joist. Insert galvanized 3/8 inch diameter by 3 inch lag screws with washers into the pilot holes and screw into the building until the ledger is held snugly against the building (Figure 29–1).

If the ledger is to be attached to concrete block, concrete foundation, or masonry veneer, use a masonry bit to prepare 3/8 inch pilot holes every 16 inches. Anchoring must be done using galvanized expanding bolts rather than lag screws.

Attachment to an Existing Slab or Footing

Connecting a structure to an existing slab requires either drilling and setting an expanding anchor bolt with a galvanized post anchor or connecting to the slab using an angle brace. In cases where wind may lift the structure, consider using both techniques. Setting an anchor bolt requires marking the center line location of the post or beam connection. Using a **hammer drill** with a **masonry bit,** drill a hole the same width and depth as the expanding bolt. Connect a galvanized post anchor to the top of the expanding bolt. Drive the anchor bolt into the hole and, using a wrench, twist until the bolt fits securely in the concrete. Set the post or beam in the post anchor and tighten.

Angle braces also use expanding bolts in the concrete with lag screws into the post or beam. Angle braces are installed by placing the post or beam in position and marking the location of the brace on the post or beam and concrete surface. Remove the brace

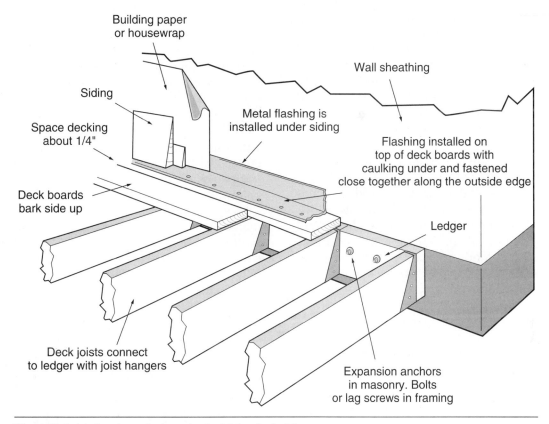

Building paper or housewrap

Siding

Space decking about 1/4"

Deck boards bark side up

Wall sheathing

Metal flashing is installed under siding

Flashing installed on top of deck boards with caulking under and fastened close together along the outside edge

Ledger

Deck joists connect to ledger with joist hangers

Expansion anchors in masonry. Bolts or lag screws in framing

Figure 29–1 A ledger is used where the deck joins the building.

and drill pilot holes in the post or beam for the lag screws. Drill the pilot holes in the concrete the same depth and diameter as the expanding anchor bolt. Place the angle brace over the expanding bolts and attach the washers and nuts loosely. Install expanding anchor bolts in the concrete and twist with a wrench until the bolt fits securely in the concrete. Install and tighten the lag screws into the post or beam. Tighten the nuts on the expanding bolts.

Cast-in-Place Concrete Footings

Cast-in-place concrete footings sunk below **frost depth** provide a stable support for any type of structure that has a substructure framework. Structures are anchored to this footing using a galvanized post anchor set in the top of the footing. Structures that rely on the footing for lateral stability do not work with this type of support (Figure 29–2). This type of footing works for structures that have multiple posts with bracing between the posts that will hold the structure square.

To construct a concrete footing, excavate a 12 inch diameter round hole to frost depth centered on the footing location marking. The hole should be as vertical as possible with straight sides. A properly excavated hole does not require forms, but if desired, a piece of

Figure 29–2 Galvanized post anchor placed in concrete footing.

tubular paper form may be installed in the hole. For installations that require a neat appearance at grade level, a piece of paper form or a box form constructed from 2 × 6s may be used to form the top 6 inches of the pour. Before pouring, recheck measurements to verify that the footing is located in the proper position and elevation. If not, expand the hole to cover the location required or fill the hole and move to the proper location. Overexcavated holes may require that a paper form be installed in the proper location and the hole backfilled around the form.

If reinforcing is desired, cut two #4 rerods 8 inches longer than the depth of the hole. Insert the rerods into the hole, pushing the extra length into the ground at the bottom of the hole until the top ends of rerods are below the top of footing. Place 2–3 inches of gravel into the bottom of the hole for drainage. Mix and pour concrete into the hole and tamp gently until settled. Fill to the top of the hole and smooth the top with a wood float. Before the concrete hardens, insert a galvanized post anchor into the center of the footing. When the concrete is hard, forming material used for the top may be removed.

If posts are placed on footings or piers, they should be braced and held in correct position until beams and joists are installed. Tack two 1 × 6s, placed at right angles to each other, to a post. Connect each 1 × 6 to stakes anchored in the ground (Figure 29–3). Each subsequent post can be braced in two directions in a similar manner, or braced against posts anchored previously. Verify that the installation is square before placing any beams or joists.

Direct Burial of Posts

For residential applications in locations with stable subsoils, treated support posts may be buried directly into the ground to provide adequate support for landscape structures. This type of support provides structural support and lateral support, and if prepared to the proper depth, can provide frost heave protection. To extend the life of a direct burial post, wrap the post with asphalt building paper (roofing felt) to prevent contact with soil at the ground line. Extend the **building paper** 6 inches above finish grade and 9 inches below finish grade (Figure 29–4).

To execute direct burial of a post, excavate an 8 inch diameter hole to frost depth centered directly on the footing location marking. The hole should be as vertical as possible with straight sides. Place 2–3 inches of gravel at the bottom of the hole for drainage. Place the post in the hole, backfill half of the hole, and tamp. Verify that the posts are plumb and square. Complete

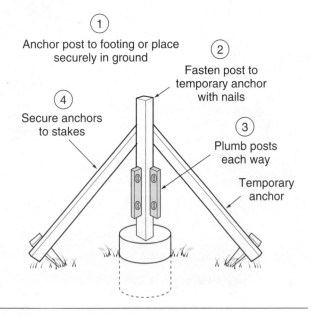

Figure 29–3 Temporary bracing for posts.

Figure 29–4 Stapling roofing felt around a direct burial post. Felt extends 6 inches above and 9 inches below the ground line.

backfilling and tamping. Brace to maintain correct position until framing begins (Figure 29–5).

Backfilling of the hole can be accomplished with several materials. Backfilling with soil excavated from the hole provides stable support of the post if the hole has not been oversized. Compact the soil with the handle of a shovel after every 8 inches of fill has been placed. Crushed aggregate can also be used for backfill, but does not provide as much stability as other choices. Place aggregate and compact after every 8 inches of gravel has been placed. Another choice for filling around a post is to place premixed concrete in the void around the post. While this method provides a very stable installation, research has shown that post replacement with this method is extremely difficult. A fourth choice is to place dry, premixed concrete in the hole for backfill. When the hole is nearly full, soak the dry concrete mix with water and mix using a long piece of rerod in a churning motion. This creates a "soft" concrete backfill that has stability slightly better than that of soil.

Whichever choice of backfill is selected, constantly check the post for plumb as the hole is filled. Correct leaning posts by wiggling the post and applying pressure in the direction the post needs to move. Recompact after alignment is corrected. If concrete is selected for backfill, smooth the surface of the hole with a wood float, tapering the surface down away from the post so that water will drain away from the post.

Precast Pier Blocks

A choice for support when frost protection is not an issue is the use of precast **pier blocks.** These units are tapered, concrete blocks with a square opening in the top designed for the placement of a treated 4 × 4 post (Figure 29–6). Pier blocks provide no lateral support for a structure and are recommended only for supporting decks or freestanding structures where frost and stability are not problems.

To install precast pier blocks, locate the center of the post location. In a 12 × 12 inch square centered on the

Figure 29–5 Setting and leveling posts for deck construction.

Figure 29–6 Precast concrete pier holding a support post.

post location, clear away all vegetative growth. If the topsoil is unstable, excavate away any unsuitable soil until a solid subgrade is reached. Replace the excavated area with granular base material, compacting in 4 inch layers. Place the block under the center of the post location and level in both directions. The pier block may be moved slightly to accommodate any errors in layout. Posts should be placed in the blocks and extended up to the deck structure. Attach posts to the structure using carriage bolts and trim flush with deck joists. Posts will require temporary bracing until deck structure is installed.

Wood Sleepers

When deck material is to rest directly on grade, wood **sleepers** are installed to provide a connecting point for surfacing. Wood sleepers are decay-resistant, dimensioned lumber, typically 4 × 4s or 6 × 6s, which are placed level on a compacted base material. Sleepers must be positioned under all edges of the surfaced area and spaced no more than 2 feet apart. Granular material is filled between and to the tops of the sleepers to hold them in place. Deck structure or surfacing

is then attached directly to the sleepers in the same manner as attachment to wood structural framing.

SUBSTRUCTURE CONSTRUCTION

Two basic choices are available when building the substructure for a deck—traditional post and beam framing (also called drop beam or plank and beam) (Figure 29–7), or platform framing (also termed box or flush beam) (Figure 29–8). Post and beam utilizes a system of posts to support heavy construction beams. These beams in turn support structural joists onto which the surfacing material is applied. While very durable and time-tested, the post and beam construction methods require enough clearance between the surfacing and ground to accommodate the structural components. Sites that are relatively flat may be difficult to adapt to a post and beam construction method.

Platform framing consists of a rim joist box that runs around the outside of a structure inside which the joists are hung. This platform is connected directly to the posts, bypassing the requirement for beams. Platform framing requires fewer structural members

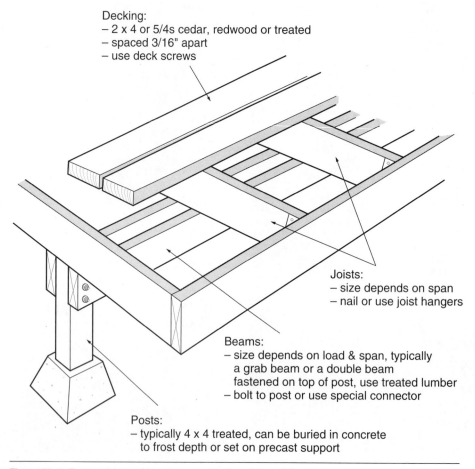

Decking:
– 2 x 4 or 5/4s cedar, redwood or treated
– spaced 3/16" apart
– use deck screws

Joists:
– size depends on span
– nail or use joist hangers

Beams:
– size depends on load & span, typically a grab beam or a double beam fastened on top of post, use treated lumber
– bolt to post or use special connector

Posts:
– typically 4 x 4 treated, can be buried in concrete to frost depth or set on precast support

Figure 29–7 Post and beam (drop beam or plank and beam) framing diagram.

to complete the substructure, but because those joists are carrying the same loads as would a beam, the members must be larger and/or spaced closer together. From a structural soundness standpoint, each method can provide adequate support, so the choice may depend on space and clearance.

Post and Beam (Plank and Beam or Drop Beam) Framing

If the deck is connected to a building, install a ledger plate at the correct location and elevation. Allow space for surfacing material between the finish elevation and top of the ledger. No beam needs to be installed at the ledger location because the ledger carries the load typically carried by a joist and beam. Hangers for joists will be fastened directly to the ledger (see Figure 29–1). Locate post locations and install as directed in previous paragraphs. Post and beam framing requires that all posts be installed and stabilized before beams are connected.

Beams may rest on top of posts or be attached to the sides of posts as single or grab beams (double beams, one attached to each side of the post). If the beam is to rest on top of the posts, the posts must be trimmed to the correct height before installing the beams. If

beams are attached to the side, the posts must be marked but may be trimmed after the beams are connected.

To set the post heights when using a ledger plate, a level stringline must be extended from the top of the ledger to a post on the edge of the project. The ledger plate functions as a rim joist and the position of all structural members must be set in accordance with this stringline. From the stringline elevation, the contractor must subtract any cross-slope desired for the deck (typically a 1/4 inch per foot fall away from structures), the depth of the joist material, and if the beam is to rest on top of the post, the depth of the beam.

Use a **speed square** to mark the post. If placing the beam on top of the post, trim the post using a circular saw. The cut must be square and level for the beam to be positioned properly. Position and attach the beam using a saddle or metal T connector on both sides of the beam (Figure 29–9).

Attaching the beams to sides of the post provides a stable installation. This also allows the use of grab beams, constructed of two smaller pieces of lumber rather than one large piece (two 2 × 10s rather than one 4 × 10) (Figure 29–10). One section of the beam is attached to each side of the post. This provides the same structural support as one large beam, but makes

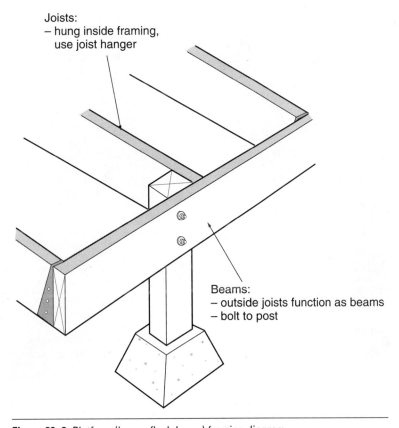

Joists:
– hung inside framing, use joist hanger

Beams:
– outside joists function as beams
– bolt to post

Figure 29–8 Platform (box or flush beam) framing diagram.

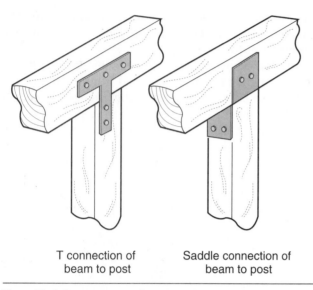

T connection of
beam to post

Saddle connection of
beam to post

Figure 29–9 Common post to beam connectors.

construction more cost-effective. Run the stringline as described in the previous paragraphs and mark the proper location for the top of the beam. Be certain to include the depth of the joists and slope of the deck in the measurement. Since the beams will be attached to the sides of the post, this mark indicates where the top of the beam(s) should set. Attach the beam(s) by drilling and inserting at least two 1/2 inch diameter bolts (Figure 29–11). Larger, structural installations may require more or heavier bolts based on the design specifications. After the beams have been installed, the top of the post may be trimmed flush with the top of the beams using a reciprocating saw.

If no ledger plate is used, mark the **finish floor elevation** on one post. Subtract the floor surfacing dimension to determine the top-of-joist elevation. Remark the post with the top-of-joist elevation and extend a level stringline to an opposite post, using this stringline to establish elevations for all posts.

Figure 29–10 Grab beam connections used for seating framing.

Figure 29–11 Post to joist connection using carriage bolts. At this corner, bolts are installed from both sides. Markings indicate top and bottom of deck surfacing.

Joists rest on top of the beams and are hung from the ledger using joist hangers, regardless of the method used for beam/post connection. Recheck the installation to assure level and square before placing any joists. Mark the locations of all joists along the outside beam and the ledger. Install joist hangers along the ledger at each location (Figure 29–12). Beginning at one side of the deck, place the first joist in the joist hanger and on the beam at the correct marking. If a joist is too high, it may require notching at the end or moving the hanger down. If the joist is too low, the hanger should be moved up or cedar shims placed under the joist to raise its level. Attach the joist to the joist hanger using joist hanger nails.

Secure the joist to the beam by toenailing (see Chapter 7) with 16d galvanized nails. Proceed with placement of joists along the entire length of the deck. If joists do not extend to the outside edge of the deck and splicing is required, attempt to center the splices over a beam. Beams that extend beyond the joists

Figure 29–12 Installing joist hangers. The block positions the hanger at the correct height. The prepunched tabs are driven in to hold the hanger until nails can be installed.

should be trimmed. The correct lengths for all joists that extend beyond the edge of the deck can be marked using a chalkline. From that chalkline, mark a vertical line using a speed square and trim with a circular saw. Nail a 2 × 4 diagonally across the surface to hold the joists in position until surfacing is installed.

Platform (Box or Flush Beam) Framing

Platform framing also begins with the placement of ledger plate and posts for the deck. Install a ledger plate at the correct location and elevation, remembering to allow space for surfacing material between the finish elevation and top of the ledger. The top of the ledger sets the elevation for the top of the posts and joists. If posts are direct buried, verify that they are straight, plumb, and square and proceed with connecting the rim joist. If posts are placed on footings or piers, they should be braced and held in position until all the joists are installed. Verify that the installation is square before placing any joists.

The outside, or rim, joist performs the same function as the beams in post and beam construction. Begin platform framing by using a joist hanger to connect a rim joist at one side of the deck to the ledger plate. Holding the joist against an outside post, use a carpenter's level to adjust the joist until any required slope away from the ledger is reached. Connect the joist to the post using two 1/2 inch diameter bolts. Repeat this process for the rim joist at the opposite side of the deck. Note that if the posts are not set at the outside edges of the platform it may be necessary to install two interior joists and the end rim joist before installing the side rim joists. After these initial joists are in place, install the rim joist across the end by endnailing into the perpendicular joists. Angle connectors provide a stronger connection than does end nailing. Posts that extend above the platform frame can be trimmed with a reciprocating saw to the same elevation as the top of the beams.

After the platform rim joists have been framed and securely attached to the posts, the interior joists may be placed. It is recommended that the level and square of the platform be checked before proceeding. Mark the location of all interior joists and install hangers for connections at both ends of all joists. Measure and cut joists, install in the hangers, and nail in place.

Cantilever

Joists from post and beam framing and platform framing may be designed to extend beyond outside support posts (Figure 29–13). This concept, called **cantilever,** requires extensive calculations by a design professional to determine how far beyond the posts structural members can be extended. When framing

Figure 29–13 Cantilever of deck platform 4 feet beyond outside support post.

post and beam structures with a cantilever, no special construction techniques need to be employed. Cantilevers with platform framing are constructed in the same manner as structures without cantilevers, except that rim joists are not located in the outside corners. Actual location of the posts is determined by the design of the cantilever.

Multilevel Deck Structural Preparation

Dramatic aesthetic effects can be achieved with little additional structural preparation when decks are designed as multilevel structures. Multilevel decks can be constructed as separate platforms, each standing as a separate raised deck, or as shared structures. Shared structures create a single network of posts on which deck platforms are constructed at different elevations. Connecting the different deck levels of either method with broad stairs improves the useability of the outdoor space.

When multilevel decks are constructed, bracing should be installed connecting the structural mem-

bers between levels. Angle or cross bracing, described in the following paragraphs, prevents structural problems that can occur from platforms shifting.

Additional Structural Preparation

With each of these framing methods, it is beneficial to complete additional structural preparation before the surfacing is installed. Additional structural preparation may be necessary for stairs, railings, seating, and skirting. While these structural items can be added later, installation at this point will save cutting and notching surface material and provide easier access to connections.

Bracing

Wood construction may require bracing to supplement the stability of the structure. Several types of bracing are available depending on the circumstances. Typical uses include bracing posts for installations that have a high clearance above grade or between joists to maintain spacing and straightness. Cross or angle bracing is recommended if the deck structure is more than 4 feet above finish grade or is carrying loads heavier than pedestrian traffic and deck furniture. When bracing the structure, select a system that applies braces in two directions at right angles to each other.

Cross bracing is used in post applications where lateral stability is required. Most decks or structures that have tall supporting posts require cross bracing to prevent collapse. While cross bracing is very effective, it is not particularly attractive. Locating bracing under decks will minimize the visual impact, but it will be difficult to hide in above grade situations such as roof structural reports. Cross bracing is installed by selecting sound boards, usually 2 × 6s or larger, and securely attaching then to posts in a diagonal manner (Figure 29–14). When bracing two posts, one brace connects from near the bottom of one and near the top of the second. The second brace runs the opposite diagonal direction between the two posts. The braces should be connected in the center to strengthen the brace. Braces should set flush against the surface of the post being braced and should be fastened with at least two bolts at each post connection point. If the braces are on opposite sides of the post in order to obtain a flush fit, place blocking between the two braces at the crossing point. This type of bracing should be placed between each pair of posts to stabilize an installation. Angle and W bracing is installed in a similar manner to cross bracing, with the exception that only one brace runs between each pair of posts.

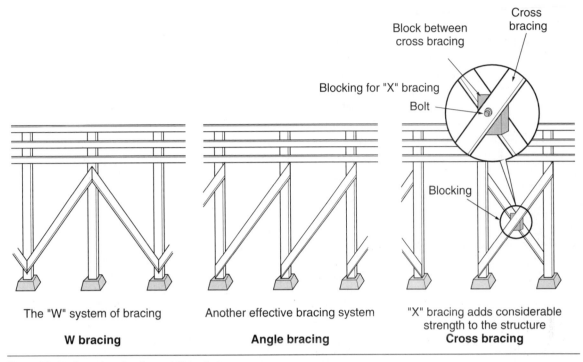

Figure 29–14 Cross bracing for stabilizing below structures.

Bridging, Blocking, and Cleats

Maintaining spacing between joists and rafters requires the installation of bridging or blocking. When joist or rafter span length exceeds 10 feet, bridging or blocking should be considered at the midpoint of the span. **Bridging** is constructed with treated 1×4s placed diagonally between each pair of joists/rafters from the top of one joist/rafter to the bottom of the other. To install bridging, measure the diagonal distance from the top of one joist/rafter to the bottom of the next. Bevel cut two 1×4s and nail into position using 5d box nails. Bridging should be placed in pairs running in opposite directions.

Blocking is the placement of solid wood blocks between joists/rafters to maintain spacing. Select blocking material of treated lumber with the same dimensions as the joist/rafter. Measure the distance between joists/rafters and square cut the blocking material. Place between the joists/rafters and end nail in place (Figure 29–15). Stagger every other piece of blocking to allow the material to be end nailed. Locate blocking on top of a beam or in any location where the joist/rafter is twisted. If located in the rafters, consider spacing blocking or bracing so it looks like a design element.

Cleats are small pieces of dimensioned lumber that are attached to the structural framework in locations where posts interrupt the joist/rafter pattern. Cleats provide support for surfacing that runs into posts. To

Figure 29–15 Blocking between joists for stability and to anchor vertical post.

install cleats, cut a piece of treated lumber the same width as the post and nail to the post where necessary (Figure 29–16). Verify that the top of the cleat is the same elevation as the joist/rafter. Pilot holes may be necessary to prevent cleats from splitting. Temporary cleats can also be installed for supporting structural pieces that are being installed. Cut a piece of dimensioned lumber and tack to a post at the proper elevation. Rest the structural piece on top of the cleat while attaching and then remove the cleat when connections are completed.

Extending Supports for Overhead Roofs, Railings, and Seating

Decks that are designed to include overhead canopies, railings, or built-in seating require posts that extend above the surface of the deck to support these elements (Figure 29–17). When ordering materials and installing the foundations and substructure, be certain to take this extra dimension and load into consideration. Extending posts that support the deck upward

to support the overhead structure provide the best lateral stability. Carefully select the posts that will perform this function for straightness, minimal knots, and no twisting.

A second choice is to attach posts for roof framing to the substructure constructed for the deck. (See Figure 30–9.) This method requires temporary bracing until the roof is complete and should have permanent bracing to prevent lateral movement. Additional strength can be gained by positioning posts at the intersection of two joists. This allows the post to be bolted in two directions.

Insect Screening

Before surfacing the deck, any desired insect screening should be installed over the structural framework to prepare the deck for future enclosure. If it is not planned to enclose the outdoor use area, failing to install screening material at this point will require that deck surfacing be removed to attach screen if enclosure is ever considered. Roll screen material over the

Figure 29–16 Cleat on which decking material will be fastened.

Figure 29–17 Post extensions through deck platform for seating supports.

joists and staple into the top of the joists. Overlap where necessary to avoid gaps in coverage.

DECK SURFACING

Covering the surface of the deck can be accomplished with a variety of patterns and materials. Before selecting, review the structural and aesthetic requirements of the project.

Deck surfacing must be applied at an angle to the joists for maximum support. Right angles provide maximum support, but angles up to 45 degrees are structurally possible. Selecting a beginning point for installing deck materials depends on the pattern and desired edge treatments. Beginning at the outer edge and working inward prevents the problem of a partial piece being installed at the outer edge, but may require the last piece next to the building to be rip cut to fit the space left.

Begin attaching surfacing at the selected location and work toward the opposite edge of the deck. Examine the cross section of each piece of decking before installation. Orientation of the grain to prevent cupping will be necessary to prevent water retention on the deck surface. Decking must be installed with the bark side up (Figure 29–18). If one piece of decking material will not cover the entire deck length, butt a second piece against the first, straddling one of the joists (Figure 29–19). Deck material that hangs over the outside edge of the deck will be trimmed off when surfacing is complete. Align the deck material and fasten to the joists.

Continue placing material using a shim or nail as a tool to keep spacing consistent throughout the surfacing. Typical spacing for 2 × 4s is 3/16 inch, but 2 × 4s

that are still wet from preservative treatment and 5/4s cedar should be spaced closer to allow for shrinkage. Most surfacing materials require two nails or deck screws at each joist location placed 1 inch in from each edge of the surfacing (Figure 29–20). One fastener per joist can be placed while the material is being initially laid and the rest of the connectors installed after all surfacing is in place. At edges where the decking was left overhanging the structure, snap a chalkline at the location of the outside edge of the structure and trim by running a circular saw from one end to the other. For a straighter edge, attach a trimming jig, or guide, which will hold the saw in the proper location along the entire cut.

If the deck is attached to a building, protection from moisture must be provided. To prevent moisture from entering between the ledger and the sheathing, a piece of galvanized Z **flashing** should be inserted under the building siding material and attached on the top of the deck material. Restore the building exterior surfacing over the structure (see Figure 29–1).

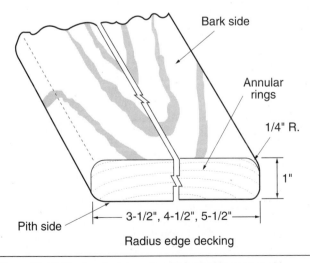

Figure 29–18 Placing deck boards with bark side up to avoid cupping.

Figure 29–19 Deck surfacing is installed with end joints staggered.

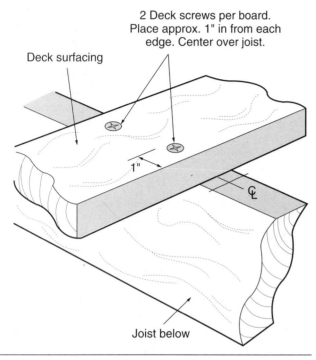

Figure 29–20 Deck surfacing connecting locations.

Figure 29–21 Trimming of deck railing.

TRIMMING

Providing the finishing touch to a deck can be done by trimming unsightly structural pieces and connections. Trimming should wait until all carpentry work, including roofing, railing, and seating, is in place. Trimming is usually done using 1 × lumber that is wider than the structural piece being covered. Bevel cut at any trim pieces that abut and attach using 8d galvanized box nails. Nailing in a straight pattern improves the finished appearance. Space fasteners 1 foot apart or less to combat the high potential for warping in 1 × trim boards. Use of wood screws also reduces the potential for warping.

Particular areas that benefit from trim pieces include the front and back sides of railing caps, posts that are exposed above the deck surface, edges between two sheets of material (such as skirting), rim joists, stair risers, and the edges of seating (Figure 29–21).

APPLYING FINISHES

Surface finishes can be treated with a variety of **sealers,** stains, and paints. Methods of application used for exterior finishes include hand brushing, swabbing with rollers, and spray application. Apply finishes using smooth, even strokes. Take the time to mask areas that are not to be finished. Use of drop cloths and covers reduces cleanup time and avoids potentially costly mistakes.

> **CAUTION**
> Follow manufacturers' safety instructions when utilizing any wood finish or wood treatment substance.

Chapter 30

Wood Stairs, Railings, Seating, and Skirting

Decks and exterior wood projects often include a variety of related amenities that aid in the safe use and enjoyment of the project. While the construction of these additions requires construction techniques similar to those of other wood components, each has special approaches and requirements, which are outlined in this chapter. Special techniques are described here for wood stairs, railings, seating, and skirting. Methods and techniques described in this chapter would be superceded by any construction documents prepared for a project.

RELATED INFORMATION IN OTHER CHAPTERS

Information provided in this chapter is supplemented by instructions provided elsewhere in this text. Before undertaking activities described in this chapter, read the related information in the following chapters:

- Safety in the workplace, Chapter 4
- Construction math, Chapter 5
- Basic construction techniques, Chapter 7
- Materials for exterior carpentry, Chapter 28
- Wood decks and platforms, Chapter 29
- Wood overhead structures, Chapter 31

FOUNDATIONS AND STRUCTURAL SUPPORTS

The material on supporting a structure in Chapter 29 presented typical methods for preparing ledgers, footings, posts, and piers for a wood deck. While most of the items listed in this chapter do not require separate foundation support, the support posts used for decks may be part of the structural system. Review of the framing methods used for wood decks will also provide valuable information regarding installation and connection of support mechanisms for deck elements. The actual method for support should be implemented based on the design for the project.

STAIRS

In most instances, it is necessary to construct some type of stairs in order to get onto or down from a deck. Basic components of a stair system include the **carriages,** or the large lumber pieces, usually notched, which support the stairs; the risers, or the vertical portion of the stair; and the tread, or the horizontal portion of the stair on which we step.

Stairs can be constructed in a straight run or wrapped around inside or outside corners of a deck. Complex stair installations may include several sets of stairs with landings, or small decks, between sets. Construction of landings is similar to that in building decks, with platform framing the easiest method to use on landings placed between stair runs. When constructing projects with complex stairs, construct and surface all platforms before installing stairs.

Stair Supports

Locations where stairs are attached can be heavy stress points for decks. To increase the strength of these areas, the rim joist can be doubled, or bridging or blocking can be added between the rim joist and the first interior joist. In locations where the rim joist

Figure 30–1 Extra stair supports. The bottom 2 × 10 provides a solid surface to nail to and the 2 × 4 assists in supporting the bottoms of the stair carriages.

Figure 30–2 Stair landing is excavated and precast blocks are placed to support the bottom of the carriage. The stake in the front left is marked with the level required for the top of the precast blocks.

of the deck is used as the top riser in a set of stairs, the structure left for attaching the stringer may be limited. In these situations, the substructure must be modified to provide adequate support for the stair carriage. This additional structural support can be accomplished by adding a second joist directly below the rim joist at the location where stair carriages are to be attached (Figure 30–1).

Landing Area Support

Stairs that land at ground level require a stable support for the ends of the carriages. Placement of a small footing or paved area provides stability against settling and prevents erosion of the landing area at the bottom of the stairs (Figure 30–2). Stairs should be attached to the landing area using galvanized angle brackets or galvanized supports when the ground support is poured concrete. If utilizing precast slab, anchor the supports using expanding bolts placed in pilot holes.

Stair Framing

Stairs are supported by joist-like structural members termed carriages. Carriages are 2 × lumber that are either notched and placed under the treads of the stairs or placed along the sides of stairs with supports connected for the treads to set upon. Side carriages are adequate for straight run stairs, while notched carriages work for straight run stairs and are the only structural method that can be used for stairs that wrap around corners. Stairs may combine side carriages with notched carriages in the center.

Spacing of carriages is determined by the tread material that is applied. Treads that are constructed of 2 × lumber should have carriages spaced no more than 2 feet apart. Treads constructed of 5/4s decking material should have carriages spaced 16 inches on center.

Construction of Notched Carriages

Review the section of Chapter 5 related to stair calculations before constructing stair carriages. If construc-

Figure 30–3 Elevation view of stair carriage supported at the top by a 2 × 10 and at the bottom by a concrete footing.

tion drawings exist for a project, utilize the dimensions provided. Construction of notched carriages requires selection of lumber that is approximately 2 feet longer than the diagonal measurement that must be covered by the stairs. 2 × 12s are a typical choice for runs with 5–10 stairs (Figure 30–3). Larger dimensioned lumber or strengthening may be required for notched stairs, since the notching reduces the effective thickness of the lumber that is supporting the load on the stairs.

Long carriages (for stairs with several risers) have a significant amount of flexibility that can be reduced by attaching a 2 × 4 along the length of the carriage below the treads. This 2 × 4 should be attached with the face flat against the carriage and the bottom edges flush. Place rust-resistant deck screws every 8 inches as connectors.

Marking and Cutting Stair Carriages

Utilize the following steps to mark carriages for side or notched stairs (Figure 30–4):

=== **STEPS** ===

1. Place lumber for carriage on supports.

2. On the outside rulers of a carpenter's square, locate the riser dimension of the short side and the tread dimension on the long side (Figure 30–4, step A).

3. Beginning near one end of the board, place the square with the corner off the lumber and the riser and tread marks on the rulers setting on the edge of the lumber (Figure 30–4, Step B).

4. Trace along the long side of the square to locate the top tread location.

5. Measure along the line the dimension of the top tread.

6. From the back edge of the tread, draw a square line down (Figure 30–4, step C).

7. Flip the square so the corner is on the lumber (Figure 30–4, step D).

8. Position the square with both the riser and tread marks on the edge of the lumber. Adjust the position of the square along the edge so the riser mark aligns with the outside mark for the top tread.

9. Along both sides of the square, trace the riser and tread locations.

10. Slide the square down the lumber and set the riser and tread marks with the riser mark aligned with the previous tread mark.

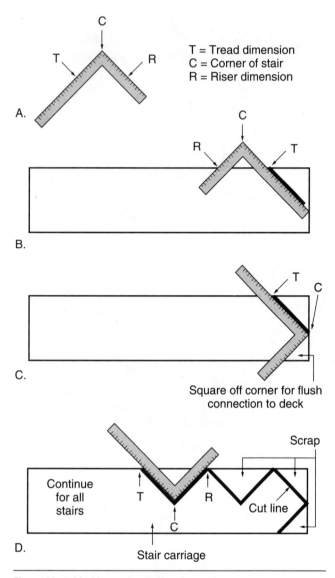

T = Tread dimension
C = Corner of stair
R = Riser dimension

A.

B.

C.

Square off corner for flush
connection to deck

Scrap

Continue
for all
stairs

T   R

Cut line

C

D.   Stair carriage

Figure 30–4 Marking stair carriage for cutting.

Figure 30–5 Framing for stairs with corners.

11. Along both sides of the square, trace the next riser and tread locations.
12. Continue this process for the number of risers and treads required.
13. At the bottom of the lowest riser, use the square to mark the bottom of the carriage. This line will be parallel to the lowest tread and square to the lowest riser.

Corner Stair Carriages

Notched carriages that are placed at the corner of stairs are placed at an angle to the other carriages and require special calculations due to their length. The riser height remains the same for corners, but the tread length is longer due to the angled placement (Figure 30–5). If stairs turn a 90 degree corner and the treads are the same width on both sides of the corner, multiply the calculated tread length by 1.41 to obtain a quick measurement of the tread length for an angled carriage. Intermediate carriages may need to be fastened to this angled carriage to support the upper treads.

Construction of Side Carriages

Review Chapter 5 for instructions on calculating stair dimensions. If construction drawings exist, utilize the dimensions provided. On stairs that are narrow, carriages can be placed along the outside of the treads. Construction of side carriages, also termed rough stringers, requires selection of lumber that is approximately 2 feet longer than the diagonal measurement that must be covered by the stairs. 2 × 10s are a typical choice for runs with 5 to 10 stairs. Mark the location of each riser and tread, subtracting the thickness of the material used for the tread marking. At each tread location, attach a 1 × 3 cleat below the mark using three rust-resistant deck screws. Pilot holes may be necessary to prevent the wood from splitting. Treads are cut to fit between the two carriages, resting on top

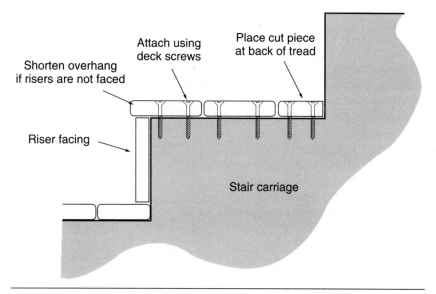

Figure 30–6 Attaching stair tread material.

of the cleats. Mark and cut the carriages as described in the previous section.

Attaching Carriages

Carriages are attached to the rim joists and/or deck framing by toenailing or by using joist hangers. Hangers are the preferred method due to their increased support capacity and ease of installation. Mark the top of the carriage locations with a chalkline to assure proper placement. Mark the location for each carriage and install the hanger. Place the carriage and check for plumb and level. Connect to the carriage. If toenailing is utilized, mark each location and hold the carriage in the proper position. Toenail from each side and from the top.

Stair Surfacing

Surfacing the stair treads can be done using 2 × lumber or 5/4s surfacing depending on the spacing of the carriages. Risers can be surfaced with 1 × lumber regardless of the carriage spacing. Risers can also be left open on any carriage arrangement if there are no concerns about tripping or seeing under the deck.

Begin tread surfacing from the front edge of the tread and work to the back. Measure and cut the first tread board the correct length. Place the first tread board overhanging the front edge of the riser by 1/4 inch. Add the thickness of the riser if risers are to be surfaced. Connect to the carriage using two rust-resistant deck screws for every carriage. Miter cut to fit the carriage for corners on wrapped stairs. Place the next tread board with a 1/4 inch gap between the first tread board and connect. Continue surfacing toward the back until the entire tread is covered. If a partial

board is required at the back of the tread, rip the dimension required and place it in the back of the tread (Figure 30–6).

Install riser surfacing after installing tread surfacing. Measure and cut risers to fit the riser dimension marked on the stairs. Rip cut the riser surface, if necessary, to match the stair riser height. Attach to the carriage with two 8d box nails at each carriage. If using side carriages, place a cleat behind each end of the riser surfacing (Figure 30–7).

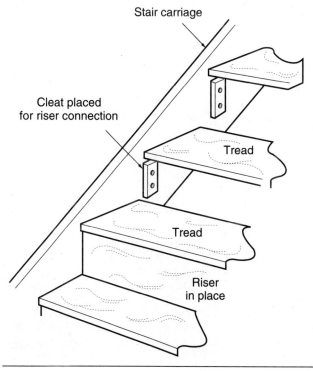

Figure 30–7 Cleats to support riser surfacing.

RAILINGS

Railings provide decks with important safety and aesthetic qualities. Decks that are above grade are candidates for railings to prevent accidental falls. Minimum railing heights are stated in codes, with 42 inches between the deck surfacing and the top of the railing a typical required railing height. Verify the railing height required with local building officials. Railings can be constructed in a variety of ways and surfaced with an even greater variety of patterns. Railings can rise up out of the deck structure, be attached to the outside of the structure, or be incorporated into seating. Horizontal boards, vertical boards, and **balusters** can all be used to create the surfacing of the fence.

Railing Supports

Railing that is placed around the edge of the deck requires some form of structural support posts. The initial framework for railing is best installed when building the substructure of the deck. Rail posts can be attached either inside or outside the rim joist (Figures 30–8 and 30–9). Space the posts according to the

design. If no design exists, the posts should be spaced no farther apart than the structural strength of the surfacing will permit. Consult a design professional to determine the correct spacing.

Mark the location of the posts and use a level when attaching to assure the posts are plumb. If attached inside the rim joist, cleats are required to support deck surfacing that abuts the post. Attach posts to the deck substructure using two 1/2 inch bolts (Figure 30–10). If posts move out of plumb when tightening, insert cedar shims between posts and joists to position the post correctly. Seating that is incorporated into decks can have structural components integrated into the substructure in the same manner. After the deck surfacing is applied, the remaining structure and facing for seating and railings may be completed.

Railing Stringers

While designs may vary, typical railing structures include a horizontal top and bottom support, termed a stringer, to support facing. Stringers run from post to post in all locations where railing is planned. Stringer

Figure 30–8 Railing support bolted to outside of deck framing.

Figure 30–9 Support bolted to inside deck framing.

material is typically a 4×4 if installed between posts, or a 2×4 or 2×6 if fastened to the face of the post (Figure 30–11). If railing surfacing material is placed vertically, these will likely be the only supports holding the surfacing. Railing surfacing material that is placed horizontally may require only the top stringer.

Measure and install the stringers level between the posts. If a stringer is placed on top of or in front of post, fasten with two deck screws. Connections on the top or face may run for several posts. If a stringer is placed between posts, use specialty connectors or angle brackets to fasten. Stringers that support railing

facing should be soundly connected to rail posts using specialty fasteners. Avoid making such connections using only end nailing or toenailing, or any of the joining techniques.

Railing Facing

Most facing material is constructed using $1 \times$ lumber or 2×2s attached horizontally or vertically using either 8d box nails or rust-resistant deck screws. Carefully plan the pattern for facing material to avoid inconsistent spacing at posts and corners. Marking the facing material location on the stringers will help in visualizing how the installation should occur. Several arrangements and combinations of facing material patterns are possible (Figure 30–12), with the three most common methods described here. Each of these methods can be modified and combined with other methods. Openings in railing facing should be reviewed carefully to assure compliance with codes

Figure 30–10 Railing post installation. Decking material has been marked and cut to place pots through deck surface to structure below.

Figure 30–11 A 2×4 stringer has been placed across the top of the posts. Balvsters are then hung from the stringer. Surfacing of 1×2s is being installed beginning at the bottom of the balvsters.

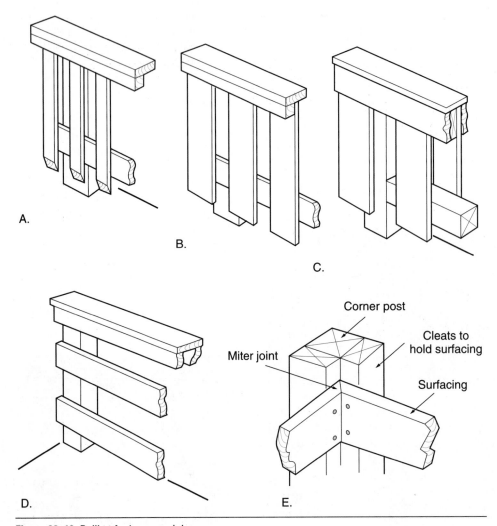

Figure 30–12 Railing facing materials.

and standards that prevent injury to children from falling through railings or placing their head into the openings.

Playground entrapment prevention standards from the Consumer Product Safety Commission should be used for calculating openings and spacing in railings. These standards state that the openings should be tested using two opening templates—a rectangular torso template that is 6.2 inches by 3.5 inches and a round head template that is 9 inches in diameter. Test openings by attempting to insert and remove the templates between the rail openings by turning the templates in all directions while holding them parallel to the surfacing. If neither template can be inserted, or if both templates can be inserted and removed, the railing should not entrap children. If the torso template can be inserted but the head template cannot be removed, there is danger of entrapment. Because of the potential for falls and entrapment, have designs

for railings reviewed by building officials and a design professional.

Horizontal facing material can be attached to posts on the front (for the strongest railing), on the back, or on both the front and back. Horizontal members are marked and cut to proper length to cover the distance between posts. One board may cover several posts, with joints bevel cut and centered on a post. Beginning at the top of the posts, place and connect the facing material to each post with two deck screws.

Vertical boards should be marked out along the top stringer before installation to determine spacing and to plan post covering. If a partial board is required to maintain even spacing for facing, place the partial board at an inside corner. Vertical boards are cut to proper length and placement should begin at a corner, working outward. Fasten to the top stringer with one fastener. Holding the board plumb, install two fasteners in the bottom stringer and the final fastener at the top stringer. As with all facing, the nail or screw pattern should be consistent in placement and spacing.

Baluster, or **spindle,** railings are 2 × 2 posts spaced closely together along the inside of the stringers. Begin placement of balusters at an outside corner by connecting the top of baluster. Holding the baluster plumb, fasten to the bottom stringer. Use the top stringer to align the tops of all balusters. Continue installation of balusters until reaching a corner. If spacing is uneven at an inside corner, use a 2 × 4 to complete one run of stringers with a 2 × 2 flush against it to start the other side.

Trimming Railings

Railings require a cap and possibly, other trim to maintain a finished appearance. Rail caps should be 1 × lumber wider than the post facing and any fascia trim planned for the inside and outside. Add 1/2–1 inch to the width to create a small overhang on both the front and back for appearance. Connect the rail cap to the top of the posts and top stringers using rust-resistant deck screws placed in pairs every 12 inches. Miter cut corners where the railing turns.

Fascia trim can be placed at the top of the facing material where it meets the rail cap. This trim piece is useful in covering the intersection of balusters and rail cap. Cut pieces of 1 × 3 or 1 × 4 and, using 8d box nails or deck screws, fasten on top of the facing material directly under the rail cap. The same installation can be done to the outside edge of the facing. The inside and outside of the bottom of the railing facing can be trimmed in a similar manner.

Exposed posts can be covered by calculating the spacing of facing material so that a piece of facing material is positioned directly over all posts. If the material or spacing does not create this effect, a piece of 1 × material can be fastened over the post using 8d box nails or deck screws. The material should be at least 1 inch wider than the post. Install the facing so it is centered on the post. Exposed corners can be trimmed using two pieces of 1 × material jointed along the corner. Miter cut the edges of the trim that will be placed together and connect using 8d box nails.

SEATING

Providing seating for decks saves space, adds convenience, and provides delineation of the edge. Seating has as many possible shapes and patterns as do railings, so it is impossible to describe the construction of every type of seat option. Covered in this section are the most common types of seating that utilize basic structural and surfacing techniques.

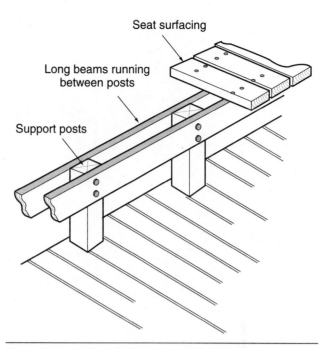

Figure 30–13 Seating framing with beams between posts.

Seating Supports

Seating supports consist primarily of posts that project up through the deck surfacing or are attached to the outside of the deck structure. (See earlier section on railing supports.) Framing is then attached to these posts to support the seat surfacing.

Seating Framing

Typical structural designs include grab beams attached to the two sides of a post or joists running between posts. Other common types of structural frameworks include seating that is supported by two posts (see Figure 30–20), or a triangle of structural pieces attached to deck framing (see Figure 30–21). Benches that have backrests require platform framing, grab beams installed on the sides of posts, or triangle framing. When 2 × lumber is used for seat surfacing, posts can be located up to 3 feet apart. When 5/4s surfacing is used, posts should be spaced no farther than 16 inches apart. Installing a structural framework under the surfacing may allow the wider spacing of posts.

Joist Framing Between Posts. Begin construction of joist supports by measuring and cutting 2 × dimensioned treated lumber that will run between each post on both the front and back side (Figure 30–13). On each post, mark the proper height for connecting the beams, subtracting the surfacing thickness from the finished seat height. Tack the beams in position and

Figure 30–14 Seating framing with grab beams.

Figure 30–15 Layout and leveling of grab beams for seating supports. Stringlines are stretched from the end supports for the bench. Remaining supports are aligned with the stringlines.

drill 2-1/2 inch diameter pilot holes through both beams and the post. Insert 1/2 inch diameter carriage bolts through the holes and connect. Repeat for each post. Where two beams meet at a post, use a butt joint centered on the post and connect each beam with two bolts. The tops of the posts can be trimmed flush with the grab beams.

Grab Beam Framing. Grab beams should be constructed of 2 × 8 or 2 × 10 treated lumber (Figure 30–14). Measure and cut two pieces for each post. The measurement should consider the width of the bench, including any extension necessary to hold a backrest support. Subtract the dimension of any trim planned for the front and back. On each post, mark the proper height for connecting the beams, subtracting the surfacing thickness from the finished seat height. Tack the beams in position, level, and drill 2-1/2 inch diameter pilot holes through both beams and the post. Insert 1/2 inch diameter carriage bolts through the holes and connect. Repeat for each post.

To maintain a straight front to the seat, properly position a beam at either end of the seat and stretch a stringline along the front between these two beams (Figure 30–15). Each subsequent beam can be positioned so that it is just touching the stringline. Trim posts to the height of the grab beams.

If a backrest support is required, cut a piece of post lumber the height of the support plus the grab beam dimension. Position the lumber between the grab beams and place at the desired angle for the backrest (Figure 30–16). Using the speed square, set at the desired angle to provide consistency for angling all posts. Tack the support in place and drill two 1/2 inch pilot holes through the support and the grab beams. Insert 1/2 inch carriage bolts and connect.

Grab beam supports can be further strengthened by adding box framing in front and in back of the grab beams (Figure 30–17). Before calculating the proper height for installing the front and back joists for box framing, determine whether the seat surfacing will overlap the framing or whether the surfacing will be

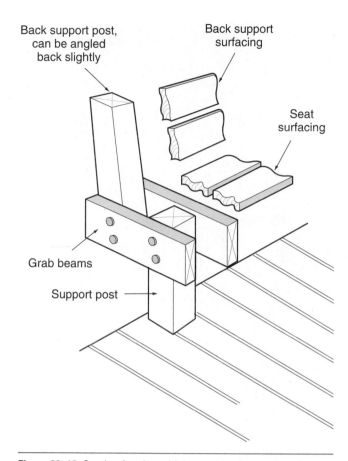

Figure 30–16 Seating framing with integral back support.

Figure 30–17 Box framing of seating.

"inset," with the seat surfacing flush with the top of the framing (Figures 30–18 and 30–19). If the surfacing overlaps, or is on top of all framing, install the framing flush with the grab beams. If the surfacing is "inset" between the front and back joists, install the front and back joists above the grab beams the same height as the thickness of the seat surfacing.

Install the front and back joists along the entire length of the seat. Attach the joists to the grab beams using two deck screws at each beam. If two joists intersect at a grab beam, use a bevel cut on the front and use two deck screws on each joist. Once the joists have been connected to the grab beams, additional seat surfacing supports can be installed between the front and back joists every 16 inches.

Double Post Framing. Seating framework that is supported by multiple posts (e.g., double posts projecting through the deck surfacing or posts supporting a roof structure) can utilize the same framing techniques described for grab beams connected to the sides of the posts or box frames that are placed around the posts.

Connections should be made using two 1/2 inch bolts at each post (Figure 30–20). Front and back joists should be connected to the posts or grab beams using two deck screws at each intersection.

Triangle Framing. Another popular version of structural support for deck benches is the triangle frame technique. Consult a design professional for sizing and positioning of materials and fasteners for this framing method. Spacing should also be determined by the design professional. Triangle framing attaches a 2 × 4 backrest support on the outside edge of the deck rim joist (Figure 30–21). This support is attached with the short dimension of the 2 × 4 flush against the rim joist, and at least three 1/2 inch bolts inserted through pilot holes, pulling the support tight against the joist. At seat height, a short 2 × 6 joist is attached to this back support, extending toward the deck. Attach the support using two 1/2 inch bolts. A third 2 × 4 support piece is attached to the front edge of this short joist through the deck surfacing to a joist below. Both connections for this support should be made with two

Figure 30–18 Seat surfacing inset at edges. Framing is set higher than beams to accommodate thickness of surfacing material. A cleat supports the framing at the end.

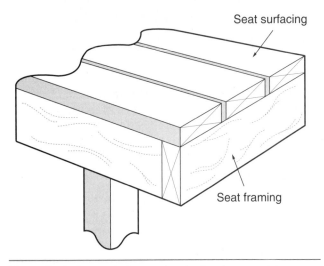

Figure 30–19 Seat surfacing overlapping framing material.

1/2 inch bolts inserted. This forms a triangular-shaped support structure using the backrest support for primary support, the short joist to hold up seat surfacing, and the angled support to hold up the front edge of the seat. An additional diagonal support is recommended between the front corner of the seat and the angled back rest to provide additional stability. Connect the support using two 1/2 inch bolts at each

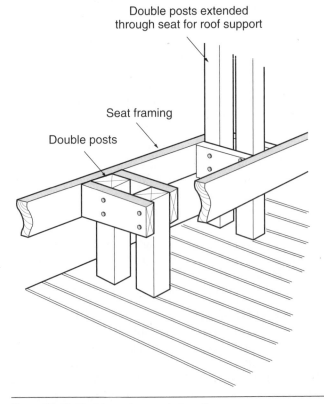

Figure 30–20 Double post support of seating.

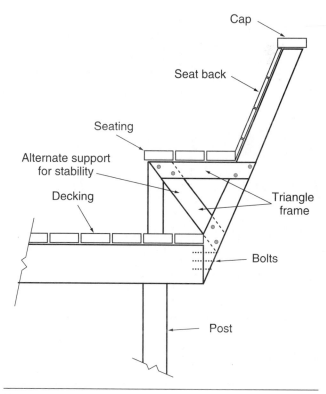

Figure 30–21 Triangle framing for seating.

end. Triangle framing is typically spaced 3 feet or closer and uses 2 × lumber for seating.

Seating Surfacing

Surfacing choices for seating range from wide 2 × lumber placed flat on the surface, 2 × 4s placed on edge, or 5/4s decking material. Benches that do not utilize a structural framework between posts require 2 × dimensioned lumber such as 2 × 6s, 2 × 8s, or wider, or 2 × 4s placed on edge. The orientation and size of this lumber provide the strength required to support loads without the structural support. Lumber that is placed flat on the surface should be applied beginning at the outer edge of the bench. Overhang the front edge 1 inch. Attach with at least two deck screws per joist and space 1/4 inch between boards. If 2 × 4s are placed on edge, place only one connector per joist. Recess the top of the fastener, drill a pilot hole through the board, and insert one 5 inch lag screw per board.

Smaller dimension lumber, such as 5/4s decking, should be applied flat over a seating structural support system. This lumber will not support a load between posts without structure. If the surfacing materials run the long direction between posts, begin attaching these boards at the outside edge of the bench. Attach with two deck screws per joist. If these seating materials run the short dimension—front to back—over grab beams, center and install the boards

at each end of the bench. Stretch a stringline between the outer edges of these two boards and align the other surfacing with the stringline. Fasten using two deck screws per joist.

Trimming Seating

Trimming can perform two important functions for seating. It can cover unsightly structural work below the seats and it can provide a safety edge that protects the legs of those who sit on the bench. Trim is typically used along the front edge of the seat, the back edge of seats that do not have a backrest, and as a cap for backrests. Typical lumber used for trim is 1 × dimensioned lumber wide enough to cover the structural pieces to which it is attached. Attach trim pieces using deck screws installed in a pattern matching that of the seat surfacing.

Before applying finishes to seating, it is advisable to sand all surfaces and edges that will potentially come in contact with occupants. Corners such as the front edge of the bench, edges of backrests, and the seat itself can be left with sharp edges or splinters that should be removed by sanding.

SKIRTING

Skirting is the screening placed under deck structures to hide structural components. Many decks are low enough that skirting is not required, but if the deck is 3–4 feet off the ground, the underside exposure may be unattractive. Decks that are viewed from a lower vantage point may also require skirting, regardless of their height.

Skirting Framing

The skirting placed under decks requires minimal framing. Skirting can usually be hung from 2 × 4 framing nailed between posts or attached directly to posts and bracing. When framing the skirting, consider framing openings that will allow future access to the space below the deck. This opening can be framed with hinges to create a doorway for spaces that require frequent access.

Skirting Facing

Materials and patterns selected for skirting should match or complement any railing pattern used on the deck. In instances where there is no dominant pattern, premanufactured lattice makes an adequate skirting material. Measure the framework openings and cut the sheets of material to fit. Attach premanufactured sheets using 5d galvanized box nails. To provide a finished look, trim the tops where skirting pieces meet.

Chapter 31

Wood Overhead Structures

Installing a roofed space creates a strong visual element with instant impact. Plants may take several years to provide shade for a patio, but a properly designed canopy provides immediate cover with a strong sense of space. Overhead structures are also an effective way of "extending" the interior spaces outdoors. Porches, sunshades, and other open or closed roofs near a house can create the feel of an additional room. Building roofs for landscape structures can be treated as an integral part of the construction of a porch, deck, or gazebo, or can be a completely independent element in the landscape. This chapter explains the techniques for framing, sheathing, and surfacing roofs used for exterior structures. Methods and techniques described in this chapter would be superceded by any construction documents prepared for a project. Connecting roofs to existing buildings may have to be done by a licensed carpenter. Consult building officials for requirements.

RELATED INFORMATION IN OTHER CHAPTERS

Information provided in this chapter is supplemented by instructions provided elsewhere in this text. Before undertaking activities described in this chapter, read the related information in the following chapters:

- Safety in the workplace, Chapter 4
- Construction math, Chapter 5
- Basic construction techniques, Chapter 7
- Materials for exterior carpentry, Chapter 28
- Wood decks and platforms, Chapter 29
- Wood stairs, railings, seating, and skirting, Chapter 30

 CAUTION Use proper ladder safety and climbing techniques when working overhead and on roofs.

FOUNDATION AND STRUCTURAL SUPPORTS

Support systems perform the important functions of holding the roof up, and holding the roof down. In roof systems composed of open framing with lightweight coverings, the most serious problem can be providing an anchor against wind. For this reason, all roofs must be securely anchored and not just rest on top of a support. Providing a roof for a landscape structure requires support from an existing structure, wall, fence, or the installation of a support system of posts or a framework. Chapter 29 includes directions for footing and post installation.

Supporting the Roof on an Existing Wall or Fence

A simple way to support a landscape structure roof is to place it upon an existing fence or landscape wall (Figure 31–1). The fence or wall must have enough structural strength to support the roof, and requires enough height to provide clearance for the roof. Wall/fence strength may be difficult to determine without knowledge of its original construction, but if the wall/fence is constructed of 4×4 posts sunk to frost depth, or masonry materials with a footing, it may provide enough strength to hold up the roof. If the height is not adequate, short posts can be installed on top of the wall/fence to provide clearance. Connect the roof framing to the existing wood structures using

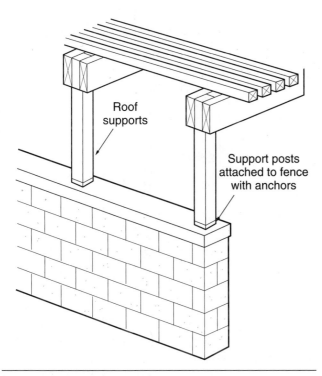

Figure 31–1 Supporting roof structure with fence framework.

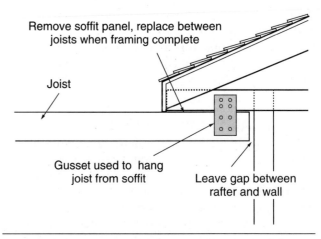

Figure 31–2 Connecting roof structure to building below soffit.

metal joist supports and to masonry structures using joist supports anchored with expanding bolts installed in pilot holes. Roof support posts can be installed on fences and walls by anchoring a post support to the fence or wall. These support posts work best if aligned with existing posts on a fence and fastened using specialty connectors. Consult with a design professional if there are questions about the ability of a wall or fence to support a roof.

Supporting the Roof from an Existing Building

In some designs, the landscape structure roof may have to connect to an existing building. If the roof is to be installed flush to a building wall, follow the instructions provided in Chapter 29 for attaching a ledger plate. Building attachments may also be accomplished by hanging the roof under a high eave. If an eave provides adequate vertical clearance for both the building's occupants and the roof structure, connection below the eave is possible. Connecting the roof below the eave will not disrupt roof drainage and minimizes future problems when reroofing. This attachment can be accomplished by hanging the roof framing under the eave using galvanized strapping that is lag screwed into the building rafters. An alternative that provides a more stable and attractive connection is to remove the **soffit,** or flat panel that covers the underside of the eave, and use gusset plates to bolt the roof framing directly to the building rafters (Fig-

ure 31–2). Install new soffit panel and trim that cover the connections, open eave area, and spaces between rafters.

Supporting the Roof from Posts and Frameworks

Frameworks that independently support landscape structure roofs require some form of post structural network. Many small posts or a few large posts perform the dual functions of support and anchoring. Post installation, either as part of a deck structure or as an independent roof system, is described in Chapter 29. When installing a roof support as part of a deck structural system, use square posts such as 4 × 4s as a minimum size.

Techniques that make structural support more attractive include using multiple posts, building up the posts, or trimming the posts. Multiple post techniques utilize two, three, or four posts at each corner to support the roof. All of these supports do not need to be placed on a footing, since typically only one is providing structural support. The space between the posts can be filled with lattice work, seating, or shelves for plants. Built-up posts add 2 × lumber to the sides of the posts to create a bulkier aesthetic and structural look. Attach 2 × 6s to opposite sides of each post with deck screws to improve the appearance. Trimming can be accomplished by **cladding,** or completely covering, all four sides of each post with 1 × lumber attached with 5d box nails or deck screws. This also creates a bulkier look with a more aesthetically pleasing appearance. An endless variety of treatments is available by combining these techniques.

Unless limited by an existing structural system, the choice for posts does not need to be limited to 4 × 4s. Built-up posts, structural timbers (6 × 6s and larger), round posts, prefabricated columns, masonry columns, or metal framework can be used to hold up the roof. Notching a flush connection area for rafters and beams on round posts makes a stronger structure. Connections for pre-fabricated components should be made using connectors recommended by the manufacturer.

ROOF BEAMS/RIDGE PLATES

Rafters used in the roof substructure need supports to which they can be connected. Rafters can be supported directly on a ledger plate or hung from eave as described in the previous section. Rafters can also be attached to a beam that is supported on posts (Figure 31–3). Attach the beams to the posts using two 1/2 inch bolts or a saddle. Verify that the system supporting the rafter and/or beams/ledger plates is square and level. Minor variations from square will make cutting rafters difficult. Adjust beams and ledgers as necessary before installing rafters.

Ridge plates are structural pieces that support the high end of rafters in peaked roofs. The ridge plate requires support to hold it in position. The ridge plate can be supported by being connected directly to a building with a joist hanger or set on top of a post (Fig-

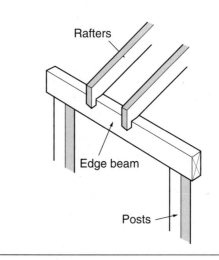

Figure 31–3 Roof beam at edge.

ure 31–4). Long structures may have several such supports on which the ridge plate is to rest. The structural support method will need to be installed before the ridge plate can be set.

To install the ridge plate, cut a 2 × 10 the length of the roof and attach it to the support. The position and

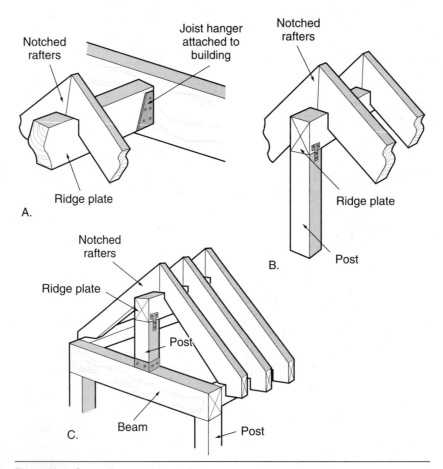

Figure 31–4 Supporting structure roof beams.

level of this ridge plate must be accurate. Fasten securely using metal straps or anchors at each post location.

ROOF SUBSTRUCTURE

Roof substructures perform the same functions as the deck substructure performs—supporting the loads of the structure. A major difference exists in structural design between the roof and deck in that the roof seldom carries as heavy a live load as the deck. Since the only live loads on a roof occur during construction and repair, the sizes of rafters are typically smaller than the normal floor joist.

Roof design can also be classified according to whether the surfacing is open or closed, and by roof styles such as shed, gabled, or hubbed. Open-surfaced roofs have no permanent solid surfacing to prevent the elements from passing through. Open-roof surfaces can be covered with lath, lattice, louvers, or dimensioned lumber with open spacing. Open roofs are prime candidates for covering with vines or fabric. Closed roofs are built with a weather-impermeable covering such as shingles, metal panels, or fiberglass panels. Closed roofs require more materials and labor, but provide protection from the elements for the occupants.

Roof styles for exterior structures typically fall into three categories. Shed roofs have a single-sloped surface running from a high point on one edge to a low point on the other. Gabled roofs have dual slopes, each running down from a central high point. The gable is the ridge that divides the two sloped sections. Hubbed roofs are a gabled roof with several facets coming together at a single high point in the center. Commonly used in freestanding structures such as gazebos, hubbed roofs can have anywhere from four to eight or more sloped sections. Hubbed roofs are the most difficult of this group to construct, but provide the highest level of aesthetics.

Shed Roof Framing

Shed roofs are framed by cutting and placing rafters along the length of the roof between supports located at both ends of the rafter, typically two beams or a beam and ledger (Figure 31–5). Before cutting rafters, verify that the supports for the roof are level and square. To make a pattern for a shed rafter which abuts a building, select a piece of rafter lumber and place it on top of the supports. Mark the angle at the top by placing a block of lumber against the building wall and transferring the angle to the rafter (Figure 31–6). Cut the angle from the rafter and use the scrap to mark and cut the same angle at the bottom of the

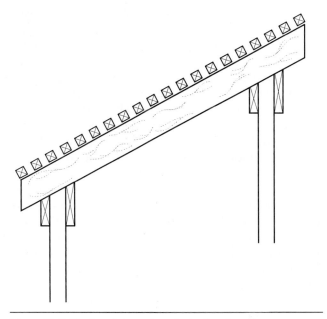

Figure 31–5 Shed roof slopes in a single direction.

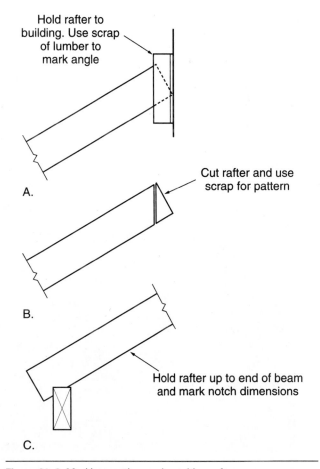

Hold rafter to building. Use scrap of lumber to mark angle

A.

Cut rafter and use scrap for pattern

B.

Hold rafter up to end of beam and mark notch dimensions

C.

Figure 31–6 Marking, cutting, and notching rafters.

rafter. When no building is available as a guide, use a carpenter's level to plumb a line from the top corner of the rafter. Return the rafter to the supports and hold it against the ends of the supports. Mark the location for a notch where the rafter will set on the support. The bottom will always be notched if resting on a beam, and occasionally the top will also be notched. Cut out this notch and check the rafter for fit at each location where a rafter will be located. If cut properly, use this rafter as a pattern for all rafters required for this section of the roof.

Begin installing rafters at spacing determined by design or calculation. Connect the rafters to the supports by toenailing with 16d nails. If attaching to a ledger, use joist hangers. For roofs that are designed with exposed rafters, blocking and bracing may be required to maintain spacing and/or to prevent twisting or toppling of rafters. For roofs that have covered ends, attach a piece of 2 × fascia across the end of the rafters at the bottom edge of the roof (and the top if not attached to an existing building). This lumber should be wide enough to cover the ends of the rafters and should be connected by endnailing with three 16d nails or two deck screws per rafter.

Gabled Roof Framing

Gabled roof framing relies on rafters that are attached to both sides of a ridge plate at the top of the gable and resting on beams at the bottom (Figure 31–7). Mark and cut rafters in a manner similar to that used for

shed roofs. Check the rafter for fit and, if properly cut, use as a pattern for all rafters in this section of the roof.

Begin installing rafters at spacing determined by design or calculation (Figure 31–8). Connect the rafters to the ridge plate using joist hangers or a lag screw on the top and by toenailing with 16d nails on each side. Attach to the beams by toenailing with 16d nails. For roofs that are designed with exposed rafters, blocking and bracing may be required. For roofs that have covered ends, attach a piece of 2 × fascia across the end of the rafters at the bottom edges of the roof. This lumber should be wide enough to cover the ends of the rafters and should be connected by endnailing with three 16d nails or two deck screws per rafter.

Hubbed Roof Framing

Roofs that have multifaceted roof segments require a central point at the top to which rafters can be connected (Figure 31–9). Depending on the number of facets the roof contains, this can be accomplished with a 4 × 4 block, a special six- or eight-sided hub block, or a metal hub kit. For each side of the hub, a rafter will run down from the hub to rest on the beams (Figure 31–10). Each rafter supports the roof surfacing for two adjacent facets. Roof support is provided by the downward pressure of all of the rafters on this hub and the support beams around the perimeter of the roof.

Angles and notching for hubbed roof rafters have to be calculated and marked from designs or measure-

Figure 31–7 Gabled roof has a long peak and two sloped sections.

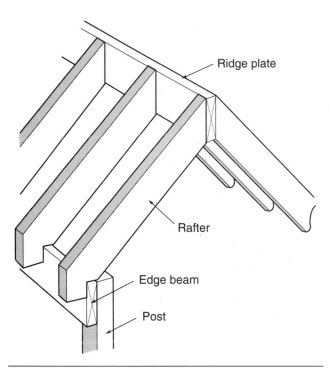

Figure 31–8 Framing a gabled roof.

Figure 31–10 Closeup showing six rafters connected to hexagonal hub.

Figure 31–9 Hubbed roof has a high point with several sloped sections, or facets.

ments provided by a hub manufacturer. Mark and cut two rafters and tack to opposite sides of the hub, then hold in place to check for fit in all directions. If the rafters are properly cut, use as patterns and cut all rafters for the project. Rafters should be connected to a wood hub using a lag screw on top and by toenailing with 16d nails on both sides. For rafters connected to metal hubs, use fasteners recommended by the hub manufacturer. To support this hub during construction, cut three rafters and attach to the hub. Place the rafters and hub on the support beams and tack in place. With the hub temporarily supported, other rafters can be attached and fastened.

For roofs that are designed with exposed rafters, block or brace as required to maintain spacing. For roofs that have covered ends, attach a piece of 1 × fascia across the end of the rafters at the bottom edges of the roof. This lumber should be wide enough to cover the ends of the rafters and should be bevel cut to match the angle between roof facets. Connect by end-nailing with three 8d nails per rafter.

Additional Structural Preparation

Because of their height, exposure to wind, and snow loads, bracing may be necessary for lateral stability required to keep roofed structures upright and square. Methods commonly used to strengthen roofed structures include bracing between the posts and beams, use of railings, or use of interior joists.

Bracing between roof posts and roof support beams is typically short blocks of wood securely fastened at an angle between these two structural members (Figure 31–11). These braces should be 1–2 feet long and fastened with 3/8 inch bolts or lag screws. Gazebos often make use of decorative trim braces for this function. Installation of railings can provide some additional structural stability to a roof structure. Securely fasten the stringers of the railing between the posts supporting the roof.

ROOF SURFACING

Choices for covering a roof include a wide range of open and closed surfacing materials. The material installations described here include the basic coverings utilized in the industry.

Open-Roof Surfacing

Open-roof surfacing involves the placement of strips of surfacing material, such as grape stakes, lath, 1 × 3s, 1 × 4s, 2 × 2s, or 2 × 4s, with an open space between strips. Open-roof surfacing can be nailed or deck-screwed directly to the structure in most applications (Figure 31–12). Begin application of surfacing

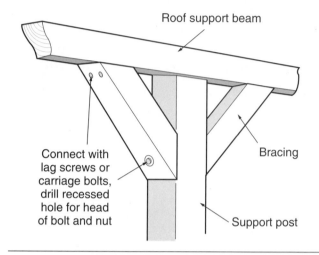

Roof support beam

Bracing

Connect with lag screws or carriage bolts, drill recessed hole for head of bolt and nut

Support post

Figure 31–11 Angle bracing of hubbed roof beams.

Figure 31–12 Open-roof shade structure.

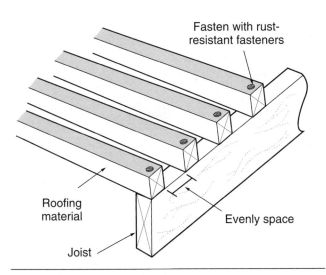

Figure 31–13 Open-roof connections.

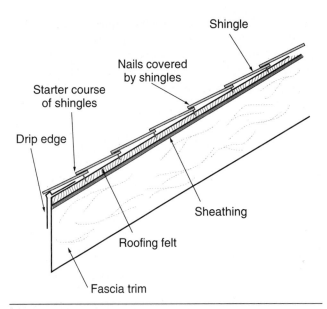

Figure 31–14 Shingled roof cross section.

materials at the top or one edge of the structure by installing the first piece and using a spacer to maintain even distances between subsequent pieces (Figure 31–13). When approaching the end of the installation, adjust the spacing slightly, if necessary, to accommodate any shortage or extra space left. Lattice work can also be directly nailed to the top surface of the structure. Cut lattice pieces so the joints butt directly over a joist and the pattern matches from one piece to the next.

Open-surfacing materials that are installed between joists can also be preassembled in panels that are connected to the joists. Cut and assemble panels using a 1 × 3 frame around the outside. Materials such as lattice or louvers can also be installed directly on cleats already installed inside the joists. Connect materials using deck screws or 8d box nails.

Open-roof surfacing may require trimming out the structure before or after surfacing is applied. Structural grade lumbers can be trimmed with 1 × stock for a more finished appearance. Joints in lattice work should also be covered with 1 × stock trim.

Fabric Roof Surfacing

Covering an outdoor roof with fabric can range in complexity based on the what the fabric is expected to do. Standard sizes of duck and vinyl drop cloths can be purchased, or a custom covering can be sewn by an awning company. Stationary fabrics can be fastened in place by inserting lag screws through **grommets** installed along the sides of a fabric. Moving fabrics require specialized framing and retraction systems that should be installed by the manufacturer. The framework placed under the fabric is important to proper functioning of the covering. The framework should not be constructed of materials that will catch

or rip the fabric and should hold the fabric taut so that moisture can drain from the roof.

Asphalt Shingle Surfacing

Three-tab asphalt shingles are a common and durable roof surfacing for many landscape structures. Available in a variety of colors and relatively easy to install, asphalt shingles are a versatile way to weatherproof an outdoor space. Shingled roofs are composed of many components installed in a particular order to maintain waterproof conditions (Figure 31–14).

Special Materials, Nails, and Techniques for Shingle Roofs. Cutting and nailing shingle roof materials require special tools and fasteners. Both roofing felt and shingles can be cut using a carpet or utility knife. Roofing felt will cut completely through, while shingles must be scored along a straightedge and bent to break through the surface. Nailing shingles requires the use of 1-1/4 inch roofing nails. Drive nails snugly into the surface without puncturing or breaking the surface of the shingle. Use four nails to fasten each shingle—spaced 1 inch in at each end and one at each shingle notch placed slightly above the black asphalt tack strip.

Sheathing. Surfacing begins with the installation of roof sheathing material. Common choices for sheathing include 1/2 inch exterior grade plywood, or **OSB.** Measure and cut sheathing material to fit the rafter spacing for the structure. Joints should be centered on rafters and each piece of sheathing should cover at least three rafters. Begin installation by placing the first piece of sheathing flush with the bottom and one edge of the roof. Nail in place with 8d ring shank box nails every 8 inches along each rafter. As the application of sheathing moves up the roof, stagger the vertical joints as much as possible. Horizontal joints for roof sheathing should be spaced 1/4 inch. Special spacing clips are available.

Subroofing. Fifteen pound roofing felt is installed next over the sheathing. Align the first piece horizontally and flush with the bottom edge of the roof and fasten using 3/8 inch staples. Overlap each subsequent piece of roofing felt 6 inches over the lower piece. Continue installing felt to the peak of the roof. At the peak of a gabled roof, install one piece of felt that overlaps both sides of the roof. Beginning at the bottom edge, install aluminum **drip edge** along all edges of the roof. Drip edge can be cut with tin snips and should be installed with 5d box nails. Marking the roof with horizontal chalk lines will help in maintaining straight shingle application. Roofing felt is also marked with lines to help keep shingle application horizontal if the roofing felt is installed horizontally.

Flashing. Flashing is required to prevent moisture leakage at valleys and in each location where items such as vents or fireplaces penetrate a closed-roof system. Flashing should be applied to valleys after any roofing felt is applied and before shingles or surfacing materials are installed. Use roll aluminum or galvanized metal to straddle the valley and extend 6 inches up each side of the roof. When a valley is adjacent to a vertical wall, run the flashing 6 inches onto the roof and 6 inches up the adjacent wall. Attach using a weatherproof glue. Metal flashing should also be installed around each roof penetration. For vents, utilize the premade flashing with a rubber boot that fits around the vent pipe. For other installations, utilize pieces of metal step flashing. Begin installation of step flashing at the bottom of the penetration, with a row of flashing placed over roofing felt and under shingles. Alternate flashing and shingles while moving up the roof, with the flashing always under the shingle. Attach flashing using a weatherproof glue.

Shingling. Shingle application begins with the installation of the starter course. First cut the colored coated tab off the bottom of enough shingles to run the entire length of the roof. Leave the black tack strip on these shingles. Cut the left 6 inches (1/2 tab) off the first starter shingle, place it so the tack strip is flush along the edge and bottom of the roof, and nail it in place (Figure 31–15). Continue placing starter shingles until the entire bottom edge is covered. Nail the first course using a full shingle placed with a 1/2 inch overhang over the left and bottom edges of the roof. Continue placing full shingles with a 1/16 inch gap between shingles along the length of the roof. To

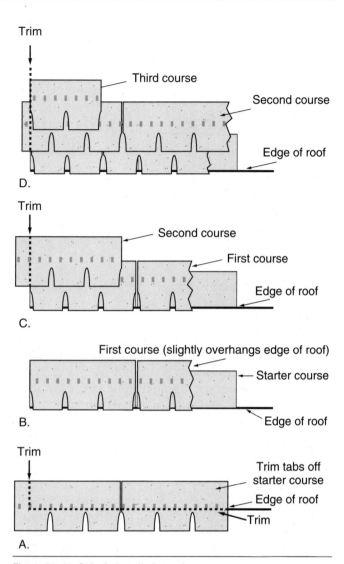

Figure 31–15 Shingle installation pattern.

install the second course, cut 6 inches (1/2 tab) off the left-hand side of a shingle and install that shingle flush with the left edge of the roof. Follow that shingle with full shingles along the entire length of the roof, trimming any excess at the end. Place each subsequent course so that 5 inches of the course below is exposed and the centers of the tabs in the second row are aligned with the gaps in the row below. To begin each subsequent row, cut an additional 6 inches off each first shingle until you begin again with a full shingle (Figure 31–16). This will stagger the joints across the roof. Continue shingle placment until the roof ridge is reached.

Ridge Treatment. Completing the shingling at the top or ridge requires cutting shingles and special installation techniques. If the roof has a ridge, cut several shingles into thirds by scoring and breaking at the tab. Bend the shingle and place, straddling the ridge with the coated tab on the left. Nail the shingle into place

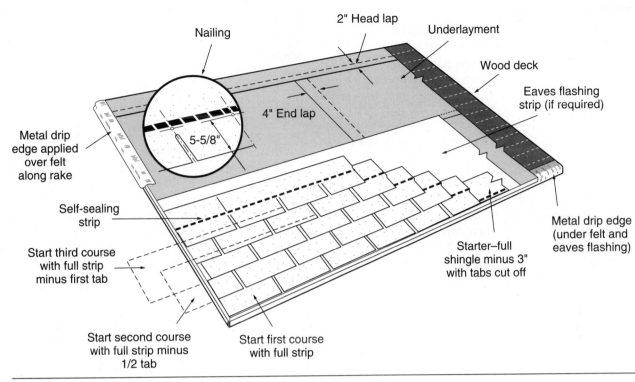

Figure 31–16 Shingled roof components.

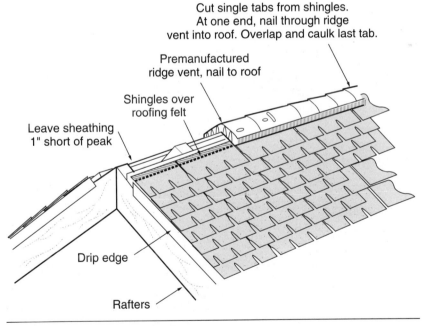

Figure 31–17 Shingled roof ridge treatment.

and lay the next shingle overlapping all but the coated tab portion of the first ridge shingle. Nail in place and continue along the entire ridge. The last shingle will require cutting off all but the coated tab. Nail this last tab over the previous shingle and caulk around the exposed nail holes. If the roof is a shed roof against a building or has no ridge, use the same process with a single row of shingles placed against the building. Roofs that are surfaced on the underside will require leaving a gap in the sheathing, installing a pre-manufactured vent, and installing ridge shingles over this vent (Figure 31–17).

Wood Shingle Surfacing

Wood shingles provide one of the most attractive roof surfacings available for landscape structures. Wood shingles are different in surface texture than wood shakes. Shingles are sawn from red cedar while shakes are split. Wood shingles for landscape structures are best installed over sheathing base, similar to the installation of asphalt shingles. Like asphalt shingled roofs, wood roofs are built of many materials installed in sequence (Figure 31–18).

Special Materials, Tools, and Techniques for Wood Shingles. Wood shingles can be sawn using a circular saw or split using a roofing hatchet. Fasteners used should be 1-1/4 inch to 2 inch 14d galvanized nails. Insert two nails per wood shingle, 3/4 inch from each side, and 1 inch up from the exposure line. The exposure line is the location where the bottom edge of the shingle placed above will rest. The exposure line varies from one-quarter to one-third the length of the shingle. The steeper the **roof pitch** the greater the expo-

sure. Nail the wood shingle firmly but do not countersink the nails into the surface of the wood shingle.

Sheathing. Sheathing installation is the same as the process described for asphalt shingles. Fifteen pound roofing felt is installed over the sheathing. Align the first piece horizontally and flush with the bottom edge of the roof and fasten using 3/8 inch staples. Subsequent courses of roofing felt are cut to an 18 inch width and installed between courses of wood shingles.

An alternative to solid sheathing is to install wood **purlins.** Purlins are wood members that run perpendicular to rafters and provide support and a nailing surface for roofing materials. Purlin attachment requires installing 1 × 4s horizontally across the roof, spaced twice the width of the shingle exposure. For wood shingle roofs, purlins should be solid for the first three courses at the bottom and the outer 1 foot at the **rakes** (Figure 31–19). Fasten the purlins with two 10d common nails per rafter. Roofing felt is not utilized if installing wood shingles on purlins.

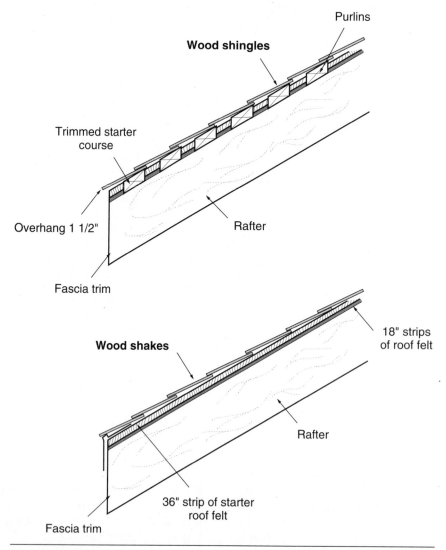

Figure 31–18 Wood shake or shingle roof crosssection.

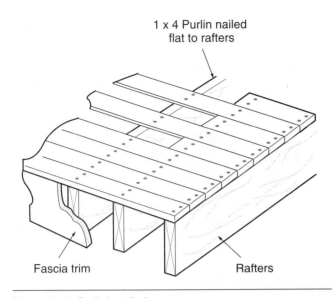

1 x 4 Purlin nailed
flat to rafters

Fascia trim

Rafters

Figure 31–19 Purlin installation.

Wood Shingle Installation. Wood shingle installation begins by trimming 2 inches off enough wood shingles to install one course the length of the roof. Nail this starter course in place with a 1-1/2 inch overhang at the bottom edge and 1/2 inch spacing between wood shingles. Install the first course of wood shingles directly on top of the starter course with the same overhang and spacing, offsetting joints between courses 1-1/2 inches. This may require splitting a wood shingle to begin the course and to finish the course. Overlap each subsequent course over the first so that the proper exposure is obtained (Figure 31–20). With shakes, between each subsequent course install an 18 inch wide piece of builder's felt, with the lower edge of the felt halfway between the next exposure line and the end of the wood shingle. Continue installing courses until reaching the ridge.

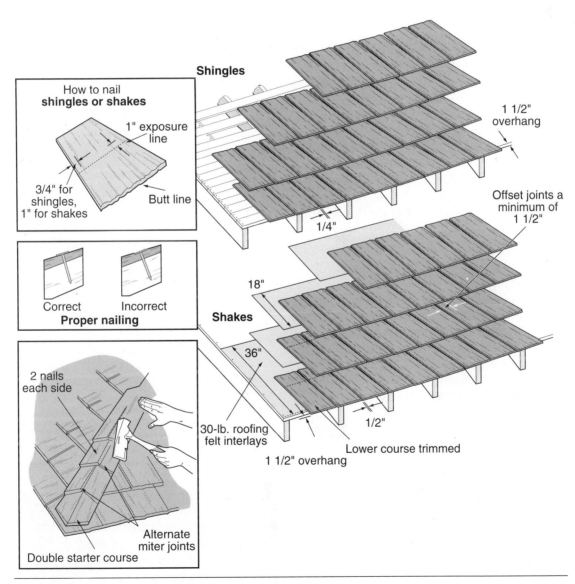

How to nail
shingles or shakes

1" exposure
line

3/4" for
shingles,
1" for shakes

Butt line

Correct   Incorrect
Proper nailing

2 nails
each side

Alternate
miter joints

Double starter course

Shingles

1 1/2"
overhang

1/4"

Offset joints a
minimum of
1 1/2"

18"

36"

Shakes

30-lb. roofing
felt interlays

1 1/2" overhang

1/2"

Lower course trimmed

Figure 31–20 Shake installation pattern.

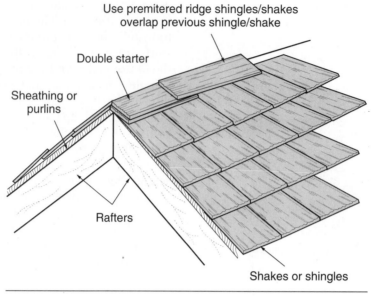

Figure 31–21 Shake roof ridge treatment.

Ridge Treatment. The top or ridge finishing course is laid perpendicular to the direction of roof wood shingles. Finish the top by splitting one shingle and placing it on one side of the ridge, butted against a split shingle on the opposite side of the ridge. Fasten with two nails for each wood shingle. Install a second pair of split shingles that overlap the first pair with double the amount of roof exposure (Figure 31–21). Continue this overlap installation along the entire ridge. Caulk around the final exposed nails. If the roof is a shed roof against a building or has no ridge, use the same process with a single row of wood shingles placed against the building.

Metal Roof Surfacing

Metal roofing provides a quick and relatively inexpensive way to provide cover for an outdoor shelter. Available in a variety of forms and colors, metal roofing is installed over 2×4 purlins attached to the roof framing or installed over a sheathed roof (Figure 31–22). Metal roofing has the disadvantage of being a rather unattractive surface when viewed from below.

Special Materials, Tools, and Techniques for Metal Roofing. Metal roofing is attached using 2 inch roofing nails with a rubber washer around the shank. Metal roofing can be cut using tin snips or a circular saw with the blade reversed. The reversed blade "grinds" a clean cut through the metal. When using a circular saw, utilize an old blade or buy an inexpensive blade. Nails must be placed along the ridges of roofing and not in the valleys. Pilot holes may be drilled or nails can be driven through the material directly. Install nails at each purlin location.

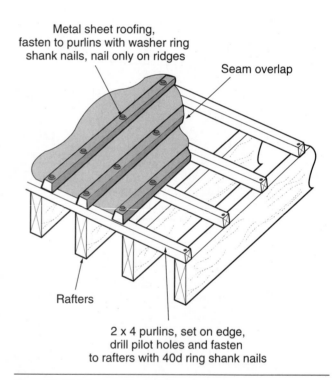

Figure 31–22 Metal roof installation cut-away.

CAUTION

Support metal and fiberglass under both sides of the cut. Wear gloves and safety glasses. Use caution when handling metal and fiberglass, particularly after it has been cut. The edges are very sharp and have small metal slivers and fiberglass shards exposed after cutting.

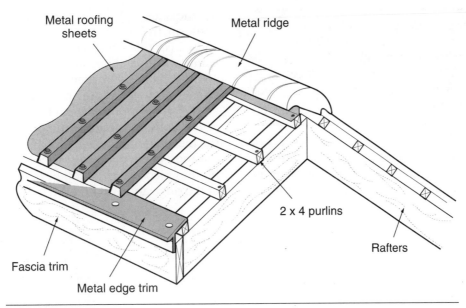

Figure 31–23 Metal roof trim installation.

Purlin Installation. Purlins are attached to the rafters by placing horizontal 2 × 4s on edge every 2 feet up the rafters. Drill pilot holes and attach the purlins to the rafters using 5 inch ring shank nails.

Trim. To finish edges and cover exposed rough cuts, premade metal trim is available that covers corners and ridges. Most trim pieces are installed before surface material is attached (Figure 31–23). Measure the area that needs to be trimmed, cut with tin snips, and install with ring shank nails.

Surfacing Attachment. Beginning at the outside edge, place the first piece of steel over the purlins aligned with the bottom and the edge and nail it in place. Most steel pieces have one edge formed to overlap and one edge formed to accept another piece. The pieces should be placed in order so each one properly overlaps the previous piece. If the pieces are not long enough to extend to the top of the roof, place the lower pieces first and overlap the bottom piece with the top by 6 inches. Fit the pieces into the previously installed trim.

Corrugated Fiberglass Surfacing

Fiberglass surfacing comes in a variety of colors and translucencies that produce an inexpensive roofing choice. Fiberglass is less expensive than other roofing materials but is also less durable. Fiberglass also requires a specialized structural system to match the material sizes (Figure 31–24).

Special Materials, Tools, and Techniques for Fiberglass Roofing. Fiberglass roofing is attached using 2 inch roofing nails with a rubber washer around the

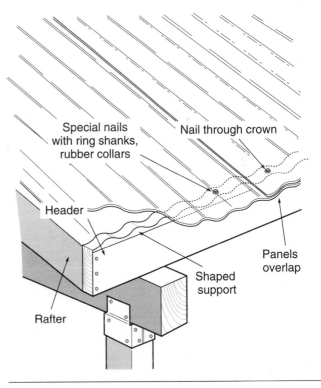

Figure 31–24 Corrugated fiberglass roof cross section.

shank. Nails must be placed along the ridges of roofing and not in the valleys. Pilot holes must be drilled using a backing block to avoid fracturing the fiberglass. Fiberglass roofing can be cut with a circular saw using an abrasive or plywood blade. Support the fiberglass on both sides of the cut and observe the safety precautions stated earlier.

Structural Support. Fiberglass sheets should be placed and nailed directly to the rafters. Blocking is placed between rafters every 2 feet to provide support for the sheets. A special formed trim is available to conform the top of the blocking to the corrugated sheets. For this reason, the rafters must be spaced the same width as the sheets of fiberglass, typically 16 or 24 inches. The pieces should be placed in order so each one overlaps the previous piece directly above a rafter. A bead of caulk can be placed along the ridge at the overlap to reduce the possibility of leakage. If the pieces are not long enough to extend to the top of the roof, place the lower pieces first and overlap the bottom piece with the top by 6 inches. Trim is placed at the ridge and around each of the edges of the roof.

A variety of wood structures is available for entertaining, functional uses, and to simply create space in the landscape. To address this variety of structure types, Section 7 presented carpentry and the construction of wood decks and landscape structures by describing the components of the construction. Covered in this section were basic materials, preparation of foundations and footings that support structures, building decks and platforms, installing functional additions such as railings, stairs, seating, and skirting, and, if the structure requires, construction of roofing and covering options. This approach allows the contractor to select à la carte from the components of wood landscape construction to fabricate any type of structure desired.

Wood structures can take many forms in the landscape. A standard for exterior use areas is the deck, but overhead structures such as screened porches, sunscreens, and arbors are also common in the industry. Gazebos combine the elements of several structure types to create a standalone space.

When planning exterior structures, consideration must be given to selection of proper materials. Due to the exposure to the elements, components must be selected that are resistant to decay, rust, and other hazards of exterior use. Additional considerations for selection of wood products include strength, grade, availability and cost, and aesthetics. Quality of wood construction is affected by the quality of the products used. Awareness of problems with warping, checks, shakes, and knots will help the contractor select appropriate materials for building. Availability of a wide range of wood products, fasteners, connectors, and premanufactured materials makes the carpenter's work more stable as well as more productive.

Preparation of the site and installation of support mechanisms for wood structures are necessary for long-term improvements. Landscape structures can be anchored to existing structures, anchored to existing slabs or footings, or have separate supporting posts installed. Posts installed to support a structure can be anchored to the top of a frost footing or buried directly. In climates where freezing occurs, these supports should extend to frost depth. For structures where movement is not a problem, or freezing conditions seldom occur, precast pier supports can be used to support structures.

Construction of decks and platforms involves the framing of a substructure on which a surfacing material is placed. Two choices for substructure are considered for landscape work—post and beam (also called plank and beam or drop beam) framing or platform (also called box or flush beam) framing. Post and beam involves the installation of posts (or other support method), attachment of beams to the posts, and setting joists on top of the beams in order to support the deck surfacing. Platform framing involves the installation of posts (or other support method), attachment of a frame of joists to the posts, and hanging joists inside the box. Surfacing is then applied over this framework. Each method has benefits and drawbacks, but proper sizing and spacing of materials will make each structurally sound. When constructing the framework, variations such as multilevel decks and cantilevered decks are possible. Support for benches, railings, and roofing must be considered before surfacing proceeds. Surfacing the framework can be accomplished using specialty decking lumber or standard dimensioned lumber. Trimming the structural components provides a finished look to the structure.

To add safety and convenience to exterior structures, adding stairs, railings, seating, and skirting to a platform should be considered. Stair configuration can take almost any form, but requires landing space and framing similar to that of the platform of the deck. Supports are carefully measured, cut, spaced, and attached and surfacing is added to provide a walking surface. Railings require the installation of structural supports that will resist the forces of people leaning against the rail. Supports may be attached to the outside edge or inside of the platform framing, and may be extensions of the posts that support the platform. Stringers are connected between supports and surfacing is applied, which creates a barrier to accidental falls from the platform. Seating for decks can be framed using supports similar to railings or specialty framing such as grab beams or the triangle frame. Surfacing is applied and, if required, backrests are installed. Skirting to hide the underside of a structure is framed between posts, with surface screening such as lattice applied to the framework.

Overhead roofing requires determination of the type of roof and the materials from which it will be constructed. Simple roof styles such as shed, gabled, and hubbed roofs are common for landscape structures. Material selections can include open-style lattice or wood, fabric, asphalt or wood shingles, metal, and corrugated fiberglass. Once choices have been made for style and material, the support for the roof is installed. Support can be from an existing structure or posts extending up from the ground or a platform. Roof framing includes installation of support beams around the perimeter of the roof that will support

rafters. Special measuring and cutting techniques are required for roof framing due to the angles and dimensions common in each of these roof styles.

When framing is complete, the selected surface material is installed. Open-roofing materials are spaced and attached according to the desired pattern. Fabric roofs may require less support, but still require sound attachment to anchor the fabric. Asphalt shingle roofs require the installation of sheathing and roofing felt over the rafters and special treatment of the edges and ridge to prevent moisture problems. Wood shingles also require sheathing or wood purlins and special ridge treatment. Both asphalt and wood shingles require starting and placing shingles in a specific pattern to assure watertight installations. Metal roofs require the installation of purlins or sheathing onto which metal sheets are attached. Special edge and ridge treatments are necessary to prevent leakage and exposure of sharp metal. Framing for corrugated fiberglass requires spacing rafters to support the edges of fiberglass sheets, and the placement of additional supports under the sheets. All roofing materials have specialty fasteners designed to effectively connect the material to the support framework.

Section 8

Fences and Freestanding Walls

INTRODUCTION

The chapters of this section describe the construction materials and installation techniques for a variety of fences and freestanding walls. The fencing chapter covers fences made of prefabricated wood panels, wood rails, wood surfacing mounted on stringers (horizontal supports), chain link, decorative metal, and vinyl. The freestanding wall chapter discusses two different fences that are comprised of basic materials: those that are stacked dry such as fieldstone, or those that are mortared together, such as stone, brick, and concrete block.

PRECONSTRUCTION CONSIDERATIONS FOR FENCING AND FREESTANDING WALLS

Before installing any fence or freestanding wall, it is advisable to have the property lines located. Do not rely on recollection or property owner descriptions for boundary locations; it is considerably cheaper to locate the property line than to relocate the fence or deed away the small strip of land the fence rests upon after an error has been made. Property lines are based on legal descriptions, and it may require a surveyor to locate the actual corner. It may also be beneficial to construct the fence 1 foot inside the property lines to

maintain a margin of error. Flagging the locations of all posts, gates, and openings before surveying will allow adjustments to be made before posts are installed.

When constructing fences, it is advisable to obtain permission from adjacent property owners for access to their land during construction. Minimal space is required for most fence or wall projects, but even a small encroachment may create problems. Always leave adjacent property clean and in good condition.

When construction documents do not dictate which surfacing must be selected for a fence, consideration should be given to using a material that is finished on both sides. A fence that is visually attractive to all affected parties will be more pleasantly accepted than one that pleases only the owner.

PRODUCTIVITY SUGGESTIONS

The following suggestions may provide general ideas from which project productivity can be improved:

- With masonry, lay out the base course to verify the number of units required for a wall. Begin work by building opposite corners and stretch a mason's line between corners to maintain alignment and elevation.

- Measure and flag post and gate locations before auguring holes.
- To improve the stability of chain link fencing placed in areas exposed to high wind, replace a line post every 50 feet with an end post set in concrete.
- Before building custom gates, check the opening for square.
- Both sides of gates should be at the same level for the gate to function properly. Adjust the grade before installing posts.

- If permanent obstructions are encountered when installing posts for fences, trim the post bottom to match the hole depth. This approach should be used only for single posts over obstructions that will not move with frost. A minimum burial depth of 18 inches should be maintained.

Chapter 32

Fencing

The variety of fence types available makes describing fence construction a difficult task. Rather than explaining every possible fence pattern and facing material, this chapter discusses fence construction in general terms for the basic materials and methods available. While this chapter does not cover every possible fencing material, it does provide information about the basic choices being used by the industry today.

Wood fencing offers the widest variety of options, with prefabricated panel fences, rail fences, fences with horizontal or vertical surfacing, and combinations of each of these styles. The first section of this chapter describes how to install posts and attach surfacing for prefabricated panel fencing. The second section covers wood fences constructed of horizontal rails. Horizontal rails are often left unsurfaced to create an open style of fence. Rail fences can also be modified by adding surfacing, such as **pickets,** boards, and other forms of vertical decoration to create a more opaque fence. The third section describes the installation of custom-built wood-surfaced stringer fences with opaque surfacing. This surfacing requires the installation of stringers, or horizontal supports between the posts, for support. Chain link, decorative metal, and vinyl fencing are described in the fourth section of the chapter. Chain link fencing is a topic by itself because the construction of this type of fence has developed into a specialized technique. Decorative metal and vinyl are also specialized applications that have unique installation techniques.

RELATED INFORMATION IN OTHER CHAPTERS

Information provided in this chapter is supplemented by instructions provided elsewhere in this text. Before undertaking activities described in this chapter, read the related information in the following chapters:

- Safety in the workplace, Chapter 4
- Construction math, Chapter 5
- Basic construction techniques, Chapter 7
- Materials for exterior carpentry, Chapter 28
- Wood decks and platforms, Chapter 29

MATERIALS FOR FENCING

Wood Prefabricated Panel Fencing

Wood prefabricated panel fencing can create an attractive perimeter for a lawn enclosure. Panel screen surfacing can either be horizontal or vertical, or composed of both patterns to reduce monotony. Surface panels can also be custom manufactured to tailor an environment visually or structurally. Compared to other fencing materials, wood panel is moderately priced and ranks as average in maintenance, requiring periodic refinishing and repairs. Installation requires basic carpentry skills and good planning. When creatively used, wood panels can traverse any slope and perform any function expected of fencing material.

Wood Rail Fencing

Rail fences are typically used to define boundaries, since they are transparent and open (Figure 32–1). Most rail fences are composed of two or three horizontal rails mounted into posts. Typical examples include **split rail,** woven board, and three-board fences. Variations of the rail fence include picket fences and other types, which add an open, vertical surfacing to improve the enclosure properties of the fence. Most rail fences are easy to install and the material is inexpensive. Aesthetics vary based on the design style selected. Maintenance includes repair and resurfacing, with periodic restraightening of posts in frost-prone areas.

Wood-Surfaced Stringer Fencing

Wood-surfaced fences are utilized primarily for privacy and screening. Similar in appearance to prefabricated panel fencing, wood-surfaced fences differ from prefabricated fencing in that they are custom-constructed using a design plan. The surfacing for stringer fences is spaced close together, rendering it more opaque than its rail counterpart (Figure 32–2). Most wood-surfaced fences are also taller than rail fencing. Surfaced stringer fences are more expensive than panel or rail fences and require more carpentry

Figure 32–1 Wood rail fencing.

Figure 32–2 Wood-surfaced stringer fencing.

skills to install. Maintenance and durability are similar to that of panel fencing.

Chain Link Fencing

Chain link fencing is a woven metal fabric that is stretched between a metal post framework (Figure 32–3). Chain link is durable for securing the perimeter of a site, but it is typically thought to be an unattractive material. Improvements have been made in appearance, however. The components of chain link can now be colored, and new decorative slattings provide some aesthetic quality. Chain link is available in a variety of heights and can be installed on flat and moderately sloped sites with minimal difficulty. Chain link fence is relatively maintenance free.

Decorative Metal Fencing

Decorative metal fencing plays an important role in historic and thematic landscape design. Metal fencing is manufactured of wrought iron or welded tubular metals such as aluminum, and is constructed by installing panels between preset posts or standards (Figure 32–4). Decorative metal fencing requires a high level of effort and expertise for installation and may also require field painting. While considered very high in aesthetic value, the fencing is also high in cost when compared to other fencing materials. Decorative metal fencing separates spaces, but does little to screen

unless covered by plantings. Metal fencing is average in maintenance, requiring periodic repainting.

Vinyl Fencing

Vinyl fencing is a plastic product available in simple residential patterns that mimic wood rail fencing (Figure 32–5). Functions performed by vinyl fencing are limited to boundary identification and possible enclosure. Costs are moderate and installation is relatively easy. Visual appeal is limited because of the available patterns and colors. Maintenance is virtually nonexistent.

PLANNING THE PROJECT

Patterns

Fence design is limited only by the imagination and the properties of the materials utilized. Figures 32–6 and 32–7 illustrate some of the common fence patterns available.

Figure 32–3 Chain link fencing.

Figure 32–4 Decorative metal fencing.

Figure 32–5 Vinyl fencing.

Fence Alignment and Wind Resistance

Fences with solid panels provide significant resistance to the wind, resulting in problems staying upright in strong winds. Adding variations in alignment to a fence increases its resistance to wind. Consideration should be given to designing the fence with **niches,** or corners, and avoiding long, straight runs of fence. While building the fence, temporary bracing should be placed for support.

Layout

Begin a fence project by installing a stringline at the fence alignment (Figure 32–8). Locate all posts along this alignment using flags or paint markings. Posts should be marked center to center. Locate any slopes where post pattern or length may have to be adjusted to step up the grade. Once end posts have been set, a stringline can be run from one end post to the next. This will provide a guide for setting the depth of posts between corners.

Fencing on Slopes

Many fence types can follow moderate slopes without losing function and aesthetic appeal, but other fence types require special planning and construction

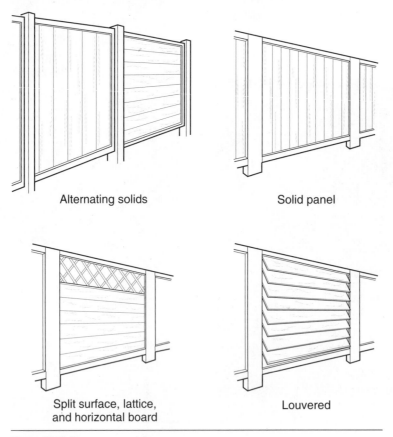

Alternating solids

Solid panel

Split surface, lattice,
and horizontal board

Louvered

Figure 32–6 Common panel fence patterns.

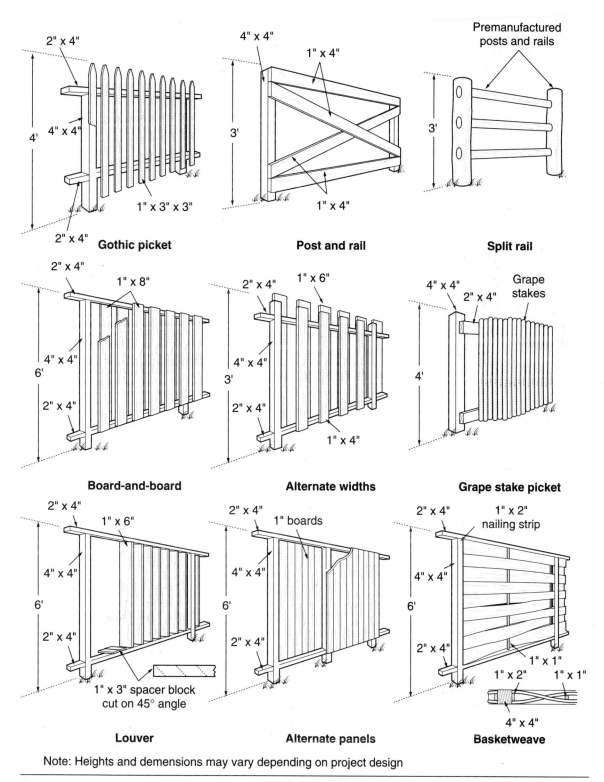

Note: Heights and demensions may vary depending on project design

Figure 32–7 Common rail and wood-surfaced stringer fence patterns.

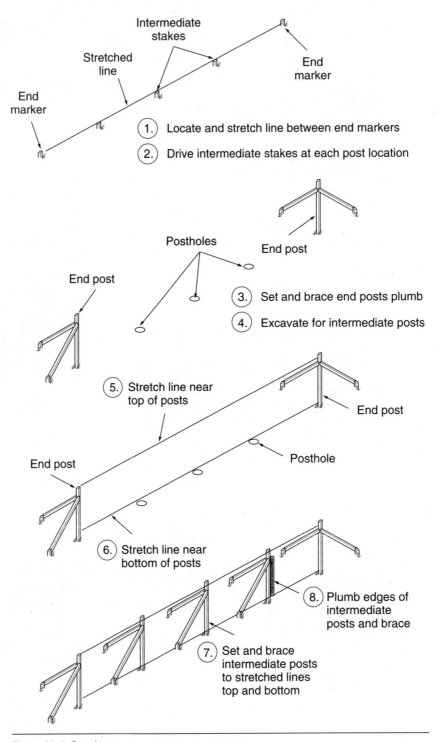

Intermediate
stakes

Stretched
line

End
marker

End
marker

① Locate and stretch line between end markers

② Drive intermediate stakes at each post location

Postholes

End post

End post

③ Set and brace end posts plumb

④ Excavate for intermediate posts

⑤ Stretch line near top of posts

End post

End post

Posthole

⑥ Stretch line near bottom of posts

⑧ Plumb edges of intermediate posts and brace

⑦ Set and brace intermediate posts to stretched lines top and bottom

Figure 32–8 Post layout.

techniques when traversing hills. Wood rail and vinyl fences are typically installed with the posts vertical and the railing parallel to the ground line. On most grades this requires little adaption of construction techniques.

Chain link fence is installed on slopes with an end post in concrete at both the bottom and top of the slope. This sloped section of the fence can be stretched separately from other fence sections. When faced with grades over 1:3 (1 foot of rise for every 3 feet of run), chain link fence can be "stepped" up slopes. To do this requires a great deal of special construction. Stepping chain link up a slope requires end posts sunk in concrete along the rise of the slope to create hori-

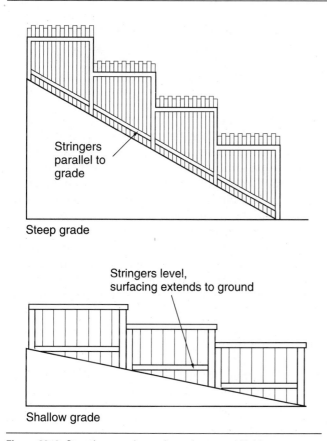

Figure 32–9 Stepping panel or stringer fence up hillsides.

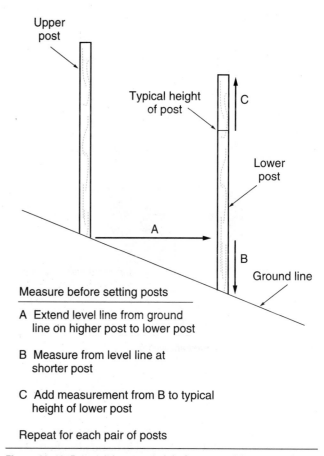

Measure before setting posts
_____

A Extend level line from ground
line on higher post to lower post

B Measure from level line at
shorter post

C Add measurement from B to typical
height of lower post

Repeat for each pair of posts
_____

Figure 32–10 Determining post height for stepped fence panels.

zontal panels. Depending on the steepness of the grade, the fence may have several short panels.

Wood panel and decorative metal fencing typically "step" along slopes of all degrees. Wood panel and metal fences are constructed with panels suspended between each pair of posts. To cover slopes, the posts are extended upward and the panels shifted up or down to match the grade. Wood-surfaced stringer fences are also constructed with stepped panels between posts. Stringers can be constructed level on shallow slopes or parallel to the grade on steep slopes (Figure 32–9). If contact with the ground is required to contain children or animals, the surfacing for each of these fence types must be extended down to meet the ground.

Posts on a hillside should also be installed plumb and should have extra height. Do the measurements for extra height before setting any posts (Figure 32–10). To determine how much extra height is required, extend a level line from the ground line on the higher post to the lower post. From that point on the lower post, measure to the ground line. Add that distance to the normal height of the lower post. This process should be used for each pair of posts on a slope until the slope is traversed.

Maintenance Strips for Fences

Maintenance around fence installations can be a time-consuming chore for the homeowner or grounds keeper. When installing fences, consider installing a **maintenance edge** under the length of the fence to reduce the time and effort required to trim and weed. The maintenance edge is a 4 inch thick by 1 foot wide strip of concrete or stone over a weed barrier (Figure 32–11). After the fence alignment is determined, use a sod cutter to remove a single strip of turf along the entire route. Install the maintenance strip surfacing after the posts have been set.

WOOD PREFABRICATED PANEL FENCE INSTALLATION

Wood, prefabricated panel fences are constructed by installing posts and mounting the panels between the posts. Panels can be solid, opaque, or a combination of designs. Most panels are set in a framework that can be fastened to the face of posts or hung between the posts.

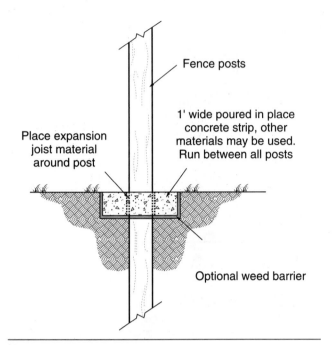

Fence posts

1' wide poured in place concrete strip, other materials may be used. Run between all posts

Place expansion joist material around post

Optional weed barrier

Figure 32–11 Fence maintenance edge.

Post Installation

The spacing of posts for panel fences depends on the panel design and how the panel is to be attached. Face attachment requires different post spacing than hanging; check the panel installation section later in this chapter before locating posts. Mark each post location carefully so no adjustment of panel length will be required during construction. Regular spacing along the entire length of the fence makes installation of the panels easier, but it may be necessary to install one smaller panel to accommodate any irregular distance left over. This small panel can be placed at either end or in the middle of the fence.

In the center of every post location, dig or auger an 8 inch diameter hole 6 inches deeper than the burial depth of the post. Install ground line protection for direct burial posts as described in Chapter 29 under the heading "Foundations for Wood Landscape Structures." Using a marker, draw a line near the base of the post that indicates the proper burial depth. Place crushed aggregate to the bottom of the hole and set posts in the holes. Adjust the post by adding aggregate until the mark is at finish grade. Begin placing backfill in the hole and tamp with a shovel handle. Recheck for height, plumb, and spacing between the posts when the hole is half filled and completely filled. An alternate method for anchoring is to attach braces to hold the posts upright until all posts are set, then backfill with concrete. To save time with height adjustments, mark and set corner posts first, then stretch a string-

line between the tops of corner posts to use as a guide for line posts (see Figure 32–8). If posts are too tall, trim with a circular saw. Refer to Figures 32–9 and 32–10 for placing posts on a hillside.

Panel Installation

Panels for this style of fence are preconstructed and typically attached between posts using metal hangers. Hold the panel in position to check for fit. If too big, trim a small amount off the side of the panel using a circular saw. Mark the top location of all panels using a **chalkline.** Mark the location of the hangers on the sides of the posts. Install the hangers using deck screws or 8d galvanized box nails. Slide the panel down between the hanger flanges. While holding the panel in the proper position and elevation, attach it to the hangers using short screws or galvanized box nails (Figure 32–12).

Another method for installing panels is to attach them to the front of the posts. This method requires spacing posts closer together during the initial phase of construction. To install the panels, hold them at proper height and attach them to the face of the posts using four deck screws per side.

Trimming

Trimming with panel fencing is usually not required. If panels are attached to the faces of posts rather than mounted between posts, consideration should be given to attaching a piece of 1×4 trim with 8d box nails over the joint between panels.

Gates

Gates are typically premanufactured for panel fences, leaving only the hardware attachment to complete. Attach the hinges and latch hardware to the gate panel and prop it in position in the gate opening. Connect the hinges to one post and check the gate for proper swing. Attach the latch hardware to the other post and adjust for proper closing.

For custom-built gates, measure the gate opening and check for square. From these dimensions, subtract hinge and latch dimensions plus an additional 1/4 inch for a safety margin. Construct a gate framework out of 2×4 material that matches the calculated dimensions. Add corner bracing or a diagonal brace made from 2×4s. Apply surfacing that matches the surfacing used for the fence. Attach the hinges and latch hardware to the gate panel and prop it in position in the gate opening (Figure 32–13). Connect the hinges to one post and check the gate for proper swing. Attach the latch hardware to the other post and adjust for proper closing.

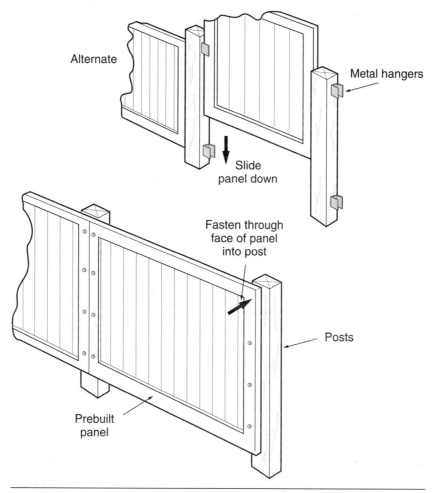

Figure 32–12 Prefabricated panel fencing, panel installation.

WOOD RAIL FENCE INSTALLATION

Wood rail fences are composed of posts with horizontal railings, such as split rail or board. Unlike prefabricated panel fences, rail fences must be custom built with each component attached separately. This requires more labor during construction, but also allows the fence to be adjusted as the installation progresses.

Post Installation

Spacing of rail fence sections is determined by the design and the strength of the material. Regular spacing along the entire length of the fence will make construction of the fence easier. If equal sections are not possible, one small section can be used to finish a run. With preengineered fence systems, the spacing is determined by the length of the rails provided.

Install posts in a manner similar to that used in panel fencing. Verify that posts are plumb (Figure 32–14). If posts are too tall, they may be trimmed with a circular saw. Except for split rail fencing, install posts along the entire fence perimeter. Split rail fencing should have only the end post set and the remaining posts left loose in the hole until rails have been attached.

Posts installed on a hillside for split rail and three-board fences should be installed plumb and at the same height above grade as posts on level ground. Railings or face boards are placed at an angle between posts parallel to the slope. Determine the correct hole location by holding one of the rails between an installed post and the next post to be dug. Posts for other rail fences installed on a hillside should be installed plumb and have extra height similar to panel fences (See Figures 32–9 and 32–10). The surfacing on this type of fence will be installed level between posts.

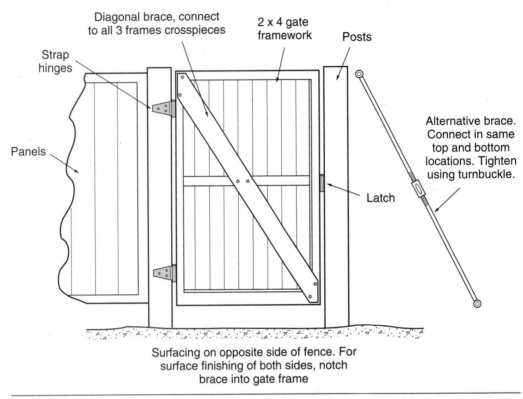

Figure 32–13 Prefabricated panel fencing, gate installation.

Figure 32–14 Rail/stringer fence post installation.

Attaching Horizontal Rails and Surfacing

Split Rail. Split rail fencing has rails with tenons on the ends that fit into notches in the posts. Rails are installed parallel to the ground line. To install split rail fencing, one post should be firmly anchored and the second post left loose so that it can be moved, allowing the rail to be installed. Holding the second post at a slight angle away from the first post, slide the bottom rail into both posts. With the rail in place pull the second post toward the first post and insert the middle rail (Figure 32–15). Again, pull the second post toward the first post and insert the top rail. When all rails are secure, pull the second post into a plumb position. Backfill and tamp around the second post. Repeat this process for the next set of rails and posts in line until all posts have been set. To improve appearance in tapered rail fences, when the rail ends inserted on one side of the post are wide, match the rail end on the other side of the post with wide ends.

Board Surfacing. A traditional board fence has three or more boards installed parallel to the ground. Board fence surfacing is typically applied to the face of the post. All posts should be installed and secured before attaching face boards. Mark the location of the tops of each board using a chalkline stretched between end posts. Space the boards so that the top board is flush with the top of the posts and the other boards are spaced evenly down the rest of the post. Attach the boards to the face of the posts using two deck screws per board. Butt the boards together with a joint centered on each post. For installations on hillsides, the board ends may need to be trimmed to match the plumb of the post. Fences with a face board located flush with the top of the posts may require a cap that overhangs the front of the face board slightly.

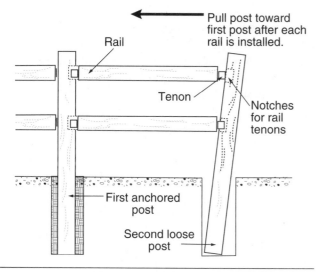

Figure 32–15 Split rail installation.

Woven Board. Woven board fencing is installed level between posts. This requires lengthening posts on hillsides (see Figure 32–10). Woven board surfacing should begin at the top of the fence. Measure from the center front of one post to the center back of the next post. Cut a surface board (typically a 1 × 6 or 1 × 8, but other 1 × lumber can be used for variation) the proper length and set in front of one post and behind the next. Lift the board into a level position and tack it in place. Attach one end to the front of the post using two deck screws and the other end to the back of the next post in the same manner. For the next lower board, reverse the location, measuring and attaching from the back of the first post to the front of the second post. These locations are also reversed for each subsequent section along the fence run (Figure 32–16). Rounded posts make attachment easier. If square posts are used, consider attaching surfacing to a cleat attached to the post.

WOOD-SURFACED STRINGER FENCING

Wood-surfaced stringer fences require the installation of horizontal stringers to support the surfacing. Post installation for this type of fencing should be completed in the same manner as that for panel fencing. Following completion of post installation, proceed with installation of the stringers and surfacing.

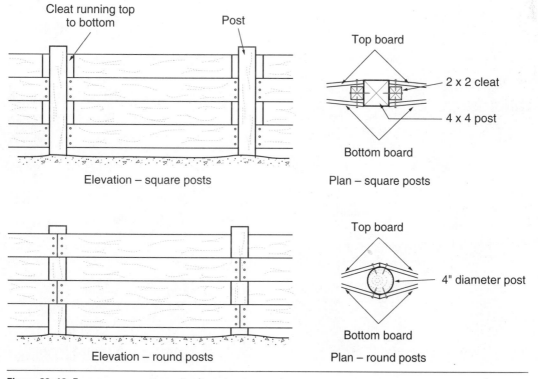

Figure 32–16 Fence weave pattern attachment.

Attaching Stringers for Vertical Surfacing

Fences that have vertical surfacing require stringers between posts to support the facing at the top and the bottom. Fences over 4 feet tall may also require a stringer placed midway between the top and bottom. Stringers attached to the face of the posts, as is the case with picket fences, utilize 2 × 4s or 2 × 6s attached flat to the posts. When stringers are hung between posts, 4 × 4s should be used.

To install stringers on the face of the posts, mark level lines between the posts using a chalkline. The top stringer should be placed at or near the top of the posts and the bottom stringer within 1 foot of the bottom. A middle stringer should be halfway between these locations. For each unit of the fence, measure the distance from the center of one post to the next. For short panel fences, a stringer may run between several posts. Cut a stringer the measured length and attach it to the posts using two deck screws per post. Joints should be centered over posts using butt joints.

Stringers can also be hung between posts as an alternative to attaching them to the face. To hang stringers, mark level lines for top, bottom, and if necessary, middle attachments, on the posts. Install metal hangers on these marks, making allowance for the thickness of the lumber that will rest on the hanger. Hold a 4 × 4 up to one of the level lines and mark the distance between the posts. Cut the stringers to this length and set them on the hangers. Fasten at all locations (Figure 32–17). Toenailing can also be used to attach the stringers to the posts, but this creates a weaker connection than using hangers.

Attaching Vertical Surfacing

Place a stringline along the fence run at the height of the top of the fencing material. Measure and cut surface material for the fence. Attach the surfacing at the top using a deck screw and check for plumb. Finish attaching the surfacing at the top and at the bottom using two deck screws per stringer (Figure 32–18). Vertical facing can be extended up to 1 foot above or below the stringers, but extensions longer than this tend to warp. Additional support may be needed to prevent warping. If vertical surfacing is designed with spaces between boards (such as alternating board

Figure 32–17 Fencing stringer attachment for vertical surfacing.

Figure 32–18 Fencing vertical surfacing attachment.

or picket fences), use a **spacer** to maintain consistent spacing along the entire fence run. Plan the spacing so that a board or picket is placed over the support posts.

Alternative vertical surfacing may require different preparation before attaching materials. **Grape stake** surfacing should have pilot holes drilled at each location where surfacing will contact a stringer. Use deck screws to attach grape stakes. Stockade fencing is typically attached without space between surfacing pieces. Louvered fencing is constructed with only a top and bottom stringer. The location and angle of each louver is marked on both the top and bottom stringer, then a spacer block of 2 × decay-resistant wood is attached with deck screws along each mark. Trim any portion of this block that extends beyond the stringers. Attach the louvers to the block at the top and bottom with two 2 inch deck screws at each location.

Lattice surfacing should be trimmed to match the opening of the panel and attached to the stringers and support posts using 6d galvanized nails. Trim must be placed over the lattice to prevent wind from dislodging the surfacing. To provide proper support for lattice, horizontal and vertical supports should be spaced no more than 2 feet in each direction. Stringers can be rip cut to a narrower width if lattice is to be inset from the face of the fence.

Trimming

To improve its appearance, 1 × trimming material can be installed at the top, posts, and corners of a fence. Measure and cut the trim using bevel cuts and attach using 8d box nails (Figure 32–19).

Gates

Gates for wood rail fences should be constructed using the same procedures described in the previous section, Wood Panel Fences.

CHAIN LINK FENCING

Chain link fencing has grown in popularity to the point that the industry manufactures a wide range of easy-to-install parts, fixtures, and accessories. These ready-made components have made the construction of chain link fencing a task that may be accomplished by the landscape contractor.

Corner/End/Gate Post Installation

Using an auger or by hand, excavate an 8 inch diameter hole to frost depth at each corner and gate post location. Mark the correct burial depth on posts using a permanent marker (Figure 32–20). Fill the holes with

Figure 32–19 Wood fence trimming.

Figure 32–20 Marking on corner post indicating proper setting depth.

a stiff concrete mix and set the corner/gate posts in the hole. If the posts sink, remove them and wait 15 minutes. When the concrete has hardened enough, the post will set in the footing and will not sink. Bounce the post down to the mark and check for plumb. Repeat for all corner and gate posts. Recheck for depth and plumb every 15 minutes for the next hour, adjusting the posts if necessary. Wait 48 hours for the concrete to completely set before continuing fence construction. At locations where posts are set in a slab, the post location should be boxed out while the slab is formed (Figure 32–21).

Line Posts

Between the tops and bottoms of corner posts, connect a stringline for the alignment of the fence. Using a fence post driver, drive a line post at each location marked along the alignment (Figure 32–22). Check often for plumb and alignment. Typical spacing for panels is 10 feet, but spacing can be reduced to maintain even panel dimensions.

An alternative to driving the line posts is to dig or auger 6 inch diameter holes and place the posts in the holes. Partially backfill the holes with gravel and adjust the height to match the stringline. Check for level in each direction and continue to backfill with soil. Compact the soil after every 6 inches of fill.

Install Framework

The framework of the fence includes the railing that runs across the tops of the posts, a **tension wire** that runs across the bottom of the fence, and, in some cases, a **mid rail** for reinforcing. The **top rail** is installed using special caps, while the mid rail and tension wire are installed using brackets with special hardware that bolt to the posts.

Install the special caps on each corner, gate, and line post. Corner post caps, which fit on top of the end posts, have openings that accommodate the end of a railing, while the line caps slip on top of line posts and have an opening through which the railing runs. Top rail has a tapered end designed to join with additional

Figure 32–21 Chain link fence post set in concrete. Concrete slab was poured earlier. A box was left out in which the fence post and footing could be placed.

sections of top rail. Beginning at one end of the fence, slide a section of top rail through the line post caps and place the non-tapered end into one corner post cap (Figure 32–23). Place a second section of top railing through the next line post caps and join with the first section. Continue placing top rail until a section passes over the corner post at the other end of the run. Hold the top rail against the corner post cap and mark where the end or the rail will meet the back of the opening in the cap. Using a reciprocating saw with a metal cutting blade, cut the top rail at the mark. Lift the corner post cap and insert the cut end of the top rail in the opening. Lower the cap back on the corner post. Continue installing top rail on all fencing runs.

Install the tension wire by first attaching the clamps around the base of each corner post. Connect the tension wire to one clamp and stretch tightly between the two corner posts. Attach it to the clamp at the base of the second corner post.

A mid rail is installed on tall fences for stability, and at each panel adjacent to corners and gates for any fence over 4 feet tall. Begin mid rail installation by bolting the rail holder fitting around the corner post at approximately mid-level. Measure and cut a piece of top rail that will fit between the corner and first rail post. The measurement must take into account the fitting on each end. Insert the cut rail into the first fitting and slip a second fitting over the other end. Clamp this second fitting around the first line post. Repeat this process for every panel that requires a mid rail.

Fabric Application

Surfacing is installed by stretching chain link fabric between corner posts and bolting it to the posts. The maximum run for stretching fabric is approximately 100 feet. It is not uncommon for fabric that is stretched tightly by hand to be stretched another 1 to 2 feet using

Figure 32–22 Chain link fencing line post driving.

Figure 32–23 Chain link fencing top rail installation.

Figure 32–24 Chain link fencing fabric layout.

Figure 32–25 Disconnecting top link.

jacks or mechanical **stretchers.** Fabric can be stretched on either side of the fence over straight runs. At curves, it should be stretched around the outside of the installation. If the fence runs up a slope, set corner posts at the top and bottom of the slope and stretch separately for this section.

Begin surfacing installation by rolling the fabric out along the fencing run. Lean the fabric against the posts to roughly determine the length needed (Figure 32–24). At one end of the run, insert a stretcher bar through the first loop of the fence. Fasten clamps around this stretcher bar and the corner post. Install one clamp for each foot of fence height. Stretch the fabric as tightly as possible by hand along the entire run. Determine where the new end to the fabric should be, after stretching, by measuring back from the unconnected end 2 inches for each 10 feet of fence.

Return the fabric to the ground to disconnect a link at the point where the fabric should end. To disconnect, straighten the bend at the top and bottom of the

link (Figure 32–25). With a twisting motion, spin the disconnected loop out of the fabric to separate the fencing (Figure 32–26). Pieces of fabric can be spliced together by overlapping loops of the fabric and placing a **stretcher bar** through this overlap (Figure 32–27). A more attractive splice is achieved by using a single link of fabric to join two pieces. Overlap the pieces and start weaving this link from top to bottom using a rotating motion. When joined, twist a bend in the top and bottom of the link to prevent unraveling.

Prepare for stretching by measuring back from the new end of the fabric 24 inches. Locate a loop at this point and insert a double stretcher bar. Attach a jack around the corner post and to the double stretcher near the center of the fabric. Jack the material toward the second corner post (Figure 32–28). When the new end of the fabric reaches the corner post, insert a single stretcher bar through the last loop in the fabric and install clamps around the corner post and stretcher bar. Install one clamp for each foot of fence height. Loosen and remove the jack and double stretcher bars from the fabric (Figure 32–29). Tall fences may require

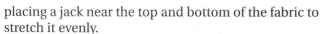

Figure 32–26 Separating links of chain link fencing.

Figure 32–27 Installing stretcher bar to join sections of fabric.

placing a jack near the top and bottom of the fabric to stretch it evenly.

Attach the fabric to the top rails, posts, and tension wires using aluminum fence **ties,** (Figure 32–30). Work from the side of the fence framework opposite the fabric. Place the hook end of the tie over a strand of the fabric. Holding the hook in place, use one finger to bend the tie around the post or rail. Push the straight end of the tie back through the fabric and bend it around a strand of fabric. Repeat every 2 feet on posts and every 4 feet along the top rail and ten-sion wire.

Gates

Gates require that a framework be measured and constructed on the ground, then "hung" from a gate post. Install hinge hardware on one gate post. Check the opening for square and measure from this hardware to the edge of the other gate post. Cut pieces of top rail to match this measurement and the gate height. Allow for hardware connections when measuring and cutting the rail. Connect the pieces using gate corner clamps. Install a mid rail in the gate for bracing. Hand stretch and attach fabric with ties over the gate framework. Install the latch and gate hardware on the gate and set it in place.

Preassembled gates are also available from many suppliers. Verify the dimensions of the gate with hardware before setting gate posts. Install hardware and hang the gate in a manner similar to installation of a custom-built gate.

Privacy Slatting

Several manufacturers make plastic **privacy slatting** that can be woven into the fabric of chain link fence for additional screening. One style uses strips of vinyl or metal that are woven diagonally into loops, using friction to hold them in place. Another style requires that a plastic channel be fed horizontally across the bottom loop of the fence. Vertical slats are then slid down each of the loops and slipped into this channel. A second channel is fed across the top of the vertical slats to lock them in position.

Figure 32–28 Stretching and connecting fabric for chain link fencing.

Figure 32–29 Chain link fencing hardware, showing top rail, post cap, stretcher bar, and connector clamps.

Figure 32–30 Connecting fabric to fence with ties.

DECORATIVE METAL FENCING

Decorative metal fencing is assembled in a manner similar to that of wood panel fencing in that posts are set in footings and preengineered panels are connected between posts.

Footing Installation

Metal fencing requires concrete frost footings for each corner post, and may require concrete footings for every post. Verify with the manufacturer or supplier the type and depth of footing required for posts with a particular fence. To install footings, auger an 8 inch diameter hole to frost depth centered at each post location. Form the top 6 inches of the footing with a box frame or tubular paper form. The spacing and level must be accurate because the fence panels cannot be adjusted to match post locations. Fill each hole with concrete and smooth the surface with a wood float. Insert the anchor bolts supplied with the fence in the top of the footing. Before the concrete hardens, verify the correct dimensions between anchor bolts in

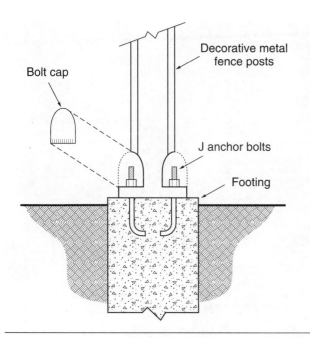

Figure 32–31 Decorative metal fencing footing-to-post connection.

each footing. Minor adjustments can be made at this point, but not after the concrete has hardened. Wait 48 hours before proceeding with fence construction.

Post Installation

Post units should fasten directly to the anchor bolts installed in the footings (Figure 32–31). Place the posts and hand tighten the nuts to allow for flexibility when installing panels. Brace posts to hold them in position until the panels are connected.

Panel Installation

Begin installation of panels at corners or gates. Connect the premanufactured panels to the posts at the top and check the panel for level. Adjust if necessary and tighten the panel connections. Complete the connections at the bottom.

Gates

Gates for decorative metal fencing should come as prefabricated units or kits. Assemble any necessary parts and install the hinge and latch hardware. Hang the gate from the hinges and adjust as necessary.

Finishing

Fencing may come from the factory as primed units that require field painting. With this approach, the damage to the finish from shipping is minimized and it allows the customer to choose custom finishes and colors. If field finishing is required, install all fencing

components and mask any items such as hinges or clasps that will not be painted. Place a drop cloth behind, below, and in front of the fence. Use an airless spray painter to apply the finish color. Follow manufacturer's instructions on type of paint and proper preparation of surface. Apply paint on a very still day when there is no expectation of rain for 24 hours. Be very cautious of any overspray and its potential to damage buildings, vehicles, and any landscaping in the immediate area.

If the fence is shipped prefinished, the only work required to finish is to touch up any damage that occurred during shipping or installation. Determine the exact paint type and color from the manufacturer before performing any touch-up work. Sand the damaged area and apply paint evenly to match the existing finish.

VINYL FENCING

Most vinyl fencing includes posts with prepunched holes and horizontal rails that are installed between posts. Prefabricated panels, which are hung between posts, are also available. Installation of vinyl fencing is similar to the installation of horizontal wood rail fences because posts must be properly spaced and the rails installed beginning from the bottom. A special crimping tool may be required to notch the ends of the rails, while other fence styles utilize connecting pins placed in predrilled holes. Vinyl fences must be custom built with each component attached singly. This requires more labor during construction, but also allows the fence to be adjusted during the course of the project.

Post Installation

Spacing of posts is determined by the length of the rail material and fence design. Regular spacing along the entire length of the fence makes construction of the fence easier. With prefabricated fence systems, the spacing is determined by the length of the rails or panels provided.

Mark each post location carefully so minimal adjustment of segment length will be required. Dig or auger an 8 inch diameter hole the burial depth of the post over every post center location. Using a marker, draw a line near the base of the post that indicates the proper burial depth. Insert posts into the holes and check the postholes for proper depth at this stage, but do not backfill or anchor posts in holes until after rail has been installed (Figure 32–32). If posts are too tall, they must be removed and the excavation deepened. If the post is set too low, granular material must be added to the hole to raise the elevation. Special posts

Figure 32–32 Vinyl fencing post installation.

Figure 32–33 Crimping vinyl fence rail with crimping tool.

are manufactured for corners and ends of runs. Verify that the proper post has been selected for each location.

On a mild slope, posts for rail fences should be installed plumb and at the same height above ground level as posts on level areas. Rails are placed at an angle between posts, paralleling the slope. Determine the correct spacing by holding one of the rails between an installed post and the next post to be dug. Abrupt changes in slope are difficult to accommodate using rail sections over 8 feet in length. Rail sections may need to be cut and an extra post installed to handle a transition from one grade to another.

Rail Installation

Long Rail Vinyl Fences. Some brands of vinyl fencing utilizes rectangular openings in a post through which the rail section can be inserted. Using the **crimping** tool, notch the rail at one end on all four sides (Figure 32–33). Slide the bottom rail through the bottom hole of the middle post and push it through until the

rail reaches the corner post (Figure 32–34). Insert the rail into the bottom opening of the corner post. Repeat this process with the middle and top rails. When the three rails are installed in the corner and middle posts, crimp the opposite end of the rail and install in the appropriate openings for the third post. The posts should be loose in the hole when performing these operations. Repeat this process for the next rail section, using the third post in the previous segment as the first post for the next unit. Crimp all ends that are inserted into posts. The rail section should not extend beyond the center of the post into which it has been inserted. If minor length adjustments are required, the rail can be trimmed with a hacksaw.

Short Rail Vinyl Fences. Begin installation for fences that utilize short railings placed between a pair of posts by crimping all sides of both ends of the rails for one section. If pins and predrilled holes are used, crimping is not necessary. Install pins after each railing end has been placed in the post. Place one end of the bottom rail in the first post. Place the other end in

Figure 32–34 Installing vinyl fence rail through post.

the second post, holding it angled away from the first post. Insert the second rail in the first post and then in the second post, pulling the second post closer to the first. Place the third rail in the first post and in the second post, pulling the second post into a plumb position. Backfill around the second post and move to the next panel.

Anchoring Posts

When all rail material has been installed, the posts can be set. To save time with height adjustments, stretch a stringline between the corner posts to use as a guide for line posts. The stringline should match the burial mark made on each earlier post. Check the post for plumb using a level on adjacent sides of the post. Backfill the holes and tamp with a shovel handle. Recheck for height and plumb when the hole is half full and completely filled. An alternative to soil backfill is to set posts in concrete. Anchor posts beginning at one end and work consecutive posts until the fence posts have been set.

Chapter **33**

Freestanding Walls

When space separation is required but the design dictates a more substantial look, a freestanding wall may be the best choice. The freestanding wall incorporates the benefits of a fence with a permanence of a landscape wall. While more expensive and time-consuming to construct, the freestanding wall adds a strong sense of permanence to a site.

Freestanding walls should not be confused with landscape retaining walls (see Section 5). Both retaining walls and freestanding walls require special construction techniques to maintain stability, but for very different reasons. Retaining walls endure pressure from the forces of soil and water behind the wall, while freestanding walls have to overcome the force of wind to remain upright. Freestanding walls do not retain soil and do not require drainage accommodation.

This chapter provides construction details for installing two common styles of freestanding walls. **Fieldstone,** or rubble, walls utilize stacked stone to recreate the look of antiquity and history in design. Masonry walls consist of manufactured materials mortared together to create visually pleasing barriers. Additionally, either wall construction can be utilized for installing piers. Piers are columns used to identify entries, to mount signs and lights, and, occasionally, to anchor posts in wood fence construction.

RELATED INFORMATION IN OTHER CHAPTERS
Information provided in this chapter is supplemented by instructions provided elsewhere in this text. Before undertaking activities described in this chapter, read the related information in the following chapters:

- Safety in the workplace, Chapter 4
- Construction math, Chapter 5
- Basic construction techniques, Chapter 7
- Mortared paving, Chapter 26

MATERIALS FOR FREESTANDING WALLS
Dry-Laid Fieldstone (Rubble) Walls
Fieldstone, or rubble, is the term for irregular-shaped round or rectangular stone that is weathered by natural forces. When used in rustic, rural settings, the fieldstone wall is one of the most attractive landscape additions available (Figure 33–1). They are primarily used for boundary identification, with separation and enclosure being additional functions. A contractor must be a skilled craftsperson to place stones in a stable, yet appealing, pattern. Maintenance requires occasional reconstruction of wall segments.

Masonry Walls and Piers
Many common building materials (stone, fieldstone, brick, block, and adobe, for instance) can be attractively used in a masonry wall and piers (Figure 33–2). Masonry installations are expensive and require a great deal of expertise to construct, but can perform all functions required of fencing with a high level of aesthetics. Maintenance is limited for masonry walls and piers.

PLANNING THE PROJECT
Beginning a masonry wall with a detailed design will reduce the number of field decisions and problems encountered during construction.

Figure 33–1 Rubble wall.

Figure 33–2 Stone masonry wall.

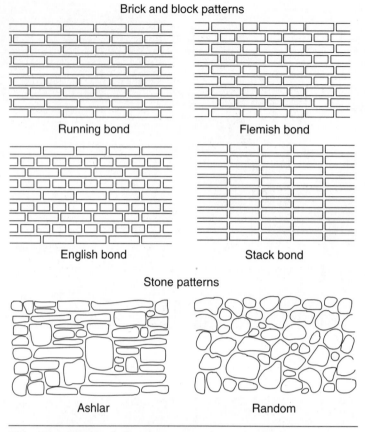

Figure 33–3 Masonry wall patterns.

Masonry Patterns

While fieldstone typically employs a random pattern, there are several common arrangements possible for masonry walls. Figure 33–3 illustrates some of the patterns available to the mason.

Masonry Wall Thickness

Most masonry walls using concrete block and wall stone are constructed the thickness of a single material, or **wythe.** A single wythe of these materials provides adequate strength and wind resistance to keep the wall upright. Brick is typically constructed using a double wythe to provide the stability necessary to keep the wall standing upright. These two wythes periodically have a unit running between the two as a bridge, or reinforcement, that ties both wythes together. This bridging course is called a header course and is common in brick wall construction. A third choice for masonry installations is a veneered wall. A veneer is a single wythe of a concrete block, or other backing material, and a single wythe of brick or stone veneer. Veneer can be placed on one or both sides of the backing block.

Freestanding Wall Shape and Wind Resistance

Because they are impermeable by wind, freestanding walls carry an extreme wind load. Strong winds can blow with enough force to topple a tall wall if precautions are not taken to counter these forces. The most common countermeasure is to avoid long, straight sections of wall. By adding corners or piers periodically, a freestanding wall gains significant wind resistance. Adjusting the alignment of a wall by adding a few niches or corners may also prevent the wall from toppling.

DRY-LAID FIELDSTONE (RUBBLE) WALLS

Base Preparation

Grade preparation for fieldstone walls requires installation of a stable granular base. Excavate a base trench 6 inches deep and 12 inches wider that the wall width

Figure 33-4 Rubble wall stone placement.

along the entire wall alignment. Fill with granular base material and compact.

Stone Placement

Stone placement for the entire wall is accomplished by hand. Because the irregularly shaped stone typically will not stack in regular patterns, fieldstone may be placed with a wide base and subsequent courses that narrow as the wall gets taller. When using rounded materials laid without mortar, utilize a tapered or pyramidal cross section. The sides may be steepened somewhat by placing larger stones at the bottom and using care to arrange the stones as they are placed so they support the next higher course. Angular, square, and rectangular materials may be stacked to heights of up to 4 feet using a rectangular cross section.

Utilize larger stones to initiate the base course and intermingle stones of varying sizes as the wall height increases. As the wall height increases, observe the pattern developing on the surface. Adjust stone placement as necessary to avoid continuous vertical or horizontal joints. Rubble walls are typically two to three wythes wide. Stones should occasionally be placed completely across the wall, front to back, to integrate the wythes. Lay out several stones, sort, and place by hand (Figure 33-4).

Stone placement requires patient sorting and fitting of materials into available openings in the wall. Avoid the use of small or irregular stone pieces by rearranging stone to place larger pieces. Small, thin shims of stone may be placed under a larger piece to level or stabilize the wall. Wall caps can be constructed of wide, flat pieces of material, if available.

MASONRY (STONE, BLOCK, AND BRICK) WALLS AND PIERS

Footing Construction

Successful masonry walls and piers require a stable base to prevent settling and movement. This is accomplished by pouring a 6 inch deep footing that is 1 foot longer and 1 foot wider than the footprint, or base dimensions, of the proposed wall (Figure 33-5). (Revisit Chapter 23 for information on forming and pouring the concrete slab that serves as the footing.) Reinforce the footing with three #4 rebars running the length of the footing. Finish the surface of the footing with a rough broom finish to provide a surface that will bond well with mortar.

Any conduit or plumbing that is required as part of the wall should be installed prior to pouring the footing. Place the conduit so that it projects above the footing in the correct location to the approximate height required for any fixtures. The mason will make final adjustments as the wall is constructed. Cover the ends of the conduit or pipe with duct tape. Pier construction requires similar preparation and installation similar to that for walls. Anchor bolts

Figure 33–5 Masonry pier footing installation, showing bond chasing and anchor bolts placed when foundation was poured.

to tie concrete block and brick walls to the footing should be installed at the correct locations before the slab sets.

Layout and Preparation

Prepare the footing to accept the mortar by scrubbing off all debris with water and a broom. When dry, use a chalkline to mark two straight lines 4 inches in front of the face of the proposed wall. These guidelines will serve as reference marks for alignment of the first course of the wall. When constructing walls using brick or concrete block, carefully measure the wall dimensions and mark beginning and ending positions on the footing. Lay out the brick or block without mortar (a process called chasing a bond) to verify dimensions before mortaring starts. Brick and concrete block cannot be cut like stone to aesthetically fit small voids, so measurements must be precise.

Prepare materials to be used in the wall construction project before beginning. Clean off any old mortar and discard broken or damaged items. When using brick, be certain there are halves available before laying begins. Lay out a variety of stone to speed the construction of a stone wall. Dip the stone in water for a short time before installation to prevent it from draw-

ing moisture too quickly from the mortar. Drawing water will speed the drying of the mortar and potentially reduce its strength.

Mixing and Placement of Mortar

After all materials are prepared, mortar mixing can begin. Typical mortar types used for exterior wall applications are types S and N. Instructions for selecting mortar are explained in Table 26–1. Refer also to mortar mixing for paving in Chapter 26.

Applying the mortar to the masonry units will take some practice. Use a pointed mason's **trowel** to apply mortar. When placing mortar on the base or the top of a masonry unit, pick up a trowel full of mortar and apply the mortar to the material with a light slapping motion. First attempts may find mortar application difficult, but practice will allow placement of mortar where planned. When put on the sides of a masonry unit, the slapping motion prevents mortar from sliding off when the unit is held upright. Apply generous amounts of mortar to fill the joints, since any extra will be squeezed out when the unit is placed.

When placing concrete blocks, be aware that there is a top and bottom. The top of the block has a wider edge than the bottom in order to hold mortar. There is no top or bottom to brick or stone. Bricks, however, may have a front and a back.

Placing Base Course

Using the reference marks as guidelines, place beads of mortar approximately 1 inch thick along the front and back locations of the wall unit. Set the first unit on these beads and check for the proper height and for level in both directions. Use the handle of the trowel to tap the unit into proper alignment and level. If the wall is constructed with a double row of wall units, both units must be level between front and back. When the unit is properly aligned, use the trowel to vertically slice the excess mortar away from the unit. Slide mortar away from the wall and do not attempt to reuse it. Mortar that has come in contact with the ground will pick up soil and other debris that will weaken the mortar if it is remixed.

If installing concrete block or brick, a stringline will help to maintain alignment and level for each. Place the blocks at both ends of the wall and install a level stringline between stakes set just beyond the ends of the wall. This stringline should just touch the top, front corner of both end blocks. As the wall grows in height, the stringline is moved up one course at a time. Clips are manufactured that allow the stringline to be hooked over the corner of the wall being built. With brick construction, it may be beneficial to build a small section of the wall four to five courses high at each end of the wall. The stringline can then be stretched between these end sections while the center part of the wall is laid. Irregular stone units must be aligned and leveled individually.

Continue laying the wall block by placing two more beads of mortar on the base for the second unit. The second unit should be "buttered" with mortar at the end that abuts the previously placed unit. Place the unit and use the trowel handle to adjust alignment against the stringline and to achieve proper spacing. Joints between units should be approximately 3/8 inch. With each unit placed, verify for level with the previous unit and check level front to back. Scrape away excess mortar. Continue placing beads of mortar and units along the entire length of the base course (Figure 33–6). If laying began from both ends, a unit will need to be placed between the two end units. This closure unit will need to be "buttered" on both ends and carefully slid between the units already in place.

Subsequent Course Placement

Place subsequent courses following the design pattern required. Joints should not be aligned vertically in a masonry wall unless it is a veneer pattern. Most patterns begin the second course using a manufactured

Figure 33–6 First course laying and leveling.

half block or brick. Special end and corner blocks are available for concrete block products. For double wythe brick walls, a brick may be placed perpendicular to the wall direction to create the half required to offset the pattern. Stone can be cut or a shorter piece selected to offset the joint pattern.

Install subsequent courses by first applying a layer of mortar to the top of the previous course and placing a block on it. Check for joint spacing, alignment, and level in both directions. Setting blocks on both corners first will allow the stringline to be moved up one course. Continue laying block in subsequent courses in the same manner as the base course (Figure 33–7).

Reinforcing should be installed every third course for brick, block, or veneer stone. The standard horizontal reinforcing used is a narrow steel ladder with angled crosspieces that is buried in the mortar placed on top of the units (Figure 33–8). No. 3 rebar is used for vertical reinforcement for tall walls. Every 32 inches, place a vertical bar into a core of the block or opening between wythes. **Grout** around the reinforcement to hold it in place. Unless the courses are laid with straight, horizontal joints, reinforcing is impractical in wall stone applications.

If the wall or pier has a mechanical fixture such as a light base, water spigot, signage anchor, or gate hardware, lay the wall up to the point where the fixture is to be located. Complete any connections to the previously placed conduit that are required, and place the fixture in the next course laid. Placement may require cutting or piecing a partial unit to make room for the fixture. Fill around the fixture with mortar and continue adding courses. Small additions such as gate

Figure 33–7 Masonry courses showing stringline used to keep blocks aligned.

hardware or anchors for signs may be embedded in the mortar between courses. Verify the dimensions and locations for all such installations before work progresses and mortar hardens. If the wall or pier has a veneer finish, small, galvanized strips of corrugated metal, called veneer ties, should be placed every other course. These ties are bent into the joints of the veneer to anchor the two wall wythes together.

Striking Joints

Joint finishing, or striking, is generally performed by running a special tool over the mortar in the joint to smooth and compress the mortar (Figure 33–9). Striking provides a more attractive finish for the mortar and makes the mortar joint waterproof. Several patterns are available, with the most common being a concave pattern. Other choices include a squared, or raked, joint, a rustic joint, an angled joint that is inset at the top and flush with the unit at the bottom, and a V joint (Figure 33–10). Each of these patterns requires a special jointing tool. Begin striking joints at the bottom course with the horizontal joints, then the vertical joints. Continue up the wall in this order until reach-

Figure 33–8 Horizontal steel reinforcement embedded in mortar.

ing the newly laid block. Timing of the striking operation is important. When the mortar begins to lose its sheen, it is time to strike the joints. Striking too soon will push mortar out of the joint, but waiting too long will allow the mortar to harden. Laying block at a slow pace may require laying one course and stopping to strike joints. The rustic joint does not remove the mortar that squeezes out between the joints, and requires no striking.

Figure 33-9 Striking joints on a masonry pier.

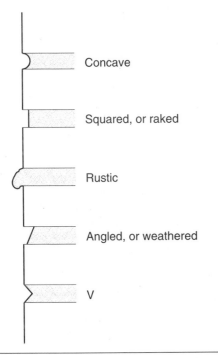

Figure 33-10 Joint patterns.

Concave

Squared, or raked

Rustic

Angled, or weathered

V

Figure 33-11 Capping a masonry pier.

until the mortar has set for 24 hours before filling the cavity with grout. Grout is a light mortar mix with double the sand content of mortar. Mix and place in a bucket, then pour into the cavity until filled.

Capping

When building walls with cavities in the center, the top of the wall must be capped to reduce the potential for moisture entering the wall. Both brick and concrete block have special capstones manufactured for this purpose. Veneer block walls may require cutting and placing a row of block perpendicular to the wall direction as a cap. Some designs select a cap unit that overhangs the wall faces slightly. Finish grouting the wall to the top of the cavity and lay beads of mortar on the top course. Set the cap course in a manner similar to laying any course, checking to maintain alignment and level both directions (Figure 33-11). If any type of structure support or fencing is to be attached to the top of the wall, insert supports or J bolts into the soft mortar between capstones and retouch the mortar. Scrape away excess mortar and strike the joints.

Grouting Cores

When building walls with concrete block, double wythe brick, or double wythe veneer block, there will be an open cavity in the center of the wall. This cavity needs to be filled to maintain structural strength and to reduce moisture infiltration inside the wall. Wait

Cleaning the Surface

> **CAUTION**
>
> When cleaning pavement using acid and caustic cleaners, follow all safety precautions identified on the label. Wear proper safety clothing. Contact and exposure to acids and caustic cleaners can cause personal injury and can harm plants.

When all mortar has hardened, clean the surface of the wall. Large pieces of mortar can be removed by rubbing a hammer over the surface. If mortar or grout has run down the wall and stained the surface, it will have to be scrubbed with a mixture of nine parts water with one part muriatic acid. Use caution for personal health and the health of surrounding plant material when cleaning with these materials.

Section 8

Summary

When a separation or definition of space is necessary, a fence or freestanding wall is often sought as the solution. In Section 8, the installation of such separators was explained. Included in the section was information pertaining to the selection of materials and preparation of the site, as well as chapters that described the specific installation techniques for wood premanufactured panel fencing, wood rail and stringer fencing, chain link fencing, decorative metal fencing, vinyl fencing, dry-laid fieldstone walls, and masonry walls. Preparation for all of these requires a consideration of alignment, slopes, surface appearance, maintenance, and access to the construction site.

Construction of wood fences requires the installation of a fencing framework, usually posts, to which additional support or panels are connected. Wood premanufactured panel fencing involves fastening panels directly to support posts or "hanging" panels between posts. Rail and stringer fencing requires the attachment of horizontal rails directly to posts, or hanging stringers between posts to which vertical facing can be connected. Gates for either fence type can be custom built or premanufactured and require an opening that is plumb and square.

Chain link fencing also requires setting posts at each end or corner of a fence section. Line posts are then driven in at regular intervals between these anchor posts. A railing is attached to the top of the posts and fabric is applied to the completed frame-work. With chain link, the fabric is stretched between end posts and anchored to the framework. Gates are custom built or prefabricated.

Decorative metal fencing necessitates a concrete base for footings. Posts are attached to anchor bolts set in these footings. Premade panels and gates are then installed between the posts.

Vinyl fencing requires direct burial of all posts along an installation. Beginning at one corner, rectangular vinyl rails are crimped and slid through openings in the posts. Both crimped ends are inserted into posts, and the process continues until all rails are installed. Posts are then plumbed, leveled, and backfilled and fitted with a cap.

Dry-laid fieldstone (rubble) walls are constructed by placing stones of irregular shape along the proposed alignment of the wall. Stone may be placed at its natural angle of repose, or it can be laid to maintain sides that are vertical.

Masonry walls and piers require a footing on which to place masonry units. Alignment is checked, mortar is mixed, and units along the base course are placed, carefully leveled, and plumbed. Each of the following courses is carefully placed on top of the previous course, with the height and spacing maintained throughout the construction. Reinforcement is placed in the mortar joints and joints are struck as the project goes up. When the wall is complete, the center cavity of the wall is filled with grout and the wall is capped.

Section 9

Site Amenities

INTRODUCTION

Sound, aesthetic design depends on the creative arrangement of major landscape elements like walls, structures, paving, plantings, and lawns. Accents and finishing touches, however, rely on more subtle site amenities. These amenities, which include everything from decorative pools to flagpoles, can enhance and enrich the overall design scheme, often making a qualitative difference in a project (Figure SI–6).

Amenities that are typically installed by the landscape contractor are discussed in this section. Each of the three chapters covers a different type of amenity: water features and bridges; premanufactured **site furniture** and play equipment; and edgings, planters, and plant protection. The first chapter on water features explains the installation of static pools created from cast-in-place concrete or flexible pool **liners,** cascades, waterfalls, fountains, and bridges. The site furniture chapter details the anchoring and assembly of benches, trash receptacles, signage, bike racks, play equipment, and other exterior furnishings. Finally, the process of edging installation for several different materials, including natural edges, plastic and metal edgings, brick, stone, precast, and concrete curbing, is explained. Also presented in the final chapter is the installation method for premanufactured landscape elements such as planters, **tree guards,** and **tree grates.**

The landscape contractor may have a small role in completing amenity installation if the scale of a project is very large. As project size increases, so does the chance that a specialized subcontractor may be necessary to finish the work. When responsibility does fall in the hands of the contractor, careful attention must be paid to the plans and specifications of the job. Many times the type, style, or detail of the amenity has been selected for a very specific reason and purpose, and the errant substitution of the wrong model number will require a costly correction. Private clients can utilize amenities to enhance a landscape, but the cost of designer benches, pools, and similar details can be very high.

SOURCES FOR AMENITIES

There are a number of places from which amenities can be ordered. Following are sources that regularly list suppliers of landscape-related products:

1. Sweets Catalog. A comprehensive listing of building product suppliers for construction. Sections 2 and 16 of the catalog have the majority of site-related products.
2. LAFile. A listing of building product suppliers for landscape-related products. LAFile is available from Kerr Publishing.

Figure SI–6 Seating for an urban plaza. (Courtesy of Margaret Sauter)

3. Journal advertisements. Most landscape-related journals provide a wide variety of advertisements from suppliers of site construction projects. Typical journals include *Landscape Architecture, Garden Design, Landscape Design, American Nurseryman,* and *Landscape Architect and Specifier News.*

4. Garden centers and related retail outlets. Many garden centers stock and sell lines of residential grade site amenities, particularly pools, fountains, edgings, and site furniture. Commercial grade selections may require searching other supply sources.

5. Manufacturer representatives. Virtually every site-related producer has sales reps who can supply information and cost data with regard to its prod-

uct. These reps can be located through direct contact with the manufacturer.

6. Trade shows. Trade shows and professional association meetings can yield numerous contacts with landscape product manufacturers. Associations that typically hold such shows include landscape contractors associations, builders associations, and nursery and landscape associations.

7. Web sites. Web sites of manufacturers and professional associations often include information related to amenity suppliers. Key search words include: landscape amenities; site amenities; site furnishing; landscape construction; and landscape materials.

Chapter 34

Water Features and Bridges

Whether a design relies on the reflections of a static pool or the dynamic sound and motion of a waterfall, a well-planned addition of water can be more provocative than any other design element on a site. New construction materials have made the introduction of water into the landscape a feasible undertaking. Traditionally, landscapes have relied on construction using concrete-lined pools with underground plumbing. While this construction technique has not disappeared, the advent of pool liners, small pumps, and plastic tubing has brought the world of water into the residential domain.

Although today's technology has allowed for water features to be easily implemented in landscapes, their high maintenance character has changed little. In all climates, the pool owner must make a commitment to regulate water quality, and in some climates, the pool may require periodic draining if the feature is to stay attractive and functional.

RELATED INFORMATION IN OTHER CHAPTERS

Information provided in this chapter is supplemented by instructions provided elsewhere in this text. Before undertaking activities described in this chapter, read the related information in the following chapters:

- Safety in the workplace, Chapter 4
- Construction math, Chapter 5
- Basic construction techniques, Chapter 7
- DC site lighting and related electrical work, Chapter 14
- Related site utility work, Chapter 15
- Cast-in-place concrete, Chapter 23
- Wood deck construction, Chapter 29
- Wood steps, railings, benches, and skirting, Chapter 30

TYPE OF WATER FEATURES AND BRIDGES

Pools and water features can take many forms in the landscape. The components of water features can be combined and shaped to follow any design concept desired. Following are the components whose installation is described in this chapter:

Pools and Water Features

Static pools. **Static pools** are water reservoirs without any moving water. Static pools can be constructed of concrete, which is common for larger or commercial applications, or with vinyl pool liners. Pool size, form, and depth vary based on design.

Cascades. Water that gently flows on the surface from a higher elevation to a lower elevation is termed a **cascade.** A shallow channel, often lined with stone or gravel, contains the water as it flows between elevations. Cascades can be constructed on steep or gradual slopes and constructed with either a concrete or vinyl channel. A water source, either a pump outlet or reservoir, and an ending reservoir and drain are required to make the cascade function properly. Water is often circulated from a lower reservoir to the top of the cascade to produce the water movement visible in the cascade.

Waterfalls. Water that drops from a higher elevation into a lower reservoir is a waterfall.

Depending on the effect the designer is attempting to achieve, the drop can be a constructed as a quiet, simple, short drop; a loud, dramatic, tall drop; or a combination of both (Figure 34–1). Waterfalls require a water source at the top and a reservoir at the bottom to collect outflow.

Fountains. Fountains are powerful pumps with special orifices that project water upward out of a reservoir. Depending on the size of the pump and type of orifice, water may project several feet into the air or simply bubble to the surface. Different types of fountains can be combined and lighting added for special effects. Fountains can be designed and built that are incorporated into walls, pools, and rock formations.

Bridges

A wide variety of prefabricated bridges is available. Many large and small bridges are constructed of **Cor-Ten® steel** and require only footing construction and placement by the landscape contractor. Smaller bridges can be constructed by building a wood platform or by placing a large slab of stone over a water feature. These simple structures may not require heavy piers, but still require careful anchoring and placement.

PLANNING THE PROJECT

Timing of Water Feature Placement

Water feature construction must occur in several phases. During utility construction (see Section 4), consideration should be given to drainage, water source, and electrical source issues. If a rigid pool construction technique is selected, the forming and pouring are natural extensions of the concrete paving process (see Chapter 23). Multi-level water features may also require landscape walls. Placement of liners, edging, pumps, and other necessary elements typically occurs between finish grading and planting operations.

Layout of Water Features

Pool locations should be located by painting the perimeter of the installation prior to excavation (Figure 34–2). Stakes within the reservoir area can mark proposed finish grades and guide filling and excavation procedures. Dynamic water features often involve reservoirs on multiple levels with connecting features such as piping, cascades, and waterfalls. Each of these features should have location and grade verified prior to construction (Figure 34–3). Preliminary layouts may reveal minor adjustments necessary to improve the performance of the feature.

Figure 34–1 Water feature in an urban plaza. (Courtesy of Margaret Sauter)

Figure 34–2 Layout of a multilevel water feature to be constructed with flexible pool liners. White lines indicate the approximate outside edge of each pool. Overlap area is where waterfall will be placed.

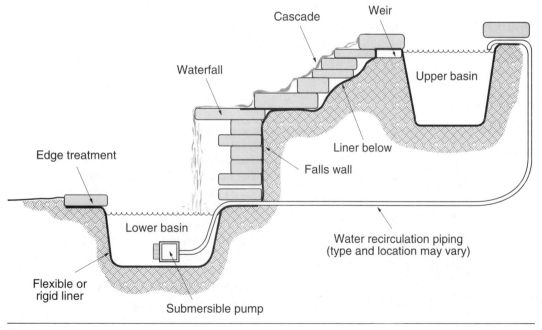

Figure 34–3 Cross section of water feature showing components.

Testing Water Feature Operation

Unlike most landscape features, visual appearance is not the final test of a proper installation. Testing of the various components as they are completed is important to avoid disruption of completed features of the site for repairs later in the project. Test the utility connections when they are installed. As component features are installed, test the liners for leaks, the pumps and tubing for proper operation, and waterfalls and cascades for flow before moving to the next stage.

Testing waterfall and cascade operation is critical. Few adjustments can be made after the landscape treatments are completed. Fill all reservoirs and turn on the pump or water supply. Verify that the basins are maintaining the proper water elevation. Adjustments in pump and weir opening size should be made if necessary to correct flow problems in these two areas.

Cascades can be tested by filling all reservoirs and operating the pump. Water should flow evenly from top to bottom down the center of the channel. Expect some of the water to flow along the sides of the decorative liner rather than over the top. If the liner pieces are not properly placed, the flow of water will miss the decorative liner completely. For a flexible liner, adjust the pieces by twisting them so the water flows down the center. For a concrete liner, add small pieces of stone, or perhaps even mortar, to divert the water back to the center.

Location of Bridges

Bridge location should be based on the bridge function, aesthetics, and water conditions of the feature. If the bridge is crossing a body of moving water, care should be given to the position of the footing. Consult a design professional if there is doubt about the stability of the footing position or about potential bridge failure.

SITE PREPARATION FOR WATER FEATURES

Site Conditions

Grade preparation is essential before pools and channels can be excavated. Because pools require level surfaces, it is easier to grade a level, compacted area prior to digging the reservoir area. Similarly, cascades and waterfalls need varying degrees of grade change, meaning the subgrade for each must be prepared before construction begins. Hours of hand excavation can be saved if the locations of features are marked before filling operations are undertaken. It is also important to maintain proper compaction in areas under and around water features, since settlement will disrupt the level conditions required to make the features work properly.

Electrical Source

Determine the electrical requirements for the feature in the early utility stages of the project. Electrical circuits that supply exterior fixtures should be ground fault circuit interrupt (**GCFI**) circuits. If the feature is to be serviced by running a power cord to an exterior receptacle, the primary concern will be that a nearby receptacle exists, and that a concealed route from the pump location to the receptacle can be located.

A large feature will usually have a circuit(s) dedicated to its operation. Burial of the electrical source should be accomplished before excavation of the feature. Conduit is required for high-voltage circuits, and if the site electrical work is not underway, the conduit may be placed and the wire pulled in at a later time. Stake the electrical source and the service point for the feature and excavate an 18 inch deep trench between the two points. In the trench, place a 2 inch or larger schedule 40 PVC conduit with radius ends that bring the conduit to the surface. Cover the ends with duct tape. A licensed electrician should perform electrical wiring.

Water Source

Water may be provided through a hose connected to an external **hose bibb** or through a separate line dedicated to the water feature. Water provided through a hose requires minimal preconstruction preparation and is usually an acceptable way to supply water for a small residential setting. Larger pools and commercial applications have a much higher demand for water, and frequent filling with a hose will prove to be a nuisance. In these cases, provide a separate underground supply line to save time and maintenance over the long run. As with the electrical source, final, above grade connections of the water source can be completed after the water feature is installed.

If a separate water supply line is provided for an exterior feature, it should be either excavated below frost depth or sloped toward the source to allow for backdraining when necessary. The source of water will most likely be inside a nearby structure where a meter is located. Inside the structure, provide a water shutoff and backdrain to allow the system to be shut down and drained to protect pipes from frost. Consideration should also be given (and may be required in some locations) to installing a backflow preventer in the water line. This will prevent accidental contamination of potable water supplies from exterior sources. The backflow preventer should be installed 2 feet higher than the high point of the water supply line. If the outflow end of the water supply never comes in contact with the water feature (air break), the backflow preventer may not be necessary. Verify water source requirements with a professional engineer or building official before beginning installations.

Drain Installation

At some point in time, the water feature will need to be drained. Consideration should be given to how water will be removed from the site. In rigid liner pools a common method is to connect a drain from the low point of the pool directly to a storm sewer or outlet. Do not connect pool drains to sanitary sewers. This drain line is best installed when subsurface utility lines are being installed, but can be installed at the same time as pool excavation. Identify the location on the storm sewer line at which the drain connection is planned, and the location of the low point of the pool. Excavate a trench between the points that maintains a downward grade from the pool to the storm sewer at a depth that is at least 8 inches below the lowest elevation of the pool.

Cut and place a T connection at the storm sewer line and connect lengths of PVC pipe to the pool location (Figure 34–4). Pipe diameter will vary, but a 2–4 inch diameter will drain most residential pool installations. Place a trap and elbow below the pool's low point, and install a vertical pipe that extends above the planned bottom elevation for the pool. Cover this end of the pipe with duct tape. This pipe can be trimmed and a drain installed while the pool is built.

Draining for flexible liner pools requires pumping water from the low point of the pool to a point where a drain exists. This is often accomplished by placing a submersible pump in the low part of the pool and running a hose to the drain location. This drain can be a storm sewer or water drain, and can be run onto a lawn area in some situations. No special construction is required for draining flexible liner pools.

Subdrain

Where soils are predominantly clay, subsurface water can lift and damage a rigid or flexible liner. A subdrain is necessary under the pool to remove subsurface water and reduce the chance of such damage. Excavate the entire pool basin 4 inches deeper than required. Excavate a 6 inch deep by 6 inch wide trench along the center of the entire length of the bottom of the pool basin. In this trench, install a 4 inch diameter socked, perforated drain tile. If a drain has been constructed for the pool, connect the tile to that drain. If no drain is present, continue the trench and tile to an outlet point. Cover the tile and entire bottom of the excavation with 4 inches of pea gravel before installing the liner or constructing the pool.

INSTALLATION OF FLEXIBLE AND RIGID LINERS FOR STATIC POOLS

Flexible and rigid vinyl liners make pool construction an easy task. While not appropriate for all installations, the versatility of liners allows for a variety of shapes and depths in pool construction.

Layout

Begin flexible pool installation by marking the proposed perimeter of the pool with paint. This marking provides a guideline for excavation of the reservoir. Rigid liners require transferring the exact pool perimeter dimensions to the location where the pool will be placed.

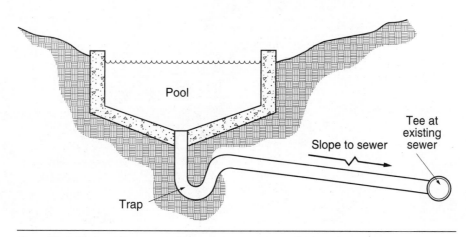

Figure 34–4 Drain and trap at pool. Do *not* connect to a sanitary sewer.

Excavation

Excavate the **reservoir basin** to the correct depth over the entire area. Very steep angles work best to create side walls (Figure 34–5), but in a sandy soil, the angle may have to be shallower to avoid sidewall collapse. The bottom of the flexible pool liner basins can be excavated to different levels if the design requires, but each terrace should be sloped slightly toward the low end of the basin. Bottom areas where fountains or pumps are planned should be excavated level to allow the pump to set flat. Corrections to the base can be made by adding sand.

Around the entire edge of the reservoir basin, excavate a level ledge for **coping material.** This ledge should be the same width and depth as the material selected to hold the pool liner in place. If edge installation is unstable, excavate the ledge deeper and install precast concrete wall caps level around the entire pool perimeter. The depth of this ledge should be deep enough that when edge material is placed, it will match the finish grade of the material that is adjacent to the pool (walk, patio, lawn, etc.). *Verify that all points of this ledge are level.* Any low point in this ledge will create an unplanned drain where water will spill out of the pool. Correct low points by filling with fine granular material.

Flexible Liner Placement

The pool liner must be sized to provide seamless coverage of the reservoir. To calculate liner length, add the maximum length of the pool plus twice the depth plus 3 feet. To calculate liner width, add the maximum width of the pool plus twice the depth plus 3 feet. For large pools, separate liners can be spliced together using bonding material available from the manufacturer. Unfold the pool liner on a flat area and trim it to the correct dimensions. Lay the liner over the reservoir basin; step carefully inside the reservoir and gently work the liner down to the bottom of the basin (Figure 34–6). Be sure the liner sets as flush as possible on the bottom of the basin. Mold the liner over terraces, tucking and folding where necessary to accommodate grade changes. Use the same process to fold the liner

Figure 34–5 Flexible liner pool excavation.

up the vertical sides and over the ledge created at the top of the pool. Fold and tuck the liner if required to adapt to corners. Wipe the liner clean before climbing out.

Fill the pool with water to settle any high points. Verify the ledge elevation is level, correcting any low points. Check to make sure the liner is snug against the sides and bottom of the basin before trimming any excess. Using a sharp carpet knife, trim the liner at the outside of the coping ledge (Figure 34–7). Hold the liner in position to cut around the outside edge. Extra liner folds can be trimmed, but do not trim beyond the inside edge of the ledge. Use caution not to puncture the liner, since even small holes create serious leaks. A damaged liner can be repaired at a tire shop.

Rigid Liner Placement

After excavation, carefully place the rigid liner into the basin and test for snug fit. If the liner sides bow, or if the liner does not sit level or completely in the excavation, remove the liner and adjust the excavation. If liner sets too low in the excavation, adjust the bottom elevation using sand or pea gravel. When correctly positioned, backfill any visible voids between the liner and the excavation with sand.

Edge Treatment

Once the flexible or rigid liner is stabilized, coping material can be placed. Select and place coping pieces

Figure 34–6 Installation of liner in flexible liner pool.

Figure 34–7 Trimming flexible liner.

Figure 34–8 Edging a flexible liner pool.

so that they fit together tightly with no liner visible between pieces. Place the coping pieces so that they overhang the edge of the pool approximately 2 inches to hide the top edge of the liner (Figure 34–8).

INSTALLATION OF CONCRETE POOLS

Pools with rigid walls are constructed of various types of concrete, some with and some without forming. Choice of a material and the construction method should be based on the shape of the pool. Pools with natural-looking surfaces cannot be created with wood forms, and should instead be constructed with a very stiff concrete mixture known as **Gunnite®.** Gunnite is a common application material for swimming pools and artificial rock features like those seen in large motels and zoos. Gunnite installations are excavated and reinforced, with the Gunnite sprayed over the reinforcing and hand finished. Because of the special skills required to install this material, Gunnite installation is not covered further in this text. It is recommended that an experienced contractor be hired to perform work for free-form pools, waterfalls, and artificial rock installations using this product.

Pools with square corners and vertical sides are best constructed by separately forming and pouring the bottom and sides of the pool. Much of the work performed for pool construction is similar to the concrete construction covered in Chapter 23.

Excavating for Concrete Pools

Excavation for rigid pools requires that the basin be over-excavated to provide enough dimension for the pool floor and walls. Examine the floor and wall thickness shown on the plans. Typical wall widths are between 4 inches and 8 inches, based on the depth and size of the pool. Excavate the basin as closely as possible to the depth required. The sides should be excavated wide enough to allow front and back forms to be constructed for the pool walls.

Forming and Pouring Concrete Pool Floors and Walls

Form the bottom of the pool in a manner similar to installation of a concrete slab. Reinforce with wire mesh and rerod for stability. Every 1 foot around the edge of the slab, an L shaped rerod should be placed

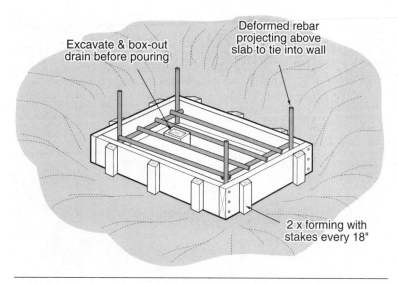

Excavate & box-out drain before pouring

Deformed rebar projecting above slab to tie into wall

2 x forming with stakes every 18"

Figure 34–9 Excavation and forming for concrete pool.

with one leg of the L projecting 6 inches above the slab. This rod serves as a connection between the floor slab and the wall. Prepare the drain pipe that was placed earlier by trimming it and attaching a drain fitting with a threaded plug. Be certain the pipe is trimmed low enough that the floor can be sloped down to the drain. Cover the drain fitting with duct tape. Pour the floor and finish with a wood float surface that maintains a slope to the drain from all directions (Figure 34–9).

After the floor forms are removed, form the sides of the pool using plywood with a 2 × 4 framework backing. Place forms for the inside of the pool wall that are set back from the edge of the floor slab the thickness of the pool wall. Brace across from the opposite side. If reinforcing is desired, place horizontal rerod around the sides and connect to the L rerod anchors projecting up from the base. Connect them by twisting 14 gauge wire around both rerods. Use this first horizontal to support vertical and horizontal rerod placed in a 1 foot by 1 foot grid around all sides of the pool. Bending rerods around the corners provides more stability than tying together the ends of two straight bars.

Any lighting, **skimmers,** water supply lines, or other treatments that need to be placed in or through the pool wall should be boxed out for future installation before the outside wall is formed. If piping, such as drains or supply lines, need to extend through to the inside wall of the pool, cut a short section of the pipe that is 4–6 inches longer than the width of the wall. Drive a nail in through the inside form at the height the pipe is to be located. Glue any fittings on the pipe,

cover the ends with duct tape, and slide the inside end over the nail. Drill a hole through the outside form of the same diameter and at the same height and slide the form over the pipe. This will hold the pipe in position during the pour and the duct tape can be cut away after the forms are removed. A connector can be glued on the outside and the piping continued to its source.

Construct the outside wall forms using the same materials as used for the inside form. The height of the exterior form should extend from the top of the pool wall to the base of the floor slab. Connect the inside and outside walls together using 1 × 4s nailed on top of the forms every 2 feet (Figure 34–10). Brace the forms on the outside using lumber braces and by piling earth against the base. Verify that the tops of the forms are level at all points and adjust if necessary using shims. Small gaps between the bottom of the forms and floor slabs can be filled with expanding foam insulation. Squirt the foam into the gap and trim on the inside of the form after the insulation hardens.

Pour the sidewalls slowly, watching the forms for movement and excess leakage. Tap forms with a rubber mallet as the pour progresses to settle the concrete and to reduce **honeycombing.** Remove the 1 × 4s immediately after the pour to allow access to the top of the wall for finishing. Smooth the tops of the walls using a float. After the walls are hardened, remove the forms (Figure 34–11). Providing a bead of silicon caulking at the joint of the wall and floor may reduce water seepage if the bond between the wall and floor is poor. Complete plumbing and related work behind the walls, and backfill with granular materials.

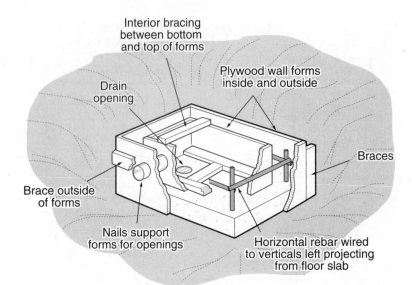

Figure 34–10 Cut-away view of completed forming for concrete pool walls.

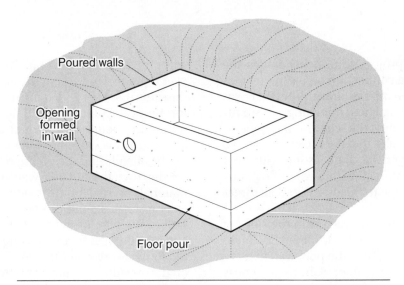

Figure 34–11 Completed pool pour.

Interior Treatment

> **CAUTION**
>
> Cleaning and painting pools require use of caustic materials. Mix, apply, and clean only according to manufacturer's instructions to avoid injury.

Pools can be filled and used without additional treatment of the inside, however, other treatments can be utilized to achieve special effects. A common treatment is to paint the interior with a waterproof paint. A dark paint provides a consistent finish that improves the reflective qualities of the pool. Painting the wall should be preceded by cleaning with a mix of nine parts water to one part muriatic acid. Priming the wall before painting reduces future maintenance.

Another treatment is to plaster the inside of the pool with a thin **spackle.** This treatment does little to improve water retention, but adds a consistent surface color and texture to the pool. Clean the inside wall with the muriatic acid mixture. When the wall is dry, mix spackle plaster with water to create a pasty solution and apply it slowly, beginning at the bottom of the pool and working up to the top of the walls. Prime and paint if desired.

Figure 34–12 Construction of a wall between flexible liner pools set at different levels. The upper pool is adjacent to the lower pool with a waterfall planned between the two.

A fourth choice is to add a vinyl liner to improve water retention. The liner is cut to fit the inside dimensions and glued into position. Any drains, supply lines, lighting, and other openings in the pool will require that the liner be trimmed and fitted around these openings.

Edge Treatment

The top edge of the pool can be treated in a number of ways. One method is to place precast wall caps or limestone paving pieces on the top of the pool wall. A more permanent installation is to place a precast concrete coping on a 1 inch bed of mortar around the perimeter of the pool.

INSTALLATION OF WATERFALLS AND CASCADES

The creation of waterfalls and cascades requires installation of basic elements that, when assembled properly, can create a dramatic landscape feature.

Lower Basin

The first element to consider for waterfall or cascade construction is the basin that will collect water at the bottom of the falls. The basin may be constructed using either the liner or concrete static pool techniques described earlier in this chapter, but the pool must be constructed with part of the basin directly under the location where the projection channel will deliver the water. The projected fall of water should land in this lower basin far enough from the edge to allow water displaced by splashing to remain in the pool. If a flexible liner is used for the lower basin, the vertical sides of the pool basin in this area should be built with concrete blocks under the liner to provide stability (Figure 34–12). This technique also provides support for the falls and any basin used above to supply the falls. The lower basin contains the pump that recirculates water to the top to supply the falls.

Water Supply/Upper Basin

The second consideration for a waterfall or cascade is the water source at the top of the falls. The falls can be supplied by water flowing out of an upper reservoir basin or directly from a water supply line recirculating water from a pump in the lower basin. Due to the need for an opening to allow water to exit the pool, rigid vinyl liners cannot be easily used for upper basins.

When an upper basin is constructed for a waterfall, the edge should be placed as close as possible to the lower basin. Concrete pools can be constructed adjacent to each other to accommodate the falls construction, but flexible liner basins require the construction of a retaining wall to support the portion of the upper basin that will supply the falls (Figure 34–12).

Weir and Pump

Both concrete and liner pools require the construction of a weir at the location where water is to drain from the reservoir to supply the falls or cascade. A **weir** is an opening in the edge of the basin with dimensions that allow only a calculated amount of water to pass (Figure 34–13). The depth and width of this opening should be taken from plans or calculated by a design professional, since the size of the opening will control the amount of water that supplies the falls. In the absence of precise weir calculations, construction of an oversized wire is recommended since the flow through the weir can be restricted but not easily increased. The weir needs to be placed lower than the perimeter of the basin to allow water to exit the upper reservoir. The opening can be constructed in a flexible liner pool by over-excavating the earthen dam to the dimensions desired. If a concrete pool is constructed, the opening will have to be blocked out before the pour or cut out after the pour.

The amount of water that flows through a waterfall or cascade is not a chance occurrence. Control of the water amount is primarily a function of the weir. Another factor that controls the water flow is the capacity of the recirculating pump. Both of these factors work together to provide either a light or a heavy flow of water through the falls channel. If the water flow from the falls does not meet design expectations, consider installing a higher volume capacity pump or using more than one pump. Flows over a waterfall can be concentrated by narrowing the weir. If the lower basin is running dry and the upper basin is running over, the weir is undersized and needs to be enlarged, or the volume of water being pumped needs to be reduced.

Recirculation Piping

Installation of waterfalls and cascades requires piping for recirculation of water between basins. Rigid piping is required in higher volume operations, while flexible tubing can be used for pumps with a capacity of less than 10 gallons per minute. This piping should be placed between supply and use locations while the basins are being constructed. In concrete pools, the piping should be formed through the wall, then connected and placed before the pool wall is backfilled. With flexible liner pools, the piping runs over the edge of the pool hidden between coping pieces. Piping between pools can be hidden by plants. To hide the installation of flexible tubing and reduce kinking of the tube, place a 3 inch diameter, nonperforated drain tile underground and pull the tubing through it.

Projection Channel

Rigid wall pools may have water spill directly from a weir into the lower basin, but many water features have a projection channel that carries the falls water from the supply out over the lower basin (Figure 34–14). This projection assures that the water will fall into the lower basin and will not miss the pool. The projection channel can be custom-cast out of concrete or can be constructed from pieces of flat stone with a slight depression in the center.

Large volumes of water require special construction to assure the water does not flow off the sides of the projection channel. Supporting the channel also requires some form of wall under the channel and anchoring at the supply side. The greater the amount of projection for the channel, the more difficult it will be to anchor. If possible, the channel should be tilted forward slightly to keep water moving downward and to help the water project farther out into the lower basin.

If water is clinging to the end of the projection channel, either tilt the channel farther forward or cut a **reglet** (Figure 34–15). A reglet is a shallow cut on the underside of the channel, running the full width of the channel. Reglets break the surface tension between water and the projection channel, allowing water to fall rather than to adhere to the surface. A reglet can be

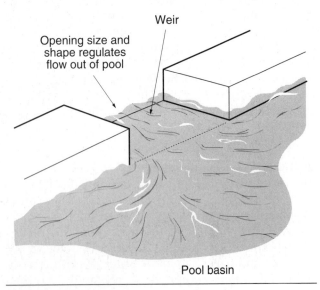

Opening size and shape regulates flow out of pool

Weir

Pool basin

Figure 34–13 Weir diagram.

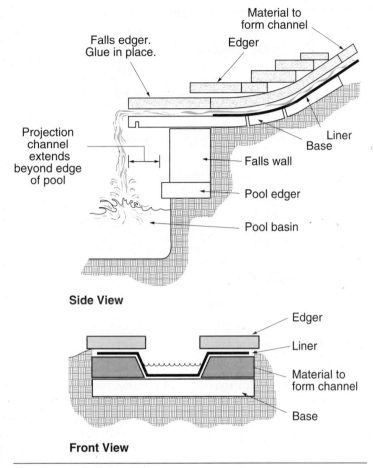

Figure 34–14 Projection channel construction for waterfall.

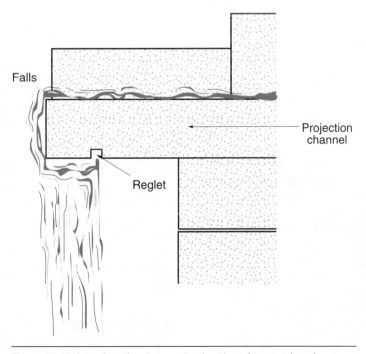

Figure 34–15 Location of reglet, used to break surface tension of water..

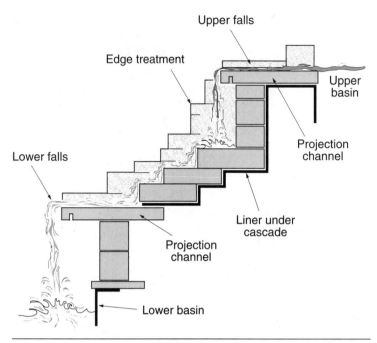

Upper falls

Edge treatment

Upper basin

Lower falls

Projection channel

Projection channel

Liner under cascade

Lower basin

Figure 34–16 Falls channel construction for multiple falls.

cut with a wet masonry saw and should be approximately 1/4 inch deep and run the full width of the channel.

Multiple Falls

Installations that call for multiple falls can be constructed using multiple basins and projection channels, or can be constructed where the projection channel is the supply for the next falls (Figure 34–16). Either arrangement requires careful positioning of channels to assure that the water falls on its target.

Cascade Channel

The channel for the cascade is a wide, shallow, U shaped channel that runs from the top weir to the lower weir. The cross section of the channel should be approximately twice the cross section of the upper weir to accommodate the water flow and decorative lining material placed in the channel. The channel may follow any alignment desired as long as the proper cross section and downward slope are maintained.

Excavate the entire route of the channel. Maintaining a consistent slope provides an even flow of water from top to bottom. Varying the slope provides a more interesting water flow, but requires that the channel be widened to accommodate heavier flow in areas. Edges of the channel need to be elevated high enough that water stays in the channel (Figure 34–17).

Waterproofing the channel can be accomplished by cutting a linear strip of pool liner material and placing it in the excavated channel. If splices need to be made

or corners must be turned, overlap two pieces of liner approximately 1 foot, placing the higher section on top of the lower section. When the waterproof lining is completed, decorative lining material should be placed. This decorative liner material is typically rock or stone and is placed in the channel to hide the waterproofing. Rock is placed along the entire length of the channel, spread from side to side to obtain complete coverage of the liner. Stone must be placed in the channel beginning at the bottom. Each succeeding piece of stone should overlap the previous piece and should have a slight forward slope. Stones should be tilted slightly toward the center of the channel.

An alternative to flexible liners for a cascade channel is pouring a concrete channel. Dimensional requirements for the cross section remain the same as those for the liner channel. Before the concrete has hardened, the decorative liner pieces should be placed. For rock, pour into the channel and seat into the concrete using a trowel. Wash lightly after the concrete has hardened slightly. To place stone, begin at the bottom of the cascade and push the decorative liner pieces into the concrete, forming the base of the channel. Continue placing the stone tilted slightly forward and toward the center of the channel until the top weir is reached.

Edge Treatment

Complete the installation by treating the edges of pools and adding edging material or plants. In falls and cascade channels, stacking stone pieces two to

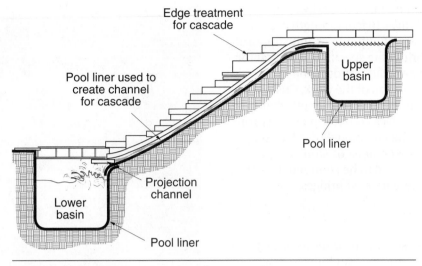

Figure 34–17 Falls channel construction for a cascade.

three high in some locations can create the illusion of water flow that has cut through several layers of stone (Figure 34–18). Careful placement of plant material can hide wiring, tubing, and other rough edges of the falls installation.

INSTALLATION OF FOUNTAINS

Fountain installation can be accomplished using a stationary pump or a floating fountain. Most fountain installations are placed within a pool or pond and utilize water from the basin of the pool for spray. Fountains that do not have pools typically include some method for water recirculation or water supply. A licensed electrician should complete electrical connections.

Stationary Pump Fountains

Select a location in the pool basin that places the fountain on a level surface at the correct depth. If necessary, a heavy block can be placed in the pool to raise the fountain pump. Lower the pump into the basin and secure in place. Bring the electrical connection out through a preplaced conduit in a concrete pool or between coping pieces for a liner pool. Complete the electrical connection by plugging into a GCFI electrical receptacle or connecting to a GCFI switched circuit. Fill the basin with water and test the fountain for proper operation.

Floating Fountains

Floating fountains require that the pool be filled before placement. Fountains that float in a pool are anchored to the sides or bottoms using cables/ropes or a weighted anchor. A maintenance rope is also run to the side and placed where it can be used to pull the

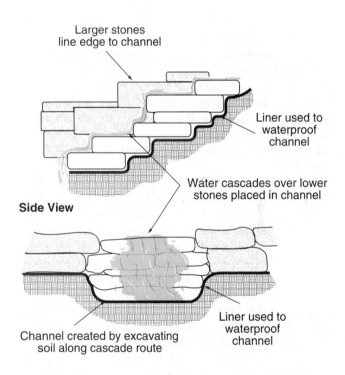

Figure 34–18 Cascade edge treatment.

fountain in for repairs. Electrical cables are attached to the fountain with waterproof connections and floated out with the maintenance cable. After anchoring and electrical connections are complete, test the fountain.

INSTALLATION OF BRIDGES

Bridge installation involves two phases—**pier** construction and placement of the structure. A specialist in concrete work should be hired to perform installation work if the bridge is over 10 feet in length or is over moving water. If concrete abutments or full supports under the ends of the bridge are specified, a contractor specializing in concrete pours should be subcontracted to perform that portion of the work. Custom or prefabricated bridges under 10 feet in length that utilize wood post or cast-in-place concrete piers are projects that might typically be completed by a landscape contractor. Truck-mounted cranes or similar types of special lifting equipment need to be contracted for the placement portion of prefabricated bridges.

Pier Construction

Prefabricated bridges rest on two or more piers that carry the load for the bridge. Carefully review the working drawings from the manufacturer to determine the dimensions of the footing, reinforcement required, and bolt pattern necessary to make connections to the bridge. Most bridge piers, regardless of size, extend below frost depth to prevent movement.

Small custom-built wood foot bridges can be placed on posts excavated to frost depth. Locate the center of each post and excavate a hole by hand or by using a post auger. Place a small amount of granular material in the hole and place the post in the hole. Plumb and trim to the elevation required for deck framing (Figure 34–19).

Piers for small bridges may also be cast-in-place concrete poured in paper tube forms. When using paper tube forms, it is critical that footings be braced against movement. There is a potential for a vertical column to tilt when filled with concrete, altering the dimensions for the bridge. Auger holes to the dimensions required for the pier and slide a piece of tubular paper, tube-forming material down the hole. Insert rerod into the form. Stabilize the rerod by driving it into the soil at the bottom of the hole. Measure and cut off the top of the tubular paper form with a reciprocating saw at the required elevation. Adjust for plumb and proper location. Pour concrete and set anchor bolts, verifying that the spacing of bolts is correct before the pour hardens.

Bridge Placement

Contracting a truck-mounted crane is the easiest way to set a prefabricated bridge. Crane operators can place the structure quickly and with minimal damage.

Figure 34–19 Use of posts installed as bridge piers.

The delivery and placement should be carefully timed to make the operation cost-efficient. Allow the crane operator access as close to the structure as possible. Wrap lift belts under the bridge and lift and place over anchor bolts. When the structure has been placed on the anchor bolts, install washers and nuts and tighten securely.

Custom Bridge Construction

Construction of custom-framed bridges is essentially deck construction with railings added for safety. Refer to the instructions given in Chapters 29 and 30. Framing can be either attached to wood posts or set and anchored to piers, as described in this section (Figure 34–20).

Figure 34-20 Deck framing for wood bridge.

Chapter 35

Site Furniture, Signage, and Prefabricated Playground Equipment

Adding furniture and site amenities to a landscape design often provides a finished touch difficult to obtain through use of other landscape elements. The aesthetic effects of pavement, walls, lawn, and plant material are enhanced by the thoughtful placement of seating and lighting. The utilization of the site is improved when trash and bikes are contained and proper signage is provided. Play equipment adds options to parks, schools, and playgrounds. Design emphasis is created by the effective use of flagpoles. While typically reserved for the commercial landscape, use of such design tools has expanded into residential design with the advent of residential grades of amenities.

Much of the furniture for a site is purchased from custom manufacturers and can be installed by a landscape contractor. While there are similarities among installation methods, be certain to review the instructions provided by the manufacturer carefully to assure warranties or guarantees are not voided.

RELATED INFORMATION IN OTHER CHAPTERS

Information provided in this chapter is supplemented by instructions provided elsewhere in this text. Before undertaking activities described in this chapter, read the related information in the following chapters:

- Safety in the workplace, Chapter 4
- Construction math, Chapter 5
- Basic construction techniques, Chapter 7
- Related site utility work, Chapter 15
- Cast-in-place concrete, Chapter 23

TYPES OF SITE AMENITIES

Benches. Site seating is available in a variety of shapes, materials, and finishes. Styles most often found include contour benches and historic benches. Seating is manufactured in wood, metal, stone, fiberglass, and precast concrete. Most seating anchors directly to concrete slabs, with some seating requiring footings or posts that are direct buried in the ground.

Park tables and grills. Heavy-duty picnic tables and charcoal grills are designed for use in park and recreation areas. Most tables are manufactured of metal and/or wood and grills are welded steel. Direct burial and concrete slab mounting are the most common installation options.

Trash receptacles. Trash receptacles are decorative enclosures into which garbage containers are placed. Materials used to construct the receptacles include wood, metal, precast concrete, or fiberglass. Most are manufactured to be anchored to a concrete slab, but some are mounted on a post that is buried in the ground or anchored to a footing. Shapes can be varied, but are typically round or square.

Signage. Identification of businesses, traffic routes, or general information is performed in the landscape using signage. Many methods of signage are available, with ground-mounted, also termed monument, and post-mounted signage most common. Many signs also require placement of electrical lines to illuminate the sign for night viewing.

Bike racks. Many designs are available for securing bikes. Most are manufactured from metal and

Figure 35–1 Bike rack anchored in a concrete slab.

Figure 35–2 Bollards used to separate traffic from pedestrians.

anchored to a concrete slab or directly buried (Figure 35–1).

Bollards. **Bollards** are short posts used to identify the edge of an area (Figure 35–2). Often used to separate pedestrian and vehicular traffic, bollards can be manufactured of wood, metal, stone, or other materials and may have lighting incorporated. Installation can be direct burial or concrete slab mounted.

Flagpoles. Flagpoles are available with little variety in shapes, but pole height, operation, and materials vary. Metal is commonly used for flagpoles, but precast concrete and wood are also used. Flagpoles require direct burial or anchoring to a footing buried to frost depth or deeper.

Light standards. The posts on which pedestrian lights are mounted are called **standards.** Styles can be found that include contemporary, historic, western, or architectural. Materials typically include metal, wood, or precast concrete. Most light standards require a footing that extends to frost depth or deeper.

Prefabricated play equipment. Like other amenities, a wide variety of play equipment choices is available. Most pieces are manufactured with a framework of wood or metal posts that are directly buried or mounted on a footing. Engineered equipment for play activities is then attached to the framework. Platforms, slides, climbers, and many other pieces of play equipment are available for the client to choose from. Custom-built wood structures follow plans prepared by the design professional and typically utilize direct burial for post installation.

PLANNING THE PROJECT

Timing of Site Furniture and Play Equipment Placement

Footings and slabs on which the furniture is anchored can be efficiently constructed when concrete for paved areas and structures is poured. Because site furniture is often shipped assembled and finished, it should be protected from potential damage by waiting for installation until major exterior construction projects are near completion. Furnishings that require assembly and/or finishing can be prepared off-site at any time and installed near the end of the project.

Manufacturers' Specifications and Shop Drawings

Materials purchased from manufacturers typically include installation and finishing specifications and **shop drawings** for proper installation. These specs and drawings should be obtained before any planning and installation begins. The contractor should also compare the specs and shop drawings with those provided by a project design professional. Any discrepancy in materials, finishes, construction, installation, or other aspects should be brought immediately to the attention of the designer.

Layout of Site Furniture, Signage, and Play Equipment

Site furniture is typically located by staking the footings used to anchor the amenity. Locate either the edges or center lines of mounting slabs. Measurements should carefully conform to drawings, since once a footing is set for an amenity, it cannot be adjusted to match the dimensions of the fixture.

Layout of play equipment is a challenge since most pieces have multiple posts. As with site furniture, obtain the working drawings from the manufacturer before proceeding. Layout using a grid with two baselines is a common method for identifying post locations. Verify that the two baselines are square to each other and double-check measurements for square often. Proper level must also be maintained between footing locations, since any variation will twist the play structure. Utilize grade stakes for each post to show the elevation and location of each footing.

ANCHORING SITE FURNITURE, SIGNAGE, AND PLAY EQUIPMENT

Site preparation for furniture, signage, and play equipment is primarily limited to the construction of footings. Three common methods of anchoring most site furniture and signage include: (1) direct burial of amenity supports, such as legs or posts, which are connected to the furniture; (2) installation of a slab onto which the furniture is anchored; or (3) construction of a frost depth footing onto which the furniture is anchored. Signage may also be anchored by attaching to an existing structure such as a building, fence,

or pole. The following sections describe the construction common for each of these techniques.

Direct Burial

For a fixture that requires direct burial, begin by locating the burial points where each support must be sunk. Verify the level between points if an amenity has several supports that must be buried. Placing a level stringline that is set at the proper burial depth and can be moved between each support location is an easy way to verify that all supports are buried to the proper depth. Burial depth can be controlled for individual supports by placing a mark on the side of the support indicating ground level.

Burial can be accomplished by either driving the support into the ground using a sledge or post driver, or by excavating a hole and placing the support. When selecting the first method, wrap the top of the support with heavy rags or place a block of wood on top of the support to avoid contact damage when driving. Using a firm stroke, drive the support into the ground. Stop periodically to check for plumb, for correct spacing between supports, and for proper orientation. Continue driving until the support is sunk to the proper depth.

If the excavation approach is selected, bore a vertical hole that is only slightly larger in diameter than the support. When the proper depth is reached, set the support and backfill. To maintain proper elevation of the support, a 1×6 cross piece can be fastened to the post at the correct burial depth. This will hold the post at the proper position in the hole until it is backfilled. This technique is beneficial when posts must be placed and equipment assembled before holes are backfilled. Fill the hole a few inches at a time and compact, rechecking to verify plumb, proper depth, and orientation. If the hole is overexcavated, it can be raised with a small amount of granular fill. Supports that are set too deep can be raised slightly by gripping the post and lifting straight up. Supports with flanges or anchors may require the use of a twisting or rocking motion. After adjusting the support, verify plumb and recompact the soil. Following support placement, the amenity may be connected.

Slab Anchoring

Furniture or monument signage that is anchored on a slab requires the forming and pouring of the slab (see Chapter 23, for instructions on pouring concrete) and setting anchors at proper dimensions before the concrete hardens. If electrical connections are required, see the section on construction of footings with electrical service later in this chapter. Verify the surface material and/or finish of the slab before pouring. The finish surface under a bench may not be the same as the adjacent paved surface, and if a special finish or allowance for pavers is required, this foundation slab must conform. To properly locate anchors, use string-lines or directly measure the placement locations. Place anchors shortly after the slab has been finished by pushing the bolt/anchor stem into the wet concrete and lifting up or wiggling slightly to settle the concrete around the stem. If the anchor sinks into the concrete, wait 15 minutes and attempt to reset it. Do not wait too long to set the anchors or they may not insert into the slab. Wait at least 48 hours before connecting the amenity (Figure 35–3).

Frost Footing

Some amenities require the stability afforded by a footing that extends below the frost line. The actual depth to avoid frost heaving varies based on geographic location. When a footing is required, excavate a vertical hole that is the diameter required by the manufacturer or the drawings. Verify that the depth meets the required depth for the installation. Amenities that have several supports may require multiple excavations. These holes do not need to be joined by a continuous footing if each extends below frost depth, but if one hole is not the proper depth, it will move up and down at a different rate than the others, causing damage to the amenity.

If steel reinforcement is required, cut rerod 8 inches longer than hole depth and secure by driving into the soil in the bottom of the footing hole. Many footings can be poured without being formed by placing the concrete and finishing the top of the footing with a trowel. If certain conditions exist, it may require forming the entire footing using a tubular paper form. These conditions include situations where the footing extends above the finish grade, where unstable soil conditions may collapse the hole, or situations in which the hole was excavated in an irregular shape. To prevent frost heaving, footing holes should have sides

Figure 35–3 Anchoring of an amenity support post to paving using an anchor bolt placed through the paving. (Courtesy of Margaret Sauter)

that are vertical. Holes that have a wider diameter at the top are subject to frost pushing up on the footing, countering the advantage of having a footing below frost depth.

Finish the top of a footing with a wood float, doming the surface slightly to create drainage away from the post. After the footing is poured, place anchors in the top of the footing by pushing the anchor stem into the concrete and lifting or wiggling slightly to settle concrete around the stem. Before the concrete hardens, verify the spacing between anchors for fixtures that have multiple supports (Figure 35–4).

Footings with Electrical Service

In some cases, the footings or slab may require additional work before the concrete can be poured. If electrical lines are to be supplied to the amenity, such as light fixtures or lighted signage, it may be necessary to install conduit through the form before the pour can take place. For slab pours, excavate the trench for conduit and place the conduit in the proper finish location. Cover both ends of the open conduit with duct

Figure 35–4 Anchoring of an amenity support on a footing poured separately from the concrete walkway.

tape. Backfill the trench with granular fill material and proceed with the pour.

To install a round lighting footing, use an auger to excavate a round hole 6 inches larger than the diameter specified and to the required footing depth. Select a tubular paper form of the specified diameter for the footing and trim to a length the same as the footing depth. Through the side of the form, cut an opening 18 inches below the surface grade. Cut and connect a conduit bend and a straight section. Place duct tape over the openings at both ends of the conduit. Place the conduit in the form, with the bend exiting the opening and the straight section extending above the top of the form. Use duct tape to position the conduit inside the form. Lower the form into the footing excavation and orient the stub of the conduit bend in the proper direction. Backfill with soil around the form while maintaining the form in a plumb position. If reinforcing is required, construct the framework and place inside the form. Pour premixed concrete into each form. Before the concrete hardens, center the conduit and insert any anchor bolts required for anchoring the fixture. When set, peel the top of the paper form away.

Anchoring Signage to Another Structure

In situations where no separate support is planned for signage, existing structures may be used to anchor signs. Signage can be connected to wood structures with the use of lag screw placed into support posts or studs, or anchored to masonry units with the use of expanding masonry bolts. If signage is attached to metal objects such as a fence, a backing strap is required. Position the sign at the proper location and insert a galvanized bolt through a hole in the sign (or through the sign's mounting bracket), through the fence, and through a hole in the mounting strap. Secure and tighten the nut using a lock washer. Various forms of mounting straps are available (Figure 35–5) that will accommodate fencing, light standards, and other types of metal landscape amenities.

INSTALLATION OF SITE FURNITURE AND PLAY EQUIPMENT

Installation of most amenities requires placing and finishing the furniture piece. In some installations, welding or special finishing may be required, and for some, the completion requires the work of an electrician or other construction specialist.

Placement and Adjustment

Depending on manufacturer's instructions, preassembly of all or part of an amenity may be required before installation. Amenities assembled piece by piece can be put together beginning at the anchors/supports and working up. Assemble any portions that are required and position over the anchors. Carefully set the fixture on the anchors/supports and install connection hardware. Secure the fixture and then begin the process of leveling and plumbing.

Key criteria for placement of an amenity are that it be placed at the proper location and elevation (Figure 35–6) and that it be placed level and plumb. To maintain these criteria connect the fixture securely but not fully tightened (termed "finger tension"). Check for level on a horizontal surface and plumb on two adjacent vertical surfaces. Adjust as necessary and complete the tightening of all fasteners. If level adjustments are required, remove the amenity and place enough galvanized washers over the anchor bolt to adjust the amenity to the proper level.

Surfacing

Resilient surfacing below play equipment is required to maintain a safe fall zone. Consult the guidelines published by the Consumer Product Safety Commission's *Playground Audit Guide* for recommended materials and depths. If granular surfaces such as mulch, sand, or gravel are being used, excavate to the proper depth and layer the material into the area with minimal compaction (Figure 35–7). If resilient paving surfacing is being used, prepare the site and pave according the instructions provided in Section 6.

Finishing

Site furniture can be shipped prefinished or requiring various degrees of **field finishing.** Selection of finished or unfinished materials is based on cost, risk of damage during shipment or installation, and the contractor's ability to field finish the product. If the contractor purchases prefinished materials, they should be covered or wrapped with a protective material until the construction project is complete. Finishing activities can range from the complete priming and painting of a fixture to touching up minor surface damage caused by transportation and installation. Most common finish activities completed by the landscape contractor include the painting of light poles, staining or painting of benches,

Figure 35–5 Attachment of signage to decorative metal fencing. (Courtesy of Margaret Sauter)

Figure 35–6 Layout of a custom-built play structure. (Courtesy of Janice Carter)

Figure 35–7 Cushioned surfacing below play structure.

installation of hardware, and installation of liner cans for trash receptacles.

One option for finishing is to perform the work after the fixture is placed. Field finishing can be difficult since the work can be disturbed by dust, inclement weather, or other construction activities. The recom- mended process is to perform any painting or staining activities in an enclosed environment, carefully trans- port the items to the site for installation, and perform only touch-up finishing in the field if necessary. Fin- ishing should be one of the last activities performed prior to final acceptance of the project by the owner.

Chapter 36

Edgings, Planters, and Plant Protection Equipment

Defining, containing, and protecting plants requires a variety of specialized construction materials and techniques. Edging is one of the more versatile materials available to the landscape designer. Edging can be used to define areas, hold pavement in place, and separate turf from planting beds. Edging materials require excavation of a trench, either by hand or machine, and placement of edging within the trench. Techniques are also available for installing concrete edging using a special curbing machine.

Planters are valuable for providing proper planting environments in paved areas. Most planters available are manufactured from wood, precast concrete, fiberglass, or composite materials, and require only placing and filling. Other installations may require that the planter be anchored to a paved surface. Tree grates allow air and moisture to reach tree roots in paved areas, while tree guards protect the trunk of the plant from damage. Tree grates and guards require preliminary work during the site paving stage, with placement and connection after planting is completed.

RELATED INFORMATION IN OTHER CHAPTERS

Information provided in this chapter is supplemented by instructions provided elsewhere in this text. Before undertaking activities described in this chapter, read the related information in the following chapters:

- Safety in the workplace, Chapter 4
- Construction math, Chapter 5
- Basic construction techniques, Chapter 7
- Concrete paving, Chapter 23

PLANNING THE PROJECT

Timing of Edging and Plant Protection Placement

Plant bed edging should be placed after the planting bed has been prepared, but before the installation of weed barrier, plantings, and mulch. Installation of the edging prior to installing any weed barrier or mulch provides a depth guide for both. Placement of edging after planting is possible if its installation will not disrupt or damage the plant material.

Planters and plant protection devices should be installed as the project nears completion. Planters should be placed after all major site construction is completed but before planting operations begin. Preliminary work for tree grates and guards is performed during the paving stages, but finishing and placement should not occur until after all plant materials have been placed.

Layout of Edging

Painting is the easiest method for identifying edging location (Figure 36–1), but a stringline may be used for straight applications. Experienced contractors may be able to locate edging as the trench is excavated, but errors require that disturbed areas be reseeded or resodded. Measurements for edging quantities can be performed by pacing or using a measuring wheel (Figure 36–2).

INSTALLATION OF LANDSCAPE EDGING

Trench preparation requires consistency and can be a laborious task. Preparation for landscape edging can be done by hand for small areas, and is best done mechanically for large installations. To edge for

Figure 36–1 Layout of edging using marking paint.

Figure 36–2 Measuring for edging with measuring wheel.

landscape installations, follow the steps listed in the following sections.

Excavation of Edging Trench

Along the entire alignment, excavate a shallow, shaped trench that is approximately 1 inch deeper than the vertical dimension of the edging (Figure 36–3). The outside edge of the trench (side nearest the lawn or paving) should be excavated with a straight, vertical edge (Figure 36–4). This allows edgings to be positioned tightly against the adjacent material. The inside edge of the trench (side toward the planting bed) should be excavated at a 45 degree angle. For curbing, stone, brick, and precast edgings, the trench should be excavated with a flat bottom that is 2 inches wider than the width material. Depending on the type of edging and what is placed on either side, the placement depth may vary. If the edging abuts an area that is seeded, the top should be flush with the finish grade. If the edging abuts an area that is sodded, the top should be placed 1 inch above the soil grade before the sod is placed. Unless otherwise specified by the

manufacturer, edging abutting paved areas should be placed flush with the top of the paved surface.

For large installations, rental of an edging machine is recommended. This piece of equipment cuts trenches with vertical and angled sides to the proper depth for plastic or metal edging (Figure 36–5). Additional blades are available that adapt the trench to wide materials or steeply angled trenches. The edging machine can be used to cut smooth radius corners and will save many hours of hand digging. To operate, mark the alignment of the edging and orient the machine with the rubber flap aimed toward the planting bed. Start the machine, and gradually lower the cutting blade onto the alignment mark. Pull the edging machine slowly along the marking.

Edging Placement

Placement of specific edgings varies depending on the material and type of installation. Installation techniques for many of the common edging materials are described in the following sections.

Figure 36–3 Proper trench for edging showing vertical excavation on lawn side.

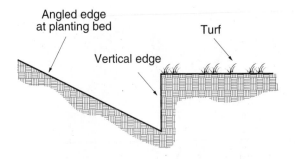

Vertical trench for plastic & metal edging

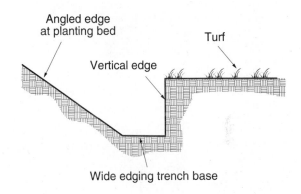

Figure 36–5 Excavation for edging using an edging machine.

Natural Edges. A simple edge that conforms to any shape and requires little special preparation is the natural edge. Created by cutting a vertical trench with a shovel, natural edges join the planting bed with the lawn without a manufactured separator. Natural edges require periodic maintenance to keep the edge clean. Respading during the year is necessary as turf encroaches into the planting bed.

Natural edges are created by using only the trench as a separation. Excavate the trench as instructed previously to a depth of 6 inches. Place the excavated material into the planting bed area. Continue this excavation along the entire length of the edging location. After the planting bed is tilled, fill the bed and trench with mulch.

Plastic. One of the most commonly used edgings is the black plastic material manufactured in rolls and long strips. This edging is flat and flexible so that it can be bent to very tight radii, is installed with metal

Figure 36–6 Staking plastic edging by driving metal stakes through channel at bottom. Edging must be held flat against edge of trench with top flush with finish grade.

stakes, and has only the rounded bead on top visible after installation. The possibility of maintenance for plastic edging is high, as the edging is often worked up by frost and equipment.

Begin installation by placing a length of the edging flat against the vertical side of the trench. If the edging has a fold of plastic, or a V channel, place the fold to the planting bed side of the trench. Verify that the top of the edging, usually the rounded bead, is at the top of the trench and is set at the proper elevation. Place an edging stake in the V channel at the bottom of the edging. Holding the stake at a slight angle toward the turf area, drive the stake through the edging into the subgrade (Figure 36–6). Place edging stakes every 2 feet along the length of the edging. Turning corners is accomplished by bending the plastic edging around the corner. If the edging buckles when turning corners, cut 2 inch slits up from the bottom to **kerf** the edging.

Join pieces of edging by cutting 6 inches of the rounded bead off the top of one end of an edging piece. Verify that both pieces are facing the same direction. Push a joining tube halfway into the rounded bead of that piece of edging (Figure 36–7). Overlap the flat sections of edging and slide them together. Slide the remaining half of the joining tube into the rounded bead on the other piece of edging. Driving a nursery nail through the overlapped pieces

Figure 36–7 Joining pieces of plastic edging using tubing joiners supplied with edging kits. Notching the edge improves the connection.

will reduce the chances of unintentional separation. If smaller pieces of edging are required, measure and cut edging with a hacksaw or tin snips (Figure 36–8). Back-fill and compact along the edge, leaving the grade on the planting bed side low enough to accommodate mulch.

Steel and Aluminum Edging. Typically reserved for commercial or large installations, a more permanent and higher priced choice for edging is steel or aluminum. While not as flexible as plastic, metal edgings can still be used to wrap around corners or curves and is less likely to be worked up by frost or equipment.

Place a piece of metal edging in the trench flat against the vertical side and with the top at the desired finish elevation. Insert a stake through the slots in the edging and drive the stake straight down into the sub-grade. Place stakes in every slot, typically every 2–3 feet. Corners are turned by bending pieces at the proper location. Most aluminum and lightweight steel edgings bend easily, but heavy steel edgings may require bending in a vise or making a cut with a hacksaw

halfway up from the bottom of the edging before bending. Measuring and bending before placement may make the process easier for heavy-duty edgings.

Metal edgings typically have notches cut from each end of the pieces that interlock for joining. Some edgings are joined by hooking these notches together, others interlock using an edging stake (Figure 36–9). Make the connection with the pieces setting in the trench and continue staking. If smaller pieces of edging are required, measure and cut using a hacksaw (Figure 36–10). Backfill and compact along the edge, leaving the grade on the planting bed side low enough to accommodate mulch.

Vertical Bricks. Vertical bricks can be used for an edging for paved surfaces or planting areas. When stood on end, or edge, the bricks can introduce earthy textures and colors into the design while matching a paving material. Installation of brick edgings is more expensive and labor-intensive than installation of most edgings, but they are relatively maintenance free when completed.

Figure 36–8 Cutting plastic edging using a hacksaw. Tin snips may also be used.

Figure 36–9 Joining pieces of metal edging using stakes provided by the manufacturer.

Figure 36–10 Cutting metal edging using a hacksaw.

To install vertical brick edging, excavate the trench along the entire perimeter of the edged area the same depth as the length of the brick. The trench must form a vertical edge on the lawn side of the trench. Place a small amount of sand in the trench and place brick vertically against the sod edge. By adding or removing sand, adjust the height of the edging so that it is set at the desired elevation. Backfill and compact leaving the planting bed side low enough to accommodate mulch. Additional stability can be obtained by backfilling with a small amount of mortar in place of soil. When hardened, cover the mortar with soil. Corners can be turned by slightly flaring the brick, leaving small gaps between bricks at one end.

Flat Brick Edge. An alternative to the vertical brick edge is to place paving bricks flat, either edge to edge or end to end. This installation requires slightly less labor and material than installation of vertical brick edgings.

Installation of a flat brick edging requires the excavation of a shallow trench the same depth as the brick and 2 inches wider than the brick length. Flat brick edgings function nicely if the trench is angled up slightly on the planting bed side, leaving the bricks with a slight upward slant after installation. Strips of landscape fabric can be placed in the trench before placing brick to reduce weed growth. Place the fabric flush with the outside edge of the trench and let it extend into the planting bed a few inches to overlap weed barrier placed in the bed. Place the bricks side by side, flat in the trench. Corners can be turned by fanning the brick around the corner, which will leave a small gap at the top or bottom between bricks. For a tighter fit around corners, diagonally trim every other brick with a wet masonry saw. If the trench is the proper depth, backfilling is usually not required.

Precast Block Edging. Manufactured concrete block can be used as an edging in almost any setting. Cast in a variety of colors and forms, these solid blocks are small enough to turn tight radii and are easy to work with. When properly installed, the precast units are low in maintenance.

Excavate a square or rectangular cross section trench with a depth the same as the block height and 2 inches wider than the width of the block. Strips of landscape fabric can be placed in the trench before placing blocks to reduce weed growth. Place the fabric flush with the lawn edge of the trench and let it extend into the planting bed a few inches to overlap the bed weed barrier. Place the blocks end to end in the trench with one side flush with the lawn edge. Turn corners by fanning the block and leaving a small gap at the front or back of the block. Special block is available with shaped ends that interlock or alternate to form smooth straight and curved sections (Figure 36–11). Backfill the edging leaving the bed side lower to accommodate mulch.

Curbing. Concrete curbing makes a durable and distinct edge for a planted area. Whether installed by hand or with the use of a curbing machine, concrete provides a low-maintenance edge that blends with many design concepts. A variety of shapes and widths is also available with this choice. Price for curbing may be expensive due to equipment rental or hand labor required for forming.

To install curbing, begin by excavating a square, cross section trench along the perimeter where the curb is to be placed. The depth should match the

Figure 36–11 Placement of precast concrete edging. This edging may be placed in straight or curved alignments.

thickness of the edging selected. Place a small amount of granular backfill in the bottom of the trench to level the base. Construct forms (See concrete installation in Chapter 23) with the top set at the desired finish elevation. Pour concrete into the forms and tap the sides to settle the mixture. Complete filling and use a float or trowel to smooth and finish the top. After the concrete has hardened, remove forms and backfill, leaving the planting bed side lower to accommodate mulch.

If a curbing machine is available, it will speed the placement of curb edging. The curbing machine requires either that a trench be excavated to the depth of the edging, or that the finish grade be built up after the edging is completed. No forms are required. A curbing machine places mixed concrete in a uniform shape along the path where the machine is operated. Finishing of the top of the edging with a trowel is usually required.

Stone Block or Flagstone. Wall stone or flagstone can also be used for edging planting beds. Using stone requires site preparation that is more extensive than preparation for plastic or metal edgings. Stone is costlier than most edgings, but is relatively maintenance free and attractive.

Stone block and flagstone are installed by excavating the trench the same depth as the thickness of the material, and 2 inches wider for block. With flagstone, excavate the trench the same width as the widest dimension of the stone. Strips of landscape fabric can be placed in the trench before placing stone to reduce weed growth. Place the fabric flush with the lawn edge of the trench and let it extend into the planting bed a few inches to overlap weed barrier in the bed. Place the blocks end to end in the trench with one side flush with the lawn edge (Figures 36–12 and 36–13). With flagstone, select the straightest edge and place that edge along the lawn edge. Flagstone leaves an irregular inner edge with interesting shapes and lines. Turn corners using smaller pieces of stone. Backfill the edging leaving the bed side lower to accommodate mulch.

Wood Edge. Relatively inexpensive and low in maintenance once installed, wood presents a natural material choice for edging beds. Wood edgers can be difficult in installations that have undulating slopes and curved edges. Only treated or decay-resistant lumbers should be used in contact with the soil.

To install wood edging, select a decay-resistant, dimensioned lumber, typically a 2×6, and cut to length for each section to be edged. Cut enough 2×2 treated stakes to place one at both ends of each edging piece and one every 2 feet along the perimeter. Excavate a vertical trench 8 inches deep along the edging alignment. Place the wood edging in the trench tightly against the vertical side and at the desired elevation. Attach one stake to each end of the edging with a single nail. Holding the stakes tightly against the edging, drive the stakes into the subgrade (Figure 36–14). Hide the tops of all stakes by setting them 1 inch below the top of the edging. Drive form nails through the stake into the edging. Adjust to the desired elevation by lifting the stakes up or driving the stakes down. After edging elevation is set, place additional stakes every 2 feet and nail to the edging. Backfill on both sides of the edge leaving the planting side low enough to accommodate mulch.

Corners are turned by butting one section of edging into another and endnailing to hold the two pieces together. Joining two sections is strengthened by nailing a treated 1×4 stake across the joint on the back side of the two pieces being joined.

An alternative to staked wood edgers is utilization of treated 4×4s or 6×6s as edging material. These edgers require a trench excavated to the same depth and width as the material. Place the material in the trench and backfill. Corners can be turned by cutting short pieces of material and placing them fanned along the curve alignment. No staking is required.

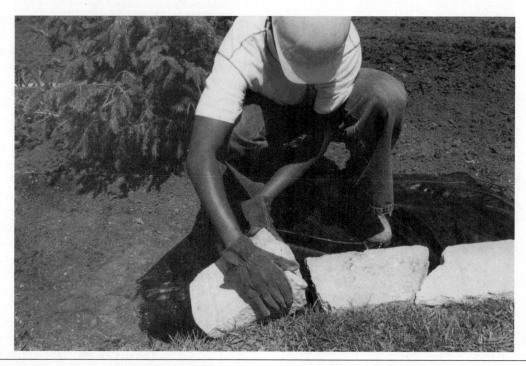

Figure 36–12 Edging with wall stone placed on weed barrier.

Figure 36–13 Edging with flagstone placed on weed barrier. Straight edge of flagstone is placed along turf edge with irregular edge facing planting bed. Place edging flush with turf and sloping up into planting bed.

INSTALLATION OF PLANTERS

Installation of plants in interior or certain exterior applications may require the use of planters. Planters are found in square, rectangular, or round shapes, with custom shapes available. Materials include wood, metal, terra cotta, fiberglass, and precast concrete. Planters typically are designed to set on a paved surface without any further anchoring, but some require anchoring to slabs. Planters can also be permanently built into projects.

Prefabricated Planters

Planter installation requires unpacking, possibly assembling, and placing at the appropriate locations throughout the site (Figure 36–15). If the planter is anchored at a specific location, anchor bolts should be placed during the paving portion of the project. When anchor bolts are preplaced, cover with Styrofoam painted a fluorescent color to reduce potential damage from traffic. Because outdoor surfaces are usually sloped for drainage, level placement of planters may require the use of cedar shims. Use a carpenter's level placed on a horizontal surface of the planter to determine the amount of shimming required. Break off a shim the correct thickness, lift the empty planter, and slide the shim underneath. Multiple shims may be required to level in all directions. Double-stick tape can be used to hold the shims at the correct location on the bottom of the planter if it must be lifted and adjusted several times. Tuck shims under the edge so they are not visible from the side.

Figure 36–14 Staking for wood edging. Place end stakes and adjust for level. Joints should overlap a single stake.

Figure 36–15 Planters in an urban plaza. Most planters are placed on pavement without anchors.

Built-in Planters

Landscape projects occasionally make use of built-in planters. Built-in planters are typically enclosed retaining walls that have been backfilled with planting soil to support plant growth. Walls for planter construction should be constructed using the same techniques described in Section 5. When constructing walls for built-in planters, accommodations for drainage may be required. To drain excess water from the planter, extend a socked, perforated drain tile around the entire interior perimeter of the wall. Connect to an underground tile or storm sewer, or drain through an opening in the wall.

INSTALLATION OF TREE GRATES AND TREE GUARDS

Tree grates are cast metal grids that are placed at openings in walkways where plant material is to be installed. Tree grates allow the water and air necessary to sustain plant growth to pass through what would otherwise be a paved surface. Tree grates have enough solid surface that they can support foot traffic. Tree guards work in conjunction with tree grates, which are metal enclosures that form a protective surrounding for tree trunks. Guards protect trees from damage from contact with vehicles, bikes, and other types of traffic. It is uncommon to install tree guards without tree grates.

Tree grate and tree guard installation requires preliminary work during the paving stage of the construction process. Tree grates are placed on a frame that is set into the concrete paving around the plant opening. The concrete used to place this frame can be part of the walkway, or a curb poured separate from the walk

Figure 36–16 Curbing with tree grate in place. Frame for tree gate can be placed in walk or placed in separate curb as shown here.

Figure 36–17 Tree grate sections are bolted together. Sections of guards are placed around the tree and bolted together, then bolted to the tree gate.

(Figure 36–16). Placement of the grate frame and a separate curb can be installed after the walk installation is complete, but will require additional forming and delivery of paving materials. The opening for the grate frame needs to be formed to the correct size and the frame placed at the proper depth for the tree grate to set flush with adjacent paving. Verify with the manufacturer the dimensions of the tree grate prior to forming the walk or curb that will support the frame.

Before placing any plant material in the tree grate opening, excavate all undesirable fill material from the hole and replace with suitable planting soil. After plant material has been installed, the tree grate may be placed on the curbing. Grates come in two or more sections to fit around the tree. If the design requires, bolt the two sections together with rust-resistant hardware. Manufacturers typically design grates so that when the opening is too small for plant material, interior rings can be cut from the grate. Use a cutoff saw with a carbide blade to remove enough inner rings to accommodate plant size. Tree guards come in two pieces and are bolted together and then to the tree grate for stability (Figure 36–17). Use touch-up paint if required to repair any finish damage.

While other elements of sitework create the major functional framework for a design, the finishing details of the hardscape are created by the site amenities. In Section 9, the installation of several common amenities was discussed. While the variety of site amenity treatments is extensive, the most common installations include use of water as a design feature, bridges, site furniture, play equipment, edgings, and plant protection devices such as tree grates and tree guards.

With the creativity and diversity of water features used in the landscape, attempting to classify uses of water can be difficult. For the purposes of this text, water-related features were grouped into flexible and rigid liner pools, concrete pools, waterfalls, cascades, fountains, and bridges. Preparation for these features requires planning the layout to accommodate the topography, electrical sources, water sources, drain installation, and recirculating piping. Bridge location also requires planning for placement, size and type of bridge, and support of bridge.

Installation of flexible and rigid liner pools is accomplished by marking the location for the pool and excavating a basin into which the liner is placed. Flexible liners can have variable levels and shapes, while rigid liners require excavation to the predetermined size and shape. Liners are placed and leveled, then edge treatments are applied to hide the top edge of the pool. Pools are filled and ready to operate. Submersible pumps are used if water is to be recirculated to other basins. Concrete pools require extensive forming of the floor and walls of the pool. Forming must include box-outs for any lighting and placement of a sump in the floor with a drain and recirculation piping. After concrete is poured, the inside of the pool is surfaced and the edges are treated with coping or edging.

Waterfalls require the construction of one or more basins into which water can fall. A lower basin to receive water is required and, in many waterfalls, an upper basin to supply water is also installed. Techniques used to construct pools can be used to create the waterfall basins. The water source and recirculating mechanism must be designed and tested to assure that the desired amount of water will circulate into the lower basin. At the top of the falls, an engineered opening called a weir controls the amount of water that passes over the falls. If multiple basins or multiple falls are part of the design, the channels between these elements must also be constructed to conduct the proper amount of water. Cascades are constructed in a manner similar to construction of waterfalls. A lower basin captures water that spills down a gently sloped channel. When waterfalls and cascades are constructed using liners, the surfaces and edges are typically covered with landscape material to hide the liner and to create a natural appearance.

Fountains can be placed in pools as stationary pumps or floated onto a pool. Each installation requires proper sizing of the pool, a pump motor, and a supply of electricity to the fountain pump. Bridge installation by the landscape contractor is typically restricted to crossings of minor water features. Both custom-built and premanufactured bridges are available to span pools, ponds, and small streams. Bridge installation begins with the accurate placement of piers for each end of the bridge. Premanufactured bridges are lifted into place using a crane. Custom bridges require framing a platform and railings, as identified in the wood construction section of this text.

Installation of site furniture and premanufactured play equipment begins with the accurate layout of the amenity. Amenities purchased may require assembly according to factory instructions and exact tolerances for placement of anchors and posts. Amenities typically have posts that are either directly buried in the ground, connected to a concrete slab using anchor bolts placed when the slab was poured, or anchored on top of a frost footing. Each anchoring method requires that the posts of the amenity resist frost action. Certain amenities may require that conduit for supplying electrical wiring or reinforcing be placed when frost footings are poured. Play equipment also requires the placement of a resilient surfacing below and around the equipment. Field finishing or touch-up of metal amenities is required by some manufacturers.

Amenities related to plant installation include edgings, planters, tree grates, and tree guards. Natural edging is accomplished by excavating a clean border between planting beds and turf. Permanent edging installation requires excavation of a trench and placement of the edging material flat against the turf side of the trench. Plastic edging is joined using short tubes and anchored using metal stakes driven through the edging. Metal edgings are overlapped and joined with a metal stake driven through a slot in both adjoining pieces. Metal stakes are driven along the length of the edging to secure it. Wood edging is anchored at each end, and at regular intervals, using wood stakes. Other edgings, including bricks, stone, and precast concrete, are installed by laying them edge to edge in the trench and backfilling. Edging can also be accomplished using a curbing machine that lays a shaped bead of concrete in an excavated trench.

Planters are typically placed in the proper location, filled with soil, and planted. In certain locations, planters may require leveling to match pavement grades. Tree grates require a cast-in-place concrete curb to support the framework for the grate. The framework is set in the curb and the grate is set in place after plant material is installed. Many frames come in two or more pieces that are bolted together. Tree guards come in two pieces and are placed around the tree, bolted together, and bolted to the tree grate.

Appendix

EXAMPLE MATERIAL AND WORKMANSHIP SPECIFICATION FOR SEEDING PROJECT

Section 02900 Seeding

Part 1—General Information

1.1 Description of work.

    A. Work included: Provide all labor, materials, equipment, and incidentals necessary to complete topsoil preparation, fertilization, seeding, and mulching where shown on the drawings, as specified herein, and as needed for a complete and proper installation.

    B. Related documents:

        1. Documents affecting work of this section include, but are not necessarily limited to, General Conditions, Supplementary Conditions, and Sections in Division 1 of these Specifications.

1.2 Quality assurance.

    A. Use adequate numbers of skilled workers who are thoroughly trained and experienced in the necessary crafts and who are completely familiar with the specified requirements and the methods needed for proper performance of the work of this section.

    B. Seed, fertilizer, and soil conditioners shall be commercial grade.

1.3 Product handling.

    A. Comply with pertinent provisions of Section 1640, Product Storage and Handling.

Part 2—Products.

2.1 Grass seed.

    A. Provide mixture consisting of the following proportions by weight:

        1. Kentucky Bluegrass—50% by weight.

        2. Redtop—10% by weight.

        3. Creeping Red Fescue—30% by weight.

        4. Annual Ryegrass—10% by weight.

    B. Lime: Finely ground agricultural limestone containing not less than 85% calcium and magnesium carbonates.

    C. Fertilizer: Free-flowing granular type of uniform composition containing by weight the following:

        1. 5% Nitrogen, with 50% of this total organic.

        2. 10% Available Phosphoric Acid

        3. 5% Potash

    D. Mulch: Oat straw.

    E. Water: Available at site.

Part 3—Execution

3.1 Preparation of topsoil.

    A. Deposit additional topsoil as may be required to correct all settlement and erosion.

    B. Apply fertilizer at an approximate rate of 10 pounds per 1,000 square feet over area.

    C. Thoroughly and evenly incorporate fertilizer with the soil to a depth of 3 inches by discing or other approved method. In areas inaccessible to power equipment, use hand tools.

    D. Prepare seed bed in accordance with acceptable horticultural practices. Leave area smooth with a loose, friable seed bed.

    E. Remove roots, sticks, clods, rocks, stones, etc. of golf ball size or larger.

3.2 Seeding.
  A. Apply seed uniformly and evenly over entire area.
  B. Seed may be drilled or broadcast. If broadcast, lightly cover seed.
  C. Apply grass seed at a rate of 3 pounds per 1,000 square feet.
  D. Roll lightly to firm seed bed and cover with mulch.
3.3 Maintenance.
  A. Soak seed bed at initial watering. Maintain damp seed bed through germination period, watering as necessary.
  B. Contractor shall be responsible for the repair and reseeding of all eroded areas and areas that fail to show a uniform stand of grass and ground cover.
  C. Reseed those areas that do not produce proper stand within first year after seeding.

End of Section 02900

EXAMPLE PERFORMANCE SPECIFICATION FOR SEEDING PROJECT

Section 02900   Seeding

Description of work.
  Contractor shall provide all labor, materials, equipment, and incidentals necessary to provide a complete and proper seeding installation. Installation shall include topsoil preparation and modification, fertilization, seeding, and mulching where shown on the drawings and in all other areas on the site disturbed by construction activities.

Work in related documents.
  Other contract documents that relate to this section include, but are not necessarily limited to, General Conditions, Supplementary Conditions, Sections in Division 1 of these Specifications, and Planting and Seeding Plans.

Quality assurance.
  Contractor shall provide a vigorous, viable, weed-free turf cover composed of no less than 100 grass plants per square foot at the time of acceptance of the project by owner. Cover crop species will not be considered as part of this assurance.

Products.
  Turf shall be composed of approximately equal parts of three species of Kentucky Bluegrass and one species of fine-textured fescue. Cover crop species may be added as required. All turf species used shall be locally adapted cultivars.

Installation requirements.
  Installation shall be coordinated with other site and building construction activities. Installation will be in accordance with recognized horticultural standards. Any soil modifications and finish grade corrections required will be the responsibility of the contractor. Any work disturbed by construction, erosion, or other means prior to final acceptance of project by owner will be repaired at contractors' expense.

Warranty.
  Contractor shall warranty the turf completely for 30 days following acceptance by owner. Contractor shall warranty trueness of species for a period of 1 year following acceptance by owner.

End of Section 02900

Appendix **B**

MEASUREMENT CONVERSIONS

As the contractor purchases, calculates, and builds with the variety of materials used in landscape construction, he or she will encounter many situations that require conversion from one standard to another. Commonly used conversions are listed here.

Linear:

12 inches	= 1 foot
36 inches	= 1 yard
3 feet	= 1 yard
5,280 feet	= 1 mile
1,760 yards	= 1 mile

Weights:

16 ounces	= 1 pound
2,000 pounds	= 1 ton

Liquid Measure:

16 fluid ounces	= 1 pint
128 fluid ounces	= 1 gallon
2 pints	= 1 quart
8 pints	= 1 gallon
4 quarts	= 1 gallon

Area:

144 square inches	= 1 square foot
9 square feet	= 1 square yard
43,560 square feet	= 1 acre
100 square feet	= 1 square (sod and shingles)
640 acres	= 1 square mile

Volume:

1,728 cubic inches	= 1 cubic foot
27 cubic feet	= 1 cubic yard

Inch/Decimal Equivalents:

1 inch	= .08 feet
2 inches	= .17 feet
3 inches	= .25 feet
4 inches	= .33 feet
5 inches	= .41 feet
6 inches	= .50 feet
7 inches	= .58 feet
8 inches	= .66 feet
9 inches	= .75 feet
10 inches	= .83 feet
11 inches	= .92 feet

Comparative Volume/Weight:

1 cubic foot water	= 64 pounds
1 cubic foot water	= 7.48 gallons
1 gallon water	= .134 cubic feet
1 cubic foot soil	= approx. 90 pounds
1 cubic foot dry sand	= approx. 100 pounds
1 cubic foot aggregate	= approx. 140 pounds
1 cubic yard soil	= approx. 2,400 pounds
1 cubic yard dry sand	= approx. 2,700 pounds
1 cubic yard aggregate	= approx. 3,780 pounds

Glossary

A

AC. Alternating current. High-voltage electrical current.

Angle of repose. Angle formed when materials are piled and allowed to slump naturally.

Ashlar. Random stone pattern composed of cut squares and rectangles of varying dimensions.

Asphalt. A mixture of petrochemicals, sand, and aggregate used to pave surfaces.

B

Baluster. Narrow post used for railing surfacing.

Base. Layer of material below the paving. Provides support for paving.

Baseline. Layout technique that locates objects using measurements made at right angles at calculated points along a straight line.

Base material. Crushed stone used as support material for most paving material. Also referred to as aggregate, gravel, roadstone, and other regional names.

Base screed. Tool used to level base material. Typically a straight 2×4 slightly shorter that the width of the forms with handles on top.

Batter. Backward lean of a retaining wall for stability.

Batterboards. L-shaped wooden framework used to mark corners, edges, and elevations of projects.

Benchmark. Fixed reference elevation.

Bentonite. Finely ground clay material that expands upon contact with water.

Berm. Mounded and shaped soil.

BFM. Bonded fiber mulch. Hydromulch with materials are chemically bonded together.

Bidding. A pricing process in which contractors compete against each other by calculating their price and submitting to owner in a sealed envelope.

Blocking. Solid wood blocks placed between joists or rafters to maintain spacing.

Bollard. Short post used to separate traffic or protect important site elements.

Brick (paving brick, pavers). Rectangular fired clay brick with typical dimensions of 4 inches × 8 inches × 2.5 inches thick. Available in a range of colors, paving bricks have no mortar holes and an average strength of approximately 5,000 PSI.

Bridging. Diagonal wood braces placed between joists or rafters to maintain spacing.

Building paper (roofing felt). Tar-impregnated paper used to waterproof roofs.

Bull float. Wide, flat metal (usually aluminum) tool with a long handle used to smooth the concrete surface after it has been screeded.

Butt stair. Stairs that have tread material that butts up against cheek walls.

Buttress. A support placed in front of or outside an object.

C

Canopy. Top vegetative portion of a plant, including trunk, branches, and foliage.

Cantilever. Overhang of a deck beyond outside supports.

Carriage. Structural support member for stairs.

Cartesian coordinates. A measurement and layout system that locates an object using measurements off of two baselines set up at right angles to each other.

Cascade. Water feature in which water spills gently from one level to the next.

Catch basin. Concrete structure with a sump into which water runoff flows.

Chalkline. Tool composed of a housing holding chalk and string. String is pulled out and snapped, leaving a chalk line used to mark locations.

Cheek wall. Side wall that runs perpendicular to the face of a retaining wall. Stairs butt or are interlocked into the cheek wall.

Circular saw. Hand-operated power saw using a round blade.

Cladding. Covering the sides of a post or column with a decorative surfacing material.

Cleat. Small wood blocks used to support surfacing materials.

Coaxial. Cable used for TV with conducting wire in center and conducting mesh wrapped around outside of center conductor.

Column. Decorative post.

Composting. Technique for piling and turning of vegetative waste that allows decomposition of waste into useable organic matter.

Concave corners. Wall corners where the radius is measured on the inside. Inside corners.

Concrete. Mixture of cement, sand, water, and aggregate (stone) that hardens to create a solid paving surface. Typical landscape concrete is ordered as 4,000 PSI.

Concrete broom. Stiff-bristled shop broom used to put a texture on concrete.

Conduit. Plastic or metal piping in which wiring is placed.

Construction joint. Joint placed at the end of a concrete pour where work is stopped with plans to continue later. This joint can be a straight edge, an edge with rebar sticking out, or an edge with a keyway.

Contour lines. A line on a drawing that connects points of the same elevation.

Contract documents. Plans, specifications, and related documents used to guide the construction of a project.

Contraction or control joint. Joint in concrete used to control the cracking that happens when concrete dries. These shallow joints are tooled into the concrete while it is still wet, or sawed in after the concrete dries.

Controllers. Automatic switches that turn a DC circuit on and off based on time or light.

Convex corners. Wall corners where the radius is measured on the outside. Outside corners.

Coping material. Edging placed around pools to cover the top edge of the pool.

Cor-ten® steel. Type of steel that has an oxidized (rusted) finish appearance.

Countersink. Driving a fastener slightly below the surface of the material being fastened.

Cover crop. Fast-germinating plant material seeded on a site to reduce erosion.

Crimping. Deforming the surface of a material.

Curb inlet. Concrete drainage structure built into a curb into which water runoff flows.

Cut and fill. The process of removing and adding soil to a site to achieve desired grades.

D

Darby float. Long, wood hand float with handle on top, used to level concrete after placement.

DC. Direct current. Low-voltage electrical current.

Deadman. Anchor that is connected to the face of a wall and buried into a hill behind the wall for stabilization.

Design professional. A registered design professional such as landscape architect, architect, engineer, or land surveyor.

Detention. Holding back of storm water.

Drip edge. Metal flashing engineered to keep water from flowing under shingles.

Drip line. Line on the ground directly below the farthest extent of a plant's canopy.

Dry-laid stone. Stone pieces that are laid without mortar joints.

Dumpy level. A tripod-mounted telescope that can be leveled and used to measure elevations.

E

ECB. Erosion control blanket. Fibers or other mulch-like materials sandwiched between layers of thin nylon netting. Also called erosion mats.

Edger. Short tool with a handle on top, flat surface on the bottom, and a 1/4 inch lip projecting down along the edge of the tool. Used along concrete forms to create rounded edges.

Edging (edge restraint). Material used along an edge to hold paving materials in place. Used with unit pavers and granular material, edgers can be concrete, stone, brick, wood, metal, or plastic. Edgers are typically anchored into the subgrade with pins or stakes. Also refers to materials placed to separate turf from planting beds.

Erosion. Removal of soil particles through the action of water and wind.

Expanding bolt. Bolt that anchors itself by expanding when the stem is turned.

Expansion joint. Joint placed where concrete intersects with a stable, permanent object. This joint protects the stable object from being moved by the force of expanding concrete.

Expansion joint material. Plastic or fibrous strips that are placed between buildings and walks to protect both when concrete expands in hot weather.

Exterior grade plywood. Plywood manufactured with special glues and woods to resist delaminating in exterior applications.

F

Fascia. Material covering structural components.

Fibermesh. Concrete with minute fiberglass strands that act as reinforcement.

Fiber optic. Communications utility line composed of a fine glass strand encased in plastic housing.

Fieldstone. Irregular sized rounded or angular shaped stone weathered by natural forces. Used in freestanding walls.

Field finishing. Painting or touching up amenity finishes after installation.

Finish floor elevation. The elevation of the finish level of a structure floor.

Fixtures. Components attached to an electrical system, such as lights.

Flagstone. Natural stone that is either cut into standard dimensions or broken into irregular pieces with a thickness of 1.5 inches to 2.5 inches.

Flashing. Metal sheeting installed between structural members to prevent moisture penetration.

Float. A broad, flat tool with a handle on top used to smooth concrete. Manufactured from wood, steel, or magnesium.

Forms. Pieces and strips of wood or metal placed to shape concrete slabs. Forms are secured to the ground and tops are set at the desired height of the concrete.

Form nails. Special nail used in form construction that has two heads to make removal from form easier.

Free-draining angular fill. Crushed stone without rounded edges or fine granular material used as fill behind a wall.

Free-out (swale high point). Low point that allows water to pass out of a drainage area.

French drain. A short, shallow trench drain filled with loose stone. Used for water storage in low areas.

Frost depth (Frost line). Average maximum depth at which frost penetrates the soil.

G

Gabions. Wire baskets that are filled with stone and stacked to create a wall.

Galvanized. Metal that has been coated with zinc to reduce rusting.

General contractor. Contractor responsible for coordinating work performed by subcontractors from several different trades.

Geogrid. Open-weave fabric used to stabilize retaining walls.

Geotextile. Interwoven structural fabric used to assist in stabilizing subgrades.

Geotextile sock. A textile covering wrapped around a tile to keep fine particles of soil from moving into the tile.

GFCI. Ground fault circuit interrupt. Circuit engineered to automatically shut off if a voltage change or short is detected.

Gill. Tool used to cultivate and smooth soil before seeding or sodding.

Granular base. Crushed stone material used to stabilize bottom course of wall.

Grape stake. Long, rounded, irregular shaped stick used for fence surfacing.

Grommets. Round metal rings which protect the openings in fabric through which cords or rope are placed.

Grout. Weak mortar mix used to fill center cavity of a masonry wall.

Gunnite®. Stiff concrete that is sprayed on a metal framework or surface and finished in place.

H

Hammer drill. A drill utilizing a bit that rotates and vibrates to improve boring through hard surfaces such as masonry or concrete.

Honeycombing. Voids or holes that occur when concrete is not properly vibrated or tamped.

Hose bibb. Threaded adapter placed on a hydrant that allows a hose to be attached.

Hydrant. Exterior water outlet, usually with a threaded connector for a hose.

Hydraulic block splitter. Table-mounted masonry cleaving tool that exerts hydraulic pressure on a bar to split materials.

Hydromulching. Application of mulch by mixing it with water and a tackifier and spraying it onto a surface.

Hydroseeding. Seed application accomplished by mixing seed with water and spraying on a surface.

Hydrostatic pressure. Pressure caused by the buildup of water behind a retaining wall.

I

Inlet. Concrete or plastic structure into which water runoff flows.

Interlocking concrete paving block. Unit paving material composed of molded, cured concrete. Pavers come in a variety of shapes and colors, are

3.5 inches thick, and have a typical average strength of 8,000 PSI.

Interlocking stair. Stairs that have treads that interlock with cheek walls.

J

Jointer. Similar to an edger, except the lip is located in the center and the tool is used to create contraction joints in a slab.

Joist hanger. Metal bracket used to connect joists to other structural elements.

Jurisdiction. Defined boundaries within which a person or organization has control over activities.

K

Kerfing. Shallow cuts along the back of a board that allow the board to be bent.

Keyway. An angled groove placed in concrete by a special form. Used to create a lip where two concrete slabs can be held together.

L

Landscape architect. A registered (in states with title laws) design professional specializing in landscape projects.

Landscape designer. An individual who possesses special skills in arranging design elements.

Landscape timber. Dimensioned lumber wall unit treated for contact with ground.

Lath. Thin, narrow piece of unfinished wood used for spacing and surface applications.

Lattice. Framework of lath or small dimensioned lumber arranged in a decorative pattern.

Ledger. Piece of dimensioned lumber attached to a structure that supports decks and roofs.

Lifts. Layers of soil or granular base.

Light standard. Post to which a light fixture is connected.

Liner. Waterproof plastic or vinyl sheet used to line water features.

Lipped wall units. Precast wall units that utilize a lip cast at the back, bottom edge of the block to align and stabilize between courses. Lips also create a set-back batter.

M

Mag (magnesium) float. Wide, flat tool with handle mounted on top used to smooth concrete surface. Made out of lightweight magnesium.

Maintenance edge. Narrow strip of weed-proof material placed under the length of a fence to reduce trimming.

Masonry drill bit. Drill bit engineered to drill through concrete and masonry materials.

Mid rail. Bracing rail placed midway between the top and bottom of a panel of chain link fence.

Mulch. Organic or inorganic ground covering placed around plants to reduce weeds and retain moisture.

N

Niche. Realignment of a wall to create a small space and/or additional stability.

Nominal dimension. Dimensions of lumber used for ordering. Represents the actual dimensions before drying and planing.

O

Orifice. Opening in a drainage grate surface.

OSB. Oriented strand board. Sheet lumber composed of chips or strands of lumber joined with special glues.

Overhead. Costs of operating a business not specifically associated with a particular project.

P, Q

Pacing. A measuring technique that utilizes walking to obtain approximate dimensions.

Phillips drive. Screw with an X pattern driving head.

Pickets. Vertical boards used for fence surfacing.

Pier. Support for bridge platform.

Pier block. Precast concrete block used to support posts.

Pinned wall units. Precast wall units that utilize a metal or fiberglass pin to align and stabilize units between courses.

Plant schedule. A listing of plant material used in a design. Often located on the planting plan.

Plumb. Placement of an object in true vertical position.

Plumb bob. A weighted tool suspended from a string. Used to determine plumb.

Pressure-treated wood. Wood that has had a preservative forced into the surface under high pressures.

Privacy slatting. Plastic or metal materials inserted into chain link fence to increase opaqueness of fence.

PSI. Pounds per square inch. Strength measurement used for paving materials

Purlin. Wood support that runs perpendicular to rafters and supports roofing materials.

R

Rakes. The end overhang of a roof.

Reglet. Small groove under a waterfall that breaks surface tension of water.

Repetitive motion. An activity in which the same motion is repeated many times in the same manner.

Rerod (rebar). Heavy metal bar used to strengthen edges of concrete slabs. Sold in long lengths that are cut with saw or bolt cutters to length needed. Deformed rebars have a rough surface.

Reservoir basin. Water pool located at the bottom of a waterfall or cascade.

Resilient surfacing. Soft surfacing placed below play structures.

Return walls. Walls that run back from the face of a wall. Typically used to end a wall into a hillside.

Ridge plate. Dimensioned lumber piece to which rafters are attached at the top.

Rim joist. A structural wood member which runs around the outside of a deck frame in the platform framing method.

Rip-rap. Large diameter stone placed on embankments and in channels to reduce erosion.

Riser. Vertical portion of a step.

Roof pitch. The slope of a roof.

Rubble walls. Walls constructed of natural stone without mortar.

S

Scale. Instrument used to measure drawings. Also refers to the percent of reduction used to fit a full-sized site onto a sheet of paper.

Schedule 40. Thick-walled PVC pipe.

Screed. Tool used to level concrete after it is placed. Typically a straight 2×4 longer than the width of the forms. Also term used for leveling of concrete after a pour.

Screening. Metal or fiberglass cloth used to keep out insects.

Sealer. Clear chemical coating that preserves wood or paving.

Segmental precast concrete wall unit. Wall building material that is manufactured by casting concrete in a form.

Setting bed. Approximately 1 inch thick layer of sand on which unit pavers are set.

Shop drawings. Construction plans provided by manufacturers of equipment and amenities.

Site furniture. Benches, trash receptacles, bike racks, tables, and other site amenities designed for exterior applications.

Skimmers. Openings around the top edge of a pool designed to capture debris floating on the water surface.

Sleeper. Wood support for decking that is laid directly on ground.

Slotted drive. Screw with a straight, grooved driving head.

Soffit. Panel placed under the eave to cover structural work.

Spacer. Block of wood used to maintain equal spacing between boards applied to a fence.

Spackle. Thin plaster applied over concrete pool surfaces.

Specifications. Written document that includes general information about a construction project and specific information about materials and installation techniques to be used.

Speed square. Small adjustable tool that can mark square and angled markings.

Spindle. Round or square railing post for porch.

Split rail. A horizontal board fence style where the horizontal pieces are split logs, or rails.

Spot elevations. Elevation of a single point on a site.

Square. Placement of two objects with a corner that forms a right angle. Also tool used to establish square.

Square-headed drive. Screw with a square pattern driving head.

Stain. Chemical coating that colors wood.

Static pool. Water feature that does not have flowing water.

Step-back batter. Batter technique in which the front of each higher course of a wall is set back slightly from the course below.

Stone set (brick set). Chisel-like tool used to cleave brick and stone.

Stretcher. Tool used to stretch chain link fencing fabric.

Stringer. Structural support for stair surfacing.

Stringline. Tool used to identify location and/or elevation of the edge of a project.

Subgrade. Soil material below paving and base.

Sump. Recessed area at the bottom of an inlet that captures debris from runoff.

Surcharge. Water that is forced back out of an inlet.

Swale. Low ditch designed to occasionally carry water runoff.

T

Tack. To temporarily nail objects together.

Tackifier. Sticky substance used to bind hydromulch or hydroseed materials to soil.

Tension wire. Wire used to anchor the bottom of chain link fabric.

Terracing. A series of retaining walls, each set back from the one below it.

Ties. Aluminum wires wrapped around chain link fabric to hold fabric next to posts.

Tile. Round tubing used to conduct drainage water. Commonly manufactured of plastic, clay, or concrete.

Toe of slope. Low point of a slope.

Toenailing. Securing pieces of lumber by nailing at an angle.

Top of slope. Upper point of a slope.

Top rail. Top piping of a chain link fence used to support top of fabric.

Tort. Illegal action that leads to damage of the property of another person.

Transformer. Electrical component engineered to reduce voltage in current.

Transit. Tripod-mounted level that can be used to measure elevations, distances, and horizontal angles on a project.

Tread. Horizontal portion of a step.

Treated woods. Woods that have been treated with preservatives to reduce the potential of decay.

Tree grate. Metal grid that allows water and air access to root systems of trees set in paved areas.

Tree guard. Metal cage placed around trunks of trees in urban areas to protect bark from damage.

Trowel. Wedge-shaped metal hand tool. Used to mix and place mortar and concrete.

Trowelling. Smoothing the surface of a material, usually concrete, using a trowel.

U

Unit pavers. Bricks, paving block, or other manmade paving materials that are placed one at a time.

V

Verticals. Ties or timbers placed with the long dimension set vertically. Usually buried in front of a wall to prevent the wall from falling forward.

Vibratory plate compactor. Metal plate with a motor mounted on top and handle for steering. Vibrates and compacts base material as the machine moves.

W, X, Y, Z

Weepholes. Openings in the face of a retaining wall that allow water to drain from behind the wall.

Weir. An opening that regulates the amount of water flowing past a point.

Welded (woven) wire mesh (WWM). Metal wires welded together in a grid pattern used to reinforce concrete. Typically 10 gauge wire with 6 inch squares.

Wet masonry saw. Table-mounted circular saw that uses water to aid in cutting masonry materials.

Wet screed. A technique used when a screed is too short or a surface needs to be warped. Performed by resting one end of screed on a form and holding the other end at a height that is approximately where the other form would be.

Wood float. Wide, flat wood tool with a handle mounted on top used to smooth concrete surfaces.

Wythe. Width of one masonry unit.

Bibliography

Alth, Max. Do It Yourself Plumbing. New York: Sterling Publishing Co., Inc., 1987.

DeCosse, Cy, ed. *Black and Decker Home Improvement Library, Landscape Design and Construction Projects.* Minnetonka, MN: Cy DeCosse Incorporated, 1992.

Everett, T. H. *Lawns and Landscaping.* New York: Grosset Good Life Books, 1976.

Feucht, James R., and Butler, Jack D. *Landscape Management.* New York: Van Nostrand Reinhold.

Glen-Gery Brickwork. *Exterior Paving with Brickwork.* Wyomissing, PA: Glen-Gery Corporation, 1992.

Home How-To Institute. *A Portfolio of Deck Ideas.* Minnetonka, MN: Cy DeCosse Incorporated, 1993.

Hughes, Dwight Jr. *Systems for Success.* Cedar Rapids, IA: Dwight Hughes Systems, Inc., 1996.

Ingels, Jack E. *Landscaping Principles and Practices*, 5th ed. Albany, NY: Delmar Publishers, 1997.

Interlocking Concrete Pavement Institute. *Tech Spec, # 2, #3, and #5,* Sterling, VA: Interlocking Concrete Pavement Institute, 1994, 1995, 1996.

Iowa One Call. *Professional Excavators Manual.* Des Moines, IA: Iowa One Call, 1995.

King's Materials. *Installation Instructions Data Sheet.* Cedar Rapids, IA: King's Materials, 1997.

Lewis, Gaspar. *Carpentry, 2nd ed.* Albany, NY: Delmar Publishers, 1994.

National Association of Home Builders Research Foundation. *Residential Concrete*, 2nd ed. Washington, DC: Home Builder Press, 1994.

Naval Education and Training Command, and Buza, John. *Builder 3 and 2, Volume 1 and Volume 2, NAVEDTRA 12520 and 12521.* Philadelphia, PA: U.S. Navy, 1994.

Physiotherapy Associates. *Protect Your Back—Prevent Injury.* Cedar Rapids, IA: Work Injury Rehabilitation Center, 1995.

Portland Cement Association. *Design and Control of Concrete Mixtures,* 12th ed. Skokie, IL: Portland Cement Association, 1979.

Sherwood, Gerald, and Stroh, Robert. *Wood Frame House Construction.* New York: Sterling Publishing Co., Inc, 1992.

Spence, William Perkins. *Construction Methods, Materials, and Techniques.* Albany, NY: Delmar Publishers, 1998.

Sunset Books and Magazines. *Decks.* Menlo Park, CA: Lane Publishing Co., 1975.

Sunset Books and Magazines. *Fences and Gates.* Menlo Park, CA: Lane Publishing Co., 1975.

Sunset Books and Magazines. *Walls, Walks, and Patio Floors.* Menlo Park, CA: Lane Publishing Co., 1975.

Sunset Books and Magazines. *Patio Roofs and Gazebos.* Menlo Park, CA: Lane Publishing Co., 1992.

U.S. Consumer Product Safety Commission. *Handbook for Public Playground Safety.* Washington, DC: U.S. Consumer Product Safety Commission, 1995.

Wing, Charlie. *Visual Handbook of Building and Remodeling.* Emmaus, PA, Rodale Press, 1990.

Index